Lecture Notes in Networks and Systems

Volume 132

The series "Lecture Notes in Networks and Systems" publishes the latest developments in Networks and Systems—quickly, informally and with high quality. Original research reported in proceedings and post-proceedings represents the core of LNNS.

Volumes published in LNNS embrace all aspects and subfields of, as well as new challenges in, Networks and Systems.

The series contains proceedings and edited volumes in systems and networks, spanning the areas of Cyber-Physical Systems, Autonomous Systems, Sensor Networks, Control Systems, Energy Systems, Automotive Systems, Biological Systems, Vehicular Networking and Connected Vehicles, Aerospace Systems, Automation, Manufacturing, Smart Grids, Nonlinear Systems, Power Systems, Robotics, Social Systems, Economic Systems and other. Of particular value to both the contributors and the readership are the short publication timeframe and the world-wide distribution and exposure which enable both a wide and rapid dissemination of research output.

The series covers the theory, applications, and perspectives on the state of the art and future developments relevant to systems and networks, decision making, control, complex processes and related areas, as embedded in the fields of interdisciplinary and applied sciences, engineering, computer science, physics, economics, social, and life sciences, as well as the paradigms and methodologies behind them.

**** Indexing: The books of this series are submitted to ISI Proceedings, SCOPUS, Google Scholar and Springerlink ****

More information about this series at http://www.springer.com/series/15179

Dharm Singh Jat · Samiksha Shukla ·
Aynur Unal · Durgesh Kumar Mishra
Editors

Data Science and Security

Proceedings of IDSCS 2020

Editors
Dharm Singh Jat
Department of Computer Science
Namibia University of Science
and Technology
Windhoek, Namibia

Aynur Unal
Stanford University
Stanford, CA, USA

Samiksha Shukla
CHRIST (Deemed to be University)
Pune Lavasa Campus
Lavasa, India

Durgesh Kumar Mishra
Computer Science and Engineering
Sri Aurobindo Institute of Technology
Indore, Madhya Pradesh, India

ISSN 2367-3370 ISSN 2367-3389 (electronic)
Lecture Notes in Networks and Systems
ISBN 978-981-15-5311-0 ISBN 978-981-15-5309-7 (eBook)
https://doi.org/10.1007/978-981-15-5309-7

This Springer imprint is published by the registered company Springer Nature Singapore Pte Ltd.
The registered company address is: 152 Beach Road, #21-01/04 Gateway East, Singapore 189721, Singapore

Preface

This volume contains the papers presented at International Conference on Data Science for Computational Security (IDSCS 2020) held on March 13–14, 2020 in Pune.

The International Conference on Data Science and Computational Security (IDSCS-2020) invited ideas, developments, applications, experiences, and evaluations in the field of Data Science, Machine Learning, Artificial Intelligence, Data Analytics from academicians, fellow students, researchers, and practitioners.

The conference deliberation included topics specified within its scope. It focused on exploring the role played by technology, Data Science, Computational Security, and related applications in enhancing and securing business processes. The conference offered a platform for bringing forward substantial research and literature across the arena of Data Science and Security. It provided an overview of the upcoming technologies. IDSCS 2020 provided a platform for leading experts to share their perceptions, provide supervision, and address participant's interrogations and concerns.

The conference was organized by CHRIST (Deemed to be University), Pune Lavasa Campus, India, on March 13–14, 2020. IDSCS 2020 received approximately 88 research submissions from 4 different countries, viz., South Korea, USA, United Arab Emirates, and Oman. The papers included topics pertaining to varied advanced areas in technology, Data Science, Artificial Intelligence, Machine Learning, and the like. After a rigorous peer review with the help of program committee members and 60 external reviewers, 40 papers were approved.

Technology is the driving force in this era of globalization for the socioeconomic growth and sustained development of any country. The influence of Data and Security in shaping the process of globalization, particularly in productivity, commercial, and financial spheres, is highly required. This is the phase of revolution that has significant implications for the current and future societal and economic situation of all the countries in the world. Data and Security plays a fundamental role in enabling people for relying on digital media for a better lifestyle. Finally, it is concluded that Data Science and Security is a significant contributor to the success of the current initiative of Digital India.

In order to recognize the ongoing research and advancement by Data Science, Security, and allied sectors and encourage universities, scholars, and students through their research work acclimating new scientific technologies and innovations, the 2-day conference had presentations from the researchers, scientists, academia, and students on the research works carried out by them in different sectors.

IDSCS-2020 is a flagship event of CHRIST (Deemed to be University), Pune Lavasa Campus.

The Conference was inaugurated by Mr. Puroshotham Darshankar, Innovation and R&D Architect, Persistent System Limited, along with other eminent dignitaries including Dr. Fr. Jossy P. George, Dr. Fr. Arun Antony, Dr. K. Balachandran, Mr. Kapil Tiwari, and Dr. Fr. Joseph Varghese.

The conference witnessed keynote addresses from eminent speakers, namely Dr. Dong S. Han, Professor, School of Electronics Engineering, Kyungpook National University, Korea; Dr. Joseph Varghese, Professor, Department of Mathematics, CHRIST (Deemed to be University), India; Dr. Balachandran K.; and Dr. Fr. Jossy P. George, CHRIST (Deemed to be University), India.

The organizers wish to thank Dr. Aninda Bose, Senior Editor, Springer Nature and Ms. Jennifer Sweety Johnson, Springer Nature, New Delhi, India, for their support and guidance. EasyChair conference management is a really wonderful tool for easy organization and compilation of conference documents.

Windhoek, Namibia — Dharm Singh Jat
Lavasa, India — Samiksha Shukla
Stanford, USA — Aynur Unal
Indore, India — Durgesh Kumar Mishra
March 2020

Contents

Editors and Contributors

About the Editors

Prof. Dharm Singh Jat received his degree Master of Engineering and Ph.D. in Computer Science and Engineering from prestigious universities in India. He is a Professor of Computer Science at Namibia University of Science and Technology (NUST). From 1990 to 2014, he was with the College of Technology and Engineering, Maharana Pratap University of Agricultural and Technology – [MPUAT], Udaipur, India. He has guided about 8 Ph.D. and 24 master research scholars.

He is the author of more than 150 peer-reviewed articles and the author or editor of more than 20 books. His interests span the areas of multimedia communications, wireless technologies, mobile communication systems, edge roof computing software-defined networks, network security and Internet of things. He has given several Guest Lecturer/Invited talk at various prestigious conferences. He has been the recipient of more than 19 prestigious awards, such as Eminent Scientist Award, Distinguished Academic Achievement, Eminent Engineering Personality, CSI Chapter Patron, CSI Significant Contribution, Best Faculty Researcher, Best Technical Staff, Outstanding University Service Award and Distinguished ACM Speaker Award. Prof. Dharm Singh is a Fellow of The Institution of Engineers (I), Fellow of Computer Society of India, Chartered Engineer (I), Senior Member IEEE and Distinguished ACM Speaker.

Dr. Samiksha Shukla is currently employed as an Associate Professor and Head, Data Science Department, CHRIST (Deemed to be University), Lavasa, Pune Campus. Her research interest includes computation security, machine learning, data science and big data. She has presented and published several research papers in reputed journals and conferences. She has 15 years of academic and research experience and is serving as a reviewer for Inderscience Journal, Springer Nature's

International Journal of Systems Assurance Engineering and Management (IJSA) and for IEEE and ACM conferences.

Dr. Aynur Unal educated at Stanford University (class of '73), comes from a strong engineering design and manufacturing tradition, majored in Structural Mechanics-Mechanical Engineering-Applied Mechanics and Computational Mathematics from Stanford University. She has taught at Stanford University till mid 80's and established the Acoustics Institute in conjunction with NASA-AMES research fellowships funds. Her work on "New Transform Domains for the Onset of Failures" received the most innovative research awards. Most recently she is bringing in the social responsibility dimension into her incubation and innovation activities by getting involved in social entrepreneurial projects on street children and ageing care. She is also the Strategic Adviser for Institutional Development to IIT Guwahati. She is a preacher of Open Innovation, and she always encourages students to innovate and helps them with her support.

Dr. Durgesh Kumar Mishra has received M.Tech. degree in Computer Science from DAVV, Indore, in 1994 and Ph.D. in Computer Engineering in 2008. Presently he is working as Professor (CSE) and Director Microsoft Innovation Centre at Sri Aurobindo Institute of Technology, Indore, Madhya Pradesh, India. He is having around 24 years of teaching experience and more than 6 years of research experience. His research topics are secure multi-party computation, image processing and cryptography. He has published more than 80 papers in refereed international/national journals and conferences including IEEE and ACM. He is a senior member of IEEE, Computer Society of India and ACM. He has played very important role in professional society as Chairman. He has been a consultant to industries and Government organization like Sales tax and Labor Department of Government of Madhya Pradesh, India.

Contributors

Thulasi Accottillam Department of Information Technology, Kannur University, Kannur, Kerala, India

Rathnakar Achary Alliance College of Engineering and Design, Alliance University, Bangalore, India

Swati Ahirrao Symbiosis Institute of Technology, Department of Computer Science and Engineering, Lavale, Pune, Maharashtra, India

Oshin Anto CHRIST (Deemed to Be University), Bengaluru, India

Krishnan Balachandran Department of Computer Science and Engineering, Christ (Deemed to be University), Bengaluru, India

Aishika Banik CHRIST (Deemed to be University), Bengaluru, India

Vandana Bhagat CHRIST (Deemed to be University), Bengaluru, India

Praveen Kumar Bhanodia School of Computer Science Engineering, Lovely Professional University, Phagwara, India

J. Chandra CHRIST (Deemed to Be University), Bangalore, India

Xavier Chelladurai CHRIST (Deemed to be University), Bangalore, India

Ginish Cheruparambil CHRIST (Deemed to be University), Bengaluru, India

Nagaraj Cholli Department of Information Science, R V College of Engineering, Bangalore, India

Sonia Maria D'Souza Cambridge Institute of Technology, Bangalore, Karnataka, India

Sunil Kumar Dasari Department of Electronics and Communication Engineering, Presidency University, Bangalore, India

Ayan Datta CHRIST (Deemed to Be University), Bengaluru, India

D. D. Diana Steffi Department of Electronics and Communication Engineering, Presidency University, Bangalore, India

Denny Dominic Department of Computer Science and Engineering, Christ (Deemed to be University), Bengaluru, India

Ankit Singh Garia CHRIST (Deemed to be University), Bengaluru, India

Jossy George CHRIST (Deemed to Be University), Bengaluru, India

Jossy P. George CHRIST (Deemed to be University), Bengaluru, India

Arpita Ghosh CHRIST (Deemed to be University), Lavasa Pune, India

Sreyan Ghosh Department of Computer Science and Engineering, Christ (Deemed to be University), Bangalore, India

Siddharth Suresh Gosavi School of C&IT, REVA University, Bengaluru, India

G. S. Gowramma Department of Information Science, R V College of Engineering, Bangalore, India

I. Sahul Hamid The Madura College, Madurai, India

J. Hilda Janice Department of Computer Science and Engineering, CHRIST (Deemed to be University), Bengaluru, India

Suraj S. Jain Department of Computer Science and Engineering, Christ (Deemed to be University), Bangalore, India

Joshua Mammen Jiji CHRIST(Deemed to be University), Bengaluru, India

Diana Jeba Jingle I Department of Computer Science and Engineering, CHRIST (Deemed to be University), Bengaluru, India

M. Joan Jose CHRIST (Deemed to be University), Bengaluru, India

Alwin Joseph CHRIST (Deemed to Be University), Bangalore, India

Solley Joseph CHRIST (Deemed to Be University), Bengaluru, India;
Carmel College of Arts, Science and Commerce for Women, Nuvem, Goa, India

Angel P. Joshy Department of Computer Science and Engineering, CHRIST (Deemed to Be University), Bangalore, India

Aditya Khamparia School of Computer Science Engineering, Lovely Professional University, Phagwara, India

Sujatha Arun Kokatnoor Department of Computer Science and Engineering, School of Engineering and Technology, CHRIST (Deemed to Be University), Bangalore, India

Balachandran Krishnan Department of Computer Science and Engineering, School of Engineering and Technology, CHRIST (Deemed to Be University), Bangalore, India

Sonal Kumar Department of Computer Science and Engineering, Christ (Deemed to be University), Bangalore, India

Joseph Varghese Kureethara CHRIST (Deemed to be University), Bangalore, India

Samden Lepcha Department of Computer Science and Engineering, Christ (Deemed to be University), Bangalore, India

Ganesh Magar P.G. Depatment of Computer Science, S.N.D.T. Women's University, Mumbai, India

Shilpa Mehta Department of Electronics and Communication Engineering, Presidency University, Bangalore, India

Mohammed Misbahuddin Center for Development of Advanced Computing, Bangalore, India

Aniket Mohan Institute of Systems Science, National University of Singapore, Singapore, Singapore

Praveen Naik Department of Computer Science and Engineering, CHRIST (Deemed to Be University), Bangalore, India

P. Nanda Cambridge Institute of Technology, Bangalore, Karnataka, India;
East Point College of Engineering, Bangalore, Karnataka, India

K. Natarajan Department of Computer Science and Engineering, CHRIST (Deemed to Be University), Bangalore, India

Shantharam Nayak Department of Information Science, R V College of Engineering, Bangalore, India

Kunj Pahuja CHRIST(Deemed to be University), Bengaluru, India

Babita Pandey Department of Computer Science and IT, Babasaheb Bhimrao Ambedkar University, Amethi, India

Pratik P. Patil Symbiosis Institute of Technology, Department of Computer Science and Engineering, Lavale, Pune, Maharashtra, India

Aishwaria Paul CHRIST (Deemed to be University), Bengaluru, India

Ambika Pawar Symbiosis Institute of Technology, Department of Computer Science and Engineering, Lavale, Pune, Maharashtra, India

Shruti Petwal Department of CSE, Anurag University (Formerly known as Anurag Group of Institutions), Hyderabad, Telangana, India

Shraddha Phansalkar Symbiosis Institute of Technology, Department of Computer Science and Engineering, Lavale, Pune, Maharashtra, India

Shynu Philip CHRIST (Deemed to Be University), Bengaluru, India

Yash Purohit CHRIST (Deemed to be University), Lavasa Pune, India

Antony Puthussery CHRIST(Deemed to be University), Bengaluru, India

Rishab Rajput CHRIST (Deemed to be University), Lavasa Pune, India

G. Raju Department of Computer Science and Engineering, School of Engineering and Technology, CHRIST (Deemed to be University), Bengaluru, India

Tanvi Rath CHRIST (Deemed to be University), Lavasa Pune, India

Sandeep Singh Rawat Department of CSE, Anurag University (Formerly known as Anurag Group of Institutions), Hyderabad, Telangana, India

Syed Raziuddin Deccan College of Engineering, Hyderabad, India

K. Satyanarayan Reddy Cambridge Institute of Technology, Bangalore, Karnataka, India

Mallamma V. Reddy Rani Channamma University Belagavi, Belagavi, Karnataka, India

Sanjana Reddy Department of CSE, R.V. College of Engineering, Bengaluru, India

K. T. V. Remya Department of Information Technology, Kannur University, Kannur, Kerala, India

Akshay Sachdeva Institute of Systems Science, National University of Singapore, Singapore, Singapore

Divyansh Sahu CHRIST (Deemed to be University), Bengaluru, India

Manjunath Sajjan Rani Channamma University Belagavi, Belagavi, Karnataka, India

Chirag Saswat CHRIST (Deemed to be University), Bengaluru, India

Ujwala Sav Vidyalankar School of Information Technology, Mumbai, India

Anshul Saxena CHRIST (Deemed to be University), Bengaluru, India

Anjana Shaji CHRIST (Deemed to be University), Bengaluru, India

G. Shanmuga Rathinam Computer Science and Engineering, Presidency University, Bengaluru, India

G. S. Sharvani Department of CSE, R.V. College of Engineering, Bengaluru, India

Sheetal Department of Computer Applications, Presidency College, Bengaluru, India

Samiksha Shukla CHRIST(Deemed to be University), Bengaluru, India; CHRIST (Deemed to be University), Lavasa Pune, India

Gopal Krishna Shyam School of C&IT, REVA University, Bengaluru, India

S. Siddharthan CHRIST (Deemed to Be University), Bangalore, India

Nikitha Srikanth Department of CSE, R.V. College of Engineering, Bengaluru, India

K. Sunny John Department of CSE, Anurag University (Formerly known as Anurag Group of Institutions), Hyderabad, Telangana, India

Amrita Tamang CHRIST (Deemed to be University), Bengaluru, India

Tessy Tom CHRIST (Deemed to be University), Bengaluru, India

Aynur Unal Digital Monozukuri, Stanford, USA

N. Usman Aijaz VTU RRC, Brindavan College of Engineering, Bangalore, India

Vaishali S. Vairale CHRIST (Deemed to be University), Bengaluru, India

K. A. Venkatesh Department of Mathematics and Computer Science, Myanmar Institute of Information Technology, Mandalay, Myanmar

Gorla Vikas Department of CSE, Anurag University (Formerly known as Anurag Group of Institutions), Hyderabad, Telangana, India

Farha Fatina Wahid Department of Informtion Technology, Kannur University, Kerala, India

An Approach to Predict Potential Edges in Online Social Networks

Praveen Kumar Bhanodia, Aditya Khamparia, and Babita Pandey

Abstract With the advent of the Internet, online social networks are furiously growing and influencing our daily life. In this work, we have worked upon the problem of link prediction across nodes within these growing online social networks. Prediction of link within the social network is pertaining to missing and future links in the network in future. This could be attained by topological attributes or measures that collaborated with machine learning approaches. In this paper, we have tried to develop an edge prediction model which will be trained using a supervised machine learning technique. For experimental analysis, Wikipedia network has been used.

Keywords Online social network · Link prediction · Classification · Supervised learning

1 Introduction

Online social network (OSN) has highly influenced the life of humans and their day-to-day life. It is now offering avenues to build new friendships, associations and relationships that may be in the business life of social life [1]. Facebook, Twitter, LinkedIn and Flickr are a few popular names around us which have become an integrated part of our daily life. Researchers study these networks employing various mathematical models and techniques. The social networks are being crawled up by the researchers for further examination; during the collection of information, at a particular instance the network information collected is partially downloaded where

P. K. Bhanodia (✉) · A. Khamparia
School of Computer Science Engineering, Lovely Professional University, Phagwara, India
e-mail: Kumarpkb2@gmail.co

A. Khamparia
e-mail: adityakhamparia88@gmail.com

B. Pandey
Department of Computer Science and IT, Babasaheb Bhimrao Ambedkar University, Amethi, India
e-mail: shukla_babita@yahoo.co.in

D. S. Jat et al. (eds.), *Data Science and Security*, Lecture Notes in Networks and Systems 132, https://doi.org/10.1007/978-981-15-5309-7_1

certain edges between the nodes may be missing. The missing edge information is an apprehension in understanding the standing structure of the network. Apart from this, crunching the network structure attributes for the approximation of new fresh edges between the nodes is another interesting challenge to be addressed. The formal definition according to Libon Novel and Kleinberg can be studied in [6].

Identification of contributing information for determining such missing and new edges is helpful in new friendship recommendation mechanism usually used in online social networks. Several algorithmic techniques described by Chen et al. [2] have been introduced by IBM in their internal social network established for their employees to connect with each other digitally. The prediction or forecasting of such existing hidden edges or creating new fresh edges using the existing social network data is termed as the edge prediction problem. The applications of edge prediction include domains like bioinformatics for finding protein–protein interaction [3]; edge prediction can also be used to develop various recommendation systems for e-commerce websites [4]; besides, edge prediction can also assist the security systems for detecting and tracking the hidden terrorist groups or terrorist networks [5]. As to address the problem of edge prediction pertaining to answer the relevant different scenarios, many algorithms and procedures were proposed and the majority of the algorithms usually belongs to machine learning approaches. The feature estimated using neighborhood techniques like JC, AA and PA would further be used in developing a model classifier for edge prediction.

In this paper, we try to address the problem of edge prediction using a machine learning approach or classifier which will be trained using certain similarity features extracted by exploiting topological features. The proposed classifier would be experimentally evaluated using online social networking datasets (Facebook and Wikipedia).

2 Related History

There are several methods proposed for the prediction of future edges in various social networks using certain features based on the local level and global level. Typically, the methods involve computation of similarity among the node pairs and this similarity would later on be used for recognition of the potential future edges between the nodes. There are two kinds of similarities being used: first is on the basis of node features like the number of adjacent nodes or common neighbor [10], secondly, it is computed on the basis of network topological feature sets like path or distance between the connected nodes. The performance of both the methods varies as both the features are independent of each other; it also depends on the nature of the social network as well. Thus it is almost difficult to define the features which consistently identify the similarity between two unconnected nodes and consequently predict future edges in between them, due to which the performance of the methods based on similarity may vary with respect to the networks.

Table 1 Techniques exploiting node attributes

Feature extraction technique	
Common neighbors	$\Gamma(x) \cap \Gamma(y)$
Jaccard coefficient	$\mathrm{JC}_{xy} = \frac{\Gamma(x)\cap\Gamma(y)}{\Gamma(x)\cup\Gamma(y)}$
Adamic/Adar	$\mathrm{AA}_{xy} = \sum_{z\in(x)\cap\Gamma(y)}^{\infty} \frac{1}{\log\lvert\Gamma(z)\rvert}$
Preferential attachment	$\mathrm{PA}_{(x,y)} = \Gamma(x) \cap \Gamma(y)$

2.1 *Feature Extraction Techniques*

Techniques exploiting the local structure of the social network are demonstrated in this section. We are considering the node neighborhood techniques only (Table 1).

3 Classification and Classification Models

There are several algorithms available for supervised learning though the performance of each algorithm comparably varies with respect to the nature of the online social network. Some of them are reasonably fair than others or vice versa on a specific set of online social networks. The algorithms most commonly used in supervised learning are SVM, Decision tree, RBF Network and bagging, Naive Bayes and many more to explore. In this work, we have explicitly experimented with the problem of edge prediction using the ZeroR algorithm and Random forest algorithm which are known for their fast computation performance. Implementation is done using Python and Weka [9].

3.1 *The Model Proposed*

The proposed model typically works on ensemble approach exploiting the problem of edge prediction in two phases, that means two individual separate methods will work sequentially to mitigate the complexity of the problem for future edge prediction. The model proposed has been demonstrated in Fig. 1. Accordingly, the provided social network is crunched to identify the missing edges between the nodes of the graph $G = (v, e)$ where v represents the node and e is the edge to which adjacent node is connected. The node neighborhood feature has been used to generate the similarity score which is subsequently used as a score for the approximation of potential edge prediction using machine learning classification techniques. The entire dataset is sorted and arranged in ascending order. Before applying the classification algorithm, the resultant network data is preprocessed based upon the calculated similarity score wherein it has been classified into two classes: potential edges and non-potential

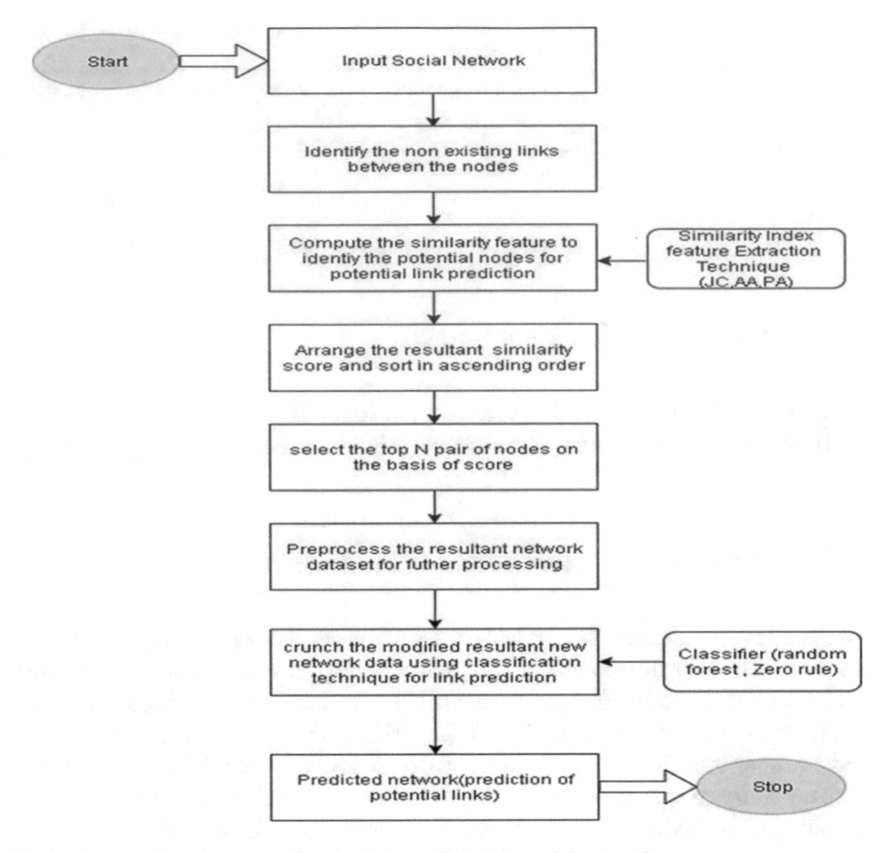

Fig. 1 Process flow demonstration for link prediction in social network

edges. Now the final moderated and modified network data is ready for crunching using classification techniques to develop a fast and effective classifier. The classification techniques used in learning and testing of the modes are Random forest and Zero Rule algorithm [7]. The model proposed exploits the advantages incorporated with both the different kinds of techniques.

3.2 Experimental Setup

In order to develop a classifier for crunching social network graphs referred to as G, the existing and non-existing edges across the graph are identified. The process is implemented in Python using the NetworkX package [8].

The crawled dataset consists of voted history data. It contains 2,794 elections (nodes) along with around 103,747 votes cast among 7,118 users including existing admins (either voted or being voted on). About half of the votes in the dataset are by existing admins, while the other half comes from ordinary Wikipedia users. The dataset is downloaded from https://snap.stanford.edu/data/wiki-Vote.html; the network nodes are users and directed edges are from node i to node j where designated user i has voted on user j. The evaluation of the model is done using two machine learning techniques; Random forest and Zero Rule algorithms.

4 Result and Discussion

The dataset created is crunched with the ZeroR algorithm for training and testing purposes using 10-fold cross-validation; the total time taken to build the model is 0.02 s which is quite reasonable. The model has correctly classified the 299 instances, that means it has predicted the probable edges with 87.172% of accuracy. The other way round the model has incorrectly classified instances too (12.828%). Random forest technique is applied to crunch the network for further edge prediction. The model has successfully predicted the probability of potential edges with cent percent (100%) accuracy. It has correctly classified all the 343 instances.

In the case of Jaccard coefficient combined with ZeroR algorithm, the model has an accuracy of 97.66% for edge prediction. To be precise the model has correctly classified 335 instances, the remaining 8 were incorrectly classified. The total time taken for the building of the model is the same, that is, 0.02 s.

Random forest classifier with JC has again proven to be cent percent accurate in the prediction of edges between nodes of the network. Although the time taken for building up the model is marginally higher than the ZeroR classifier, it is not that significant.

The network with preferential attachment feature crunched with ZeroR does not classify the probability of having a potential edge or not significant, rather the model is quite cumbersome to understand. Nevertheless, it has predicted the instances where the prediction of a future edge is negative up to the accuracy of 81.04%. This is perhaps the preferential attachment technique is quite inconsistent in generating the similarity score of the edge between the nodes u and v (Table 2).

Table 2 Detailed Accuracy of classification models

Model	TP rate	FP rate	Precision	Recall	*F* Measure	ROC area
AA+ZeroR	1.000	1.000	0.872	1.000	0.9331	0.471
AA+Randomforest	1.000	1.000	1.000	1.000	1.000	1.000
JC+ZeroR	1.000	1.000	0.977	1.000	0.988	0.396
JC+Random forest	1.000	1.000	1.000	1.000	1.000	1.000
PA+ ZeroR	1.000	1.000	0.810	1.000	0.895	0.478
PA+Random forest	1.000	1.000	1.000	1.000	1.000	1.000

5 Conclusion

Edge prediction in online social networks is an important instance of the online social network analysis for drawing decisive knowledge used in various sorts of expert and recommendation systems. Approaches for edge prediction are emphasizing the structural features and other significant features. Based on similarity these features would be computed, assuming that connection between two similar nodes across the nodes would most likely be established instead of unsimilar nodes. In order to experiment it, we have built classifiers using ZeroR and Random forest classification techniques which are applied over online social networks like Wikipedia. Moreover, the outcome is evaluated using performance parameter precision, F measure and recall. The trained and tested models are compared with each other and it has been observed that Random forest algorithm has superseded ZeroR algorithm in terms of accuracy. Multiple classifiers can also be combined to produce a combined global model but that is not our focus.

References

1. Kautz H, Selman B, Shah M (1997) Referral web: combining social networks and collaborative filtering. Commun ACM
2. Chen BJ, Geyer W, Dugan C, Muller M, Guy I (2009) Make new friends, but keep the old: recommending people on social networking sites. In: Proceedings of the 27th international conference on Human factors in computing systems, ser. CHI 2009. ACM, New York, NY, USA, pp 201–210
3. Airoldi EM, Blei DM, Fienberg SE, Xing EP (2006) Mixed membership stochastic block models for relational data with application to protein-protein interactions. In: Proceedings of international biometric society-ENAR annual meetings
4. Huang Z, Li X, Chen H (2005) Link prediction approach to collaborative filtering. In: Proceedings of the 5th ACM/IEEE-CS joint conference on digital libraries
5. Hasan MA, Chaoji V, Salem S, Zaki M (2006) Link prediction using supervised learning. In: SDM workshop of link analysis, counter terrorism and security
6. Liben-Nowell D, Kleinberg J (2007) The link prediction problem for social networks. J Am Soc Inform Sci Technol, 1019–103
7. Han J, Kamber M (2009) Data mining concepts and techniques. Morgan Kaufmann Publishers an imprint of Elsevier
8. Hagberg AA, Schult DA, Swart PJ (2008) Exploring network structure, dynamics, and function using networkx. In: Proceedings of the 7th python in science conference (SciPy 2008).
9. Witten I, Frank E (2005) Data mining: Practice machine learning tools and techniques. Morgan Kaufmann, San Francisco
10. Lin D (1998) An information-theoretic definition of similarity. In: ICML. pp. 296–304

A Novel Approach to Human–Computer Interaction Using Hand Gesture Recognition

Akshay Sachdeva and Aniket Mohan

Abstract As computers become progressively inescapable in the public arena, encouraging characteristic human–computer interaction (HCI) will positively affect their utilization. Henceforth, there has been developing enthusiasm for the improvement of new methodologies and innovations for bridging this barrier. In this project, a novel approach has been presented toward hand gesture recognition using smartphone sensor reading, which can be applied to interact with your personal computers. The method presented collects data from smartphone sensors and based on the sensor data (accelerometer and gyroscope), classifies the data using deep learning algorithms. Furthermore, once the gesture has been accurately classified, the gesture is mapped to an action. This paper also provides an analysis of a comparative study done for this area.

Keywords Hand gesture · Sensor data · Deep learning · Smartphone · Human–computer interaction (HCI) · Deep convolution neural network · Application (app)

1 Introduction

The use of hand gestures provides an attractive alternative to cumbersome interface devices for human–computer interaction (HCI). This has motivated an active research area in computer vision-based analysis and interpretation of hand gestures. HCI provides a communicative and natural way of communication between people and machines. Its easiness and intuitiveness have brought forth in investigating enormous complex information, and numerous applications like computer games, virtual reality, medicinal services, and so forth. An efficient tool for capturing hand

A. Sachdeva (✉) · A. Mohan
Institute of Systems Science, National University of Singapore, Singapore 119615, Singapore
e-mail: akshay97sachdeva@gmail.com

A. Mohan
e-mail: E0402022@u.nus.edu

D. S. Jat et al. (eds.), *Data Science and Security*, Lecture Notes in Networks and Systems 132, https://doi.org/10.1007/978-981-15-5309-7_2

gesture are magnetic sensing or electromechanical devices. These strategies make use of sensors connected to the glove that transforms finger flexions into electrical signs to decide the gesture. They convey a thorough, application-free arrangement of constant estimations of the hand. In any case, they have a few disadvantages (1) pricey for easygoing use, (2) prevent the instinctive nature of hand gesture, and (3) they require complex adjustment furthermore, and setup methodology to acquire precise estimations.

Nowadays smartphones are fitted with small different sensors such as Proximity sensor, Barometer Sensor, Accelerometer, Magnetometer, Temperature Sensor, so on. The method has been proposed to achieve this interaction by using machine learning to classify sensor data (accelerometer and gyroscope) and map the same to an action.

The paper is organized as follows. The next section includes the literature survey related to gesture recognition and Neural Networks. Section 3 elaborates on the Proposed Approach that includes data collection and preprocessing. In Sect. 4 provides the implementation details. Section 5 includes the details on approaches to classify the data. Section 6 provides the details on model performance and finally, Sect. 7 includes a discussion of our approach and conclusions.

2 Related Work

Hand gesture recognition has attracted numerous research efforts in the past decade. In this section, the goal is to present inspiration and motivation for works that aim to understand hand gestures, i.e., aim to recognize the context of this work. In [1], a deep learning framework is used to recognize hand gestures robustly. Yi et al. [2] propose a system that collects, displays sensor data using an Android smartphone. A more effective and convenient mechanism was constructed in comparison with other methods. Mylonas et al. [3] examine the feasibility of ad hoc data acquisition from smartphone sensors by implementing a device agent for their collection in Android, as well as a protocol for their transfer. Ren et al. [4] compare the performance between speed and accuracy by obtaining a difference in results between Finger-Earth Mover's Distance (FEMD) and a corresponding-based shape matching algorithm. Raptis et al. [5] present a real-time gesture classification system based on skeletal wireframe motion. Ronao and Cho [6] propose a deep convolutional neural network to perform efficient and effective HAR using smartphone sensors. Wenchao and Zhaozheng [7] assemble signal sequences of accelerometers and gyroscopes into a novel activity image. Strezoski et al. [8] conduct experiments with different types of convolutional neural networks. Gerhard [9] reviews the recent techniques in the area of user modeling research in human–computer interaction. Fragopanagos and Taylor [10] outline a neural network approach to construct an emotion recognizing system. Ibraheem and Khan [11] discuss researches done in the area of HCI based on Artificial Neural Network approaches. Guo et al. [12] provide coverage of the top approaches of semantic segmentation using deep neural network and summarize the strengths, weaknesses, and major challenges. Hasan and Abdul-Kareem [13] present a novel technique for hand gesture based on shape analysis. Chakraborty et al. [14] survey major constraints on vision-based gesture recognition. Kim et al. [15] propose an efficient hand gesture recognition algorithm, using real-time learning for different HCI applications.

3 Proposed Approach

In Fig. 1, the proposed approach has been provided. The approach has been split into two phases, i.e., Training and Testing phase. In the training phase, the sensor data has been recorded using a mobile sensor and then preprocessing on the data is applied, which has been further explained in Sects. 3.2 and 3.2.

In the testing phase, the mobile sensor is transferred through an application which has been built using AngularJS. The Python script continuously captures the values, received by the NodeJS server running on the computer system. The model built then classifies and detects the hand gestures and performs the operations such as controlling keyboard keys. These operations can be modified as per user needs.

3.1 Data Collection

The first step of the approach is to obtain the data itself. For the same, the data has been obtained from smartphone sensors (accelerometer and gyroscope) which has been saved in the form of .csv files. To record data, an open-source android app 'Sensor Record' which provides an easy interface to record sensor data when hand movements are done has been used (Figs. 2 and 3).

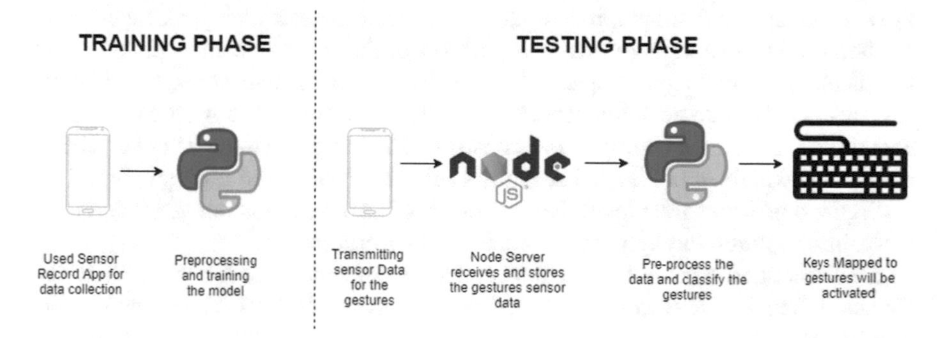

Fig. 1 Architecture diagram

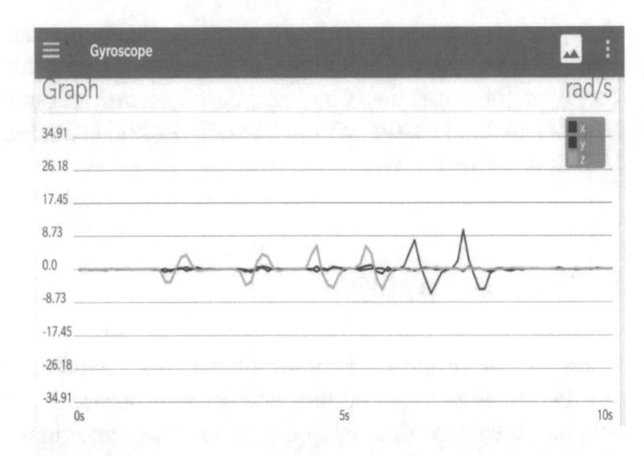

Fig. 2 Sensor data of gyroscope

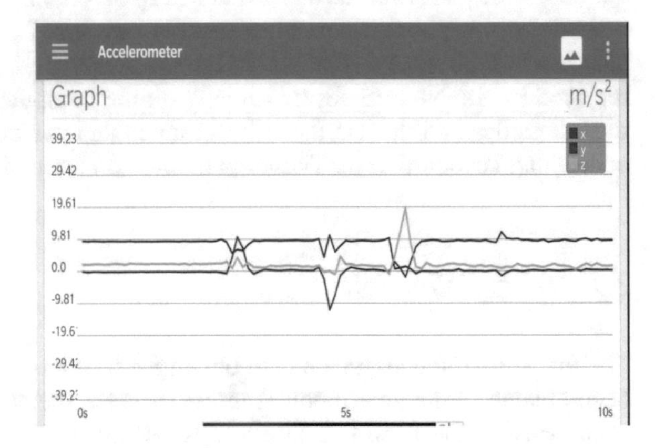

Fig. 3 Sensor data of accelerometer

3.2 Data Preprocessing and Training of the Model

Post obtaining the data, preprocessing of the accelerometer and gyroscope values fetched from the smartphone sensor is performed. For every x, y, and z parameter of the accelerometer and gyroscope, features such as mean, median, standard deviation, and median absolute deviation are obtained. The movement has been captured for 750 times for each hand gesture and consist of 36 features in total. These attributes are further explained in Fig. 4. The dataset split for training and testing is 3:1.

Figure 4 provides the model features or list of attributes that are generated from the values captured through accelerometer and gyroscope.

The generated data is trained further for classification models such as KNN, Decision Tree, Naïve Bayes, and Deep Neural Network (DNN). The trained model is used further as an input to the next step. Deep learning methodologies have emerged as a prospect for achieving these aims with their ability to capture syntactic as well as semantic features of text.

Fig. 4 Model features

acc_x_mean	Mean of accelermeter x coordinates
acc_y_mean	Mean of accelermeter y coordinates
acc_z_mean	Mean of accelermeter z coordinates
gyro_x_mean	Mean of gyroscope x coordinates
gyro_y_mean	Mean of gyroscope y coordinates
gyro_z_mean	Mean of gyroscope z coordinates
acc_x_max	Maximum of accelerometer x coordinates
acc_y_max	Maximum of accelerometer y coordinates
acc_z_max	Maximum of accelerometer z coordinates
gyro_x_max	Maximum of gyroscope x coordinates
gyro_y_max	Maximum of gyroscope y coordinates
gyro_z_max	Maximum of gyroscope z coordinates
acc_x_min	Minimum of accelerometer x coordinates
acc_y_min	Minimum of accelerometer y coordinates
acc_z_min	Minimum of accelerometer z coordinates
gyro_x_min	Minimum of gyroscope x coordinates
gyro_y_min	Minimum of gyroscope y coordinates
gyro_z_min	Minimum of gyroscope z coordinates
acc_x_mad	Mean Absolute Deviation of accelerometer x coordinates
acc_y_mad	Mean Absolute Deviation of accelerometer y coordinates
acc_z_mad	Mean Absolute Deviation of accelerometer z coordinates
gyro_x_mad	Mean Absolute Deviation of gyroscope x coordinates
gyro_y_mad	Mean Absolute Deviation of gyroscope y coordinates
gyro_z_mad	Mean Absolute Deviation of gyroscope z coordinates
acc_x_std	Standard Deviation of accelerometer x coordinates
acc_y_std	Standard Deviation of accelerometer y coordinates
acc_z_std	Standard Deviation of accelerometer z coordinates
gyro_x_std	Standard Deviation of gyroscope x coordinates
gyro_y_std	Standard Deviation of gyroscope y coordinates
gyro_z_std	Standard Deviation of gyroscope z coordinates
acc_x_y_corr	Correlation of accelerometer x and y coordinates
acc_y_z_corr	Correlation of accelerometer y and z coordinates
acc_z_x_corr	Correlation of accelerometer z and x coordinates
gyro_x_y_corr	Correlation of gyroscope x and y coordinates
gyro_y_z_corr	Correlation of gyroscope y and z coordinates
gyro_x_z_corr	Correlation of gyroscope x and z coordinates
Label	Gesture 1 or 0

The model uses Dropout of 0.5 to prevent overfitting, Binary Cross-Entropy as Loss function and Relu as an activation function for hidden layers, and Sigmoid for the output layer. The metrics used to measure the performance is ‘accuracy’.

Figure 5 provides the architecture of the two-layer dense sequential model, which has been used in the model. The neurons are grouped into three different layers—Input Layer, Hidden Layer, and Output Layer.

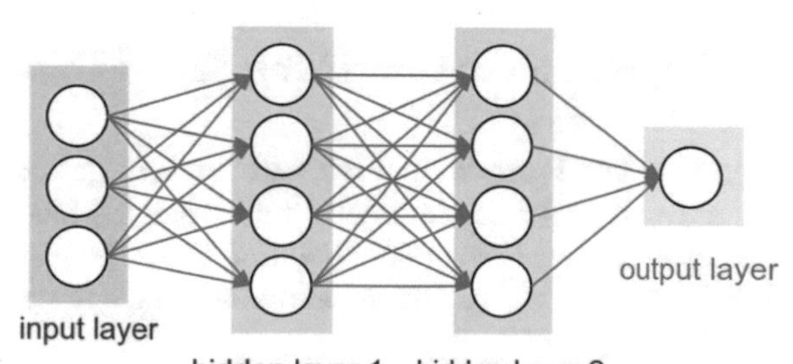

Fig. 5 Two-layer dense neural network

4 Implementation

4.1 Mobile Application

A mobile application was developed to listen for accelerometer and gyroscope values from a smartphone in which it is installed and relay to any server using the HTTP REST POST request. The app was built using React Native to reduce development time because it can be compiled to run on both android and iOS devices. It can either be a local system or on a cloud server as well. The mobile app has a simple user interface. After entering the server address, click on the big black button, perform the gesture, and release the button in order to send the data to the server.

4.2 Node.js Server

The node server built is a standard node.js server application that can run on both local and cloud servers. It uses expressJS web framework to manage the APIs and HTTP requests. The server has 2 endpoints that receive GET requests.

4.3 Python Client

The above two processes only handle relaying the data from a phone to the system. To map the gestures to actions, the Python client file needs to be run. The node.js server must be running before the client.py file can be run. The sensor values are then preprocessed, and new features are extracted (36 new features) using statistical combinations of x, y, z axes of accelerometer and gyroscope values. The processes data frame is then fitted on to the gesture classification model to be classified.

Figure 6 provides the view of the app before it is connected to the server and the gesture is captured. Once the app is connected to the server, as displayed in Fig. 7, the gestured can be performed and mapped.

5 Approaches to Classifying Sensor Data

Figures 8 and 9 display the left and right tilt gestures, respectively. The red line corresponds to the accelerometer values and the blue line to the gyroscope values. The gestures were recorded a hundred times for each gesture and stored it in csv files for preprocessing. On obtaining the data, a few approaches were researched to classify the gestures.

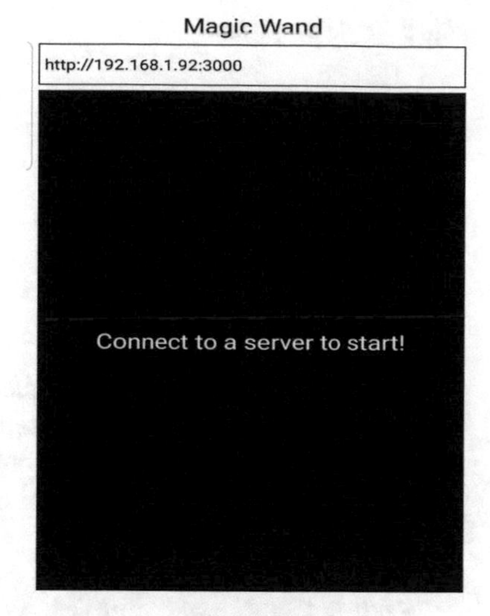

Fig. 6 IP address on the top

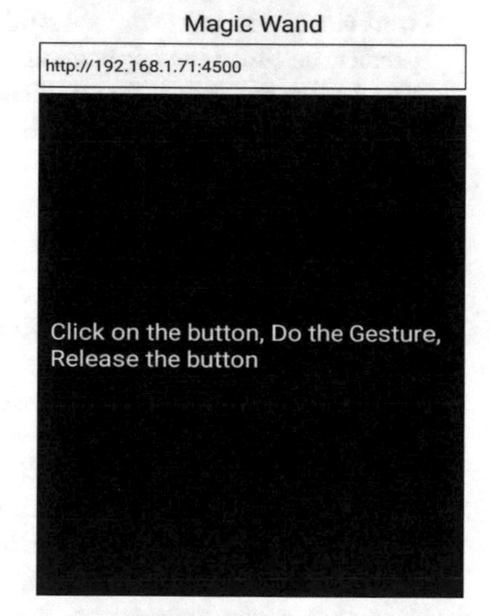

Fig. 7 Gesture and data recording

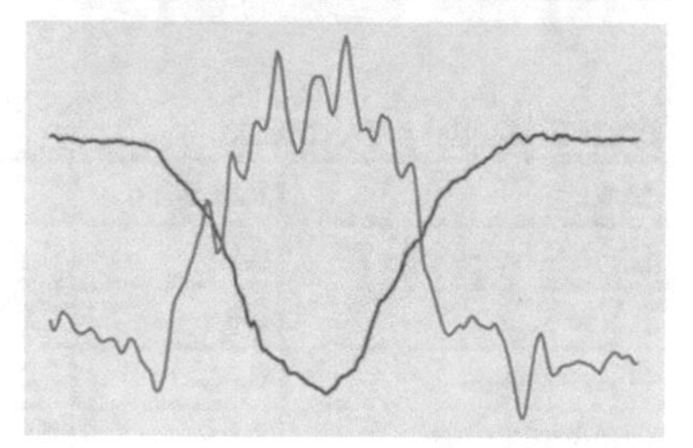

Fig. 8 Forward tilt gesture

Fig. 9 Right tilt gesture

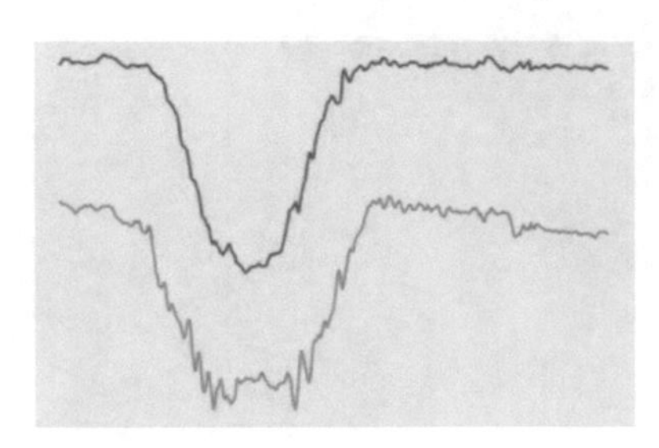

1. The approach used is the one mentioned earlier, i.e., after receiving the sensor data, new features were extracted using combinatorial statistical operations on the x, y, z axes of the accelerometer and gyroscope values to obtain 36 new columns. A 3-layer deep neural network was used to classify, and this gave us good results.
2. The other approach was to plot the gestures and save them as graph images and apply a CNN-based approach to classifying the gestures. The models worked well but it was not taken forward as it took more processing time for model prediction. Since the application was real time, there could not be any process that took more than a few milliseconds to process. Hence, this approach was discarded.

6 Model Performance

Table 1 depicts the result values obtained after generating the different models for gesture recognition.

It can be inferred that the best results have been obtained through the deep neural network model as compared to other classification models of mapping gesture to actions. The gesture system for DNN responded appropriately in over 90% of all test cases and achieved an overall classification success rate of 99% while using the model during an active gameplay environment. While there is still work to be done, the base system described here provides foundational capabilities for the design and implementation of more robust gesture recognition systems.

Table 1 Model performance

Model	Precision	Recall	F1-score	Area under curve	Accuracy
Logistic regression	0.84	0.90	0.87	0.8219	79.98
KNN	0.85	0.80	0.84	0.8092	78.02
Naïve Bayes	0.88	0.78	0.83	0.7929	76.06
Random forest	0.83	0.92	0.89	0.8512	82.55
Deep neural network	0.89	0.99	0.97	0.9985	98.90

7 Conclusions

The project has satisfied several substantive requirements and met challenges in the areas of sensor data and machine learning. This novel approach describes the implementation and testing of an accelerometer-based gesture recognition system applied to a real-time application. The principal outcome of this work is the use of accelerometers and gyroscope to define gesture windows and the classification of accelerometer and gyroscope data to achieve gesture recognition of arm movements. This application requires sensing and classifying a wide variety of arm and hand movements. Future implementations can leverage these capabilities to meet evolving requirements for natural and useful human–computer interaction.

References

1. Mohanty A, Rambhatla SS, Sahay RR Deep gesture: static hand gesture recognition using CNN. Adv Intell Syst Comput 460. https://doi.org/10.1007/978-981-10-2107-7_41
2. Yi J-W, Jia W, Saniie J (2012) Mobile sensor data collector using android smartphone. IEEE 978-1-4673-2527-1
3. Mylonas A et al (2013) Smartphone sensor data as digital evidence. Comput Secur. https://doi.org/10.1016/j.cose.2013.03.007
4. Ren Z, et al Depth camera based hand gesture recognition and its applications in human–computer-interaction. IEEE 978-1-4577-0031-6
5. Raptis M, Kirovski D, Hoppes H (2011) Real-time classification of dance gestures from skeleton animation. In: Eurographics/ACM SIGGRAPH symposium on computer animation
6. Ronao CA, Cho S-B (2016) Expert systems with applications 59:235–244
7. Wenchao J, Zhaozheng Y (2015) ACM. ISBN 978-1-4503-3459-4/15/10
8. Strezoski G et al (2016) ICT innovations. Adv. Intell. Syst. Comput. 665
9. Fischer Gerhard (2001) User modeling in human–computer interaction. User Model User-Adap Inter 11:65–86
10. Fragopanagos N, Taylor JG (2005) Emotion recognition in human–computer interaction. Neural Networks 18:389–405
11. Ibraheem NA, Khan RZ (2012) Vision based gesture recognition using neural networks approaches: a review. Int J Hum Comput Inter (IJHCI) 3(1)
12. Guo Y, Liu Y, Georgiou T et al (2018) Int J Multimed Info Retr 7:87
13. Hasan H, Abdul-Kareem S (2014) Artif Intell Rev 41:147–181
14. Chakraborty BK, Sarma D, Bhuyan MK, MacDorman KF (2018) Review of constraints on vision-based gesture recognition for human– computer interaction. IET Comput Vision 12(1):3–15
15. Kim M, Cho J, Lee S, Jung Y (2019) IMU sensor-based hand gesture recognition for human–machine interfaces. Sensors 19:3827

Insider Threat Detection Based on Anomalous Behavior of User for Cybersecurity

Ujwala Sav and Ganesh Magar

Abstract In today's competitive world, business security is essential. To secure the business processes and confidential data, organizations have to protect the system by implementing new policies and techniques to detect the threats and control it. Threats for cybersecurity are classified into two types, outsider and insider threats. Both threats are very harmful to the organization. These may convert into a severe attack on the systems upon future. Outsider threats have to take more effort to break the security system. But inside users are those who are privileged to access the system within the organization. As data form is digital, it is straightforward to transfer from one to another. Data leakage, theft, and sharing are easy for insiders. Therefore, there is a need to research in this domain. In this proposed paper, a study of insider threat detection based on the anomalous behavior of the user for cybersecurity is presented. The data processing and anomaly detection algorithms are performed for insider threat detection by researchers. This research paper presented a study on insider threat detection based on the anomalous behavior of the user for cybersecurity.

Keywords Cybesecurity · Insider threats · Anomalous behavior · Machine learning · Data leakage · Bipartite graph

1 Introduction

Insider threat detection based on the anomalous behavior of the user is required to protect the business from cyberattacks. Therefore, there is a necessity to study the anomalous behavior of users. The research paper processed the data using the machine learning algorithm and identified insider threats.

U. Sav (✉)
Vidyalankar School of Information Technology, Mumbai, 37, India
e-mail: ujwalasav@gmail.com

G. Magar
P.G. Depatment of Computer Science, S.N.D.T. Women's University, Mumbai, 49, India
e-mail: gmmagar@gmail.com

D. S. Jat et al. (eds.), *Data Science and Security*, Lecture Notes in Networks and Systems 132, https://doi.org/10.1007/978-981-15-5309-7_3

1.1 Insiders and Insider Threat

Insiders are the legal users, having trustworthy authorized access to the resources of an organization. These insiders have privileges to use and process the data. Usually, 60–70% insiders are doing their job honestly and giving their best. But, 30–40% insiders are not satisfied with their job profile, a task assigned, position, responsibilities, salary, the boss, infrastructure, coworkers, and so on. These insiders are a few but very dangerous if you do not identify them within the organization. Insiders are careless. There is a lack of awareness of security precautions. Also, orientation, induction, and training have to be given to protect the organization's confidential data.

There are many motivations for insider threats. Insider threat detection is a difficult task, as the insiders do not have to break the security layers. They are authorized to access the data. Reasons for insider threats are careless behavior, disgruntled user, the greed of users to earn more money, revenge attitude due to dissatisfaction with their job and salary, and lack of facilities availed in the organization. There are guest users—third-party users. Competitors may launched spy in your organization. Spy is working for competitors, they are leaking data, stealing data, and also sharing your important data with them.

1.2 Anomalous Behavior of User

Several methods can detect insider threats. One of the most effective ways is to monitor the anomalous behaviors of an insider. Anomalous behavior is like data outliers. Generally, 95% of insiders behave normally. Some of them are loyal do their work sincerely, and they are assets for the organization. They are dedicated ones. A few are not satisfied but work for a salary. There are very few whose behavior is anomalous, and they might be harmful to the organization badly; this impacts the image, profit, standard, and many more. Therefore, there is a requirement to monitor the behavior of the insiders and detect insider threats based on anomalous behavior analysis of insiders

2 Related Work

Business security is a significant issue in the digital era. In this digital world, it is difficult to protect the information as it is effortless to transfer from sender to receiver. Also, no checkpoint to check the confidential data transfer from the organization. With this concept, to stop data leakage and data sharing, some of the reviews have been included in the literature review.

IBAC stands for Intent Based Access Control which is recommended to detect the purposes of access based on current knowledge of purpose. It is possible to find out the risk on the basis of the study of user intention and motivation. This model helps to find out the threats based on intention, but still, there is a need to strengthen this model [1].

There is a methodology for insider threat detection based on rules of behavior of the user. This is known as BLITHE, which monitors data in the smart grid. It uses consciousness and accuracy [2].

It is observed that most of the research is based on post hoc personality and to find out the vulnerability of a person to find out the insider threats. This game-based method is used to find out the behavioral and personal differences of employees. This paper focused on facial expression, linguistic features, and personality variables [3].

Deep autoencoder and random forest classifiers are used to detect insider threats using time series activities of inside users. Data are classified and f source is computed and the result is compared with other classifiers [4].

CADS is a community anomaly detection system. Insider threats are detected using an unsupervised learning framework using logs of collaborative situations. It consists of two parts as relational pattern mining and anomaly detection, to secure organizations in wireless sensor grids, sensor nodes (SNs) must create secret collective keys with adjoining nodes [5].

There is growth in multimedia-based applications, and its usage is leading to multimedia traffic. In this paper, the hybrid deep learning algorithm for anomaly detection is proposed. It includes suspicious information of users on social media. It is evaluated by performing the experiments which are based on real time and datasets. This proves the effectiveness of anomaly detection [6].

It is a challenge to find out insider threats for the information security society. The progressing system is also functional to enhance classifier accurateness. Comprehensive representations are estimated through investigation of their connected confusion matrix along with the curve of the Receiver Operating Characteristic, and the greatest performance classifiers are combined into an ensemble classifier. "This meta-classifier has an accurateness of 96.2%, with an area below the ROC curve of 0.988". This shows the effectiveness of the confusion matrix [7].

Data leakage is an insider threat called as data leakage prevention (DLP). To quality, a low-complexity score walk algorithm is projected to control the final understanding [8]. The concept of isolation Forest enables subsampling activity to the degree, which is not there in present procedures. It generates an algorithm that needs a linear time complication with low constraints and small memory prerequisites for anomaly detection. iForest has several unrelated conditions where the working outset does not comprise any anomalies [9].

Predict the insider threats from huge and composite auditing data, and system findings can be used to outfit the anomaly detection with an ensemble of deep autoencoders. Each autoencoder in the cluster is trained using an exact group of audit data, which signifies a user's distinctive behavior accurately. Mathematical investigated using a level dataset for insider threat discovery. Outcomes that the detection system could distinguish all of the harmful insider activities with a practical false positive rate [10].

The survey work finds out three kinds of insiders: spy, masquerader, and unintended criminal. It reviews the precautionary actions from an information analytics viewpoint. When direct and indirect threats work are categorized as swarm, web, or related databases, it is possible to identify work. It is extracted from the engaged data and algorithm [11]. The finding method assesses how reliable an analyst is across dissimilar tasks and increases an awareness if any significant change is detected. A standardization process is hired, which permits us to associate across a group variation that is due to disagreeable actions [12].

In this paper, "verifying the user authorization and performing the user authentication together whenever a user accesses the devices for threat detection. The scheme computes each user role dynamically using attribute-based access control and verifies the individuality of the operator laterally with the device. Security and performance analysis display that the projected system struggles numerous insider as well as outsider threats and attacks" [13].

This research paper presents a risk of social media through insider threats in an organization. The approach is defined as Zero-Knowledge, which is leveraging the existing techniques to validate the evidence of social media without exposing organization confidential data [14].

No research is useful to identify an insider threat. Safety actions are required to find out breaches in prior. An investigation of malicious action stops it. Consequently, there is a significant theoretical shortfall in current cyber defense architecture [15]. In the relational database for classification, a framework is used for predicting and mitigating insider collusion. It monitors insider transactions, accesses, and inferences. This helps to reduce the possible collusion of attacks. This is also used to find out the collusion insider threat to reduce the potential attack [16].

3 Methodology

Insider's temporal log data is used for insider threat detection. This process helps identify the anomalous behavior of insider and its analysis is used to detect insider threats. The research flow for this process is as follows.

In the research flow, Fig. 1, firstly, insiders log data collection or data generation process is done. Once data is collected, data preprocessing, including removal of duplicate data, finding out the replacement of data, scaling, and standardization decision for data processing to work out are done. Now this preprocessed data is used to find out anomalous behavior of insiders by implementing a machine learning anomaly detection algorithm.

Fig. 1 Research flow

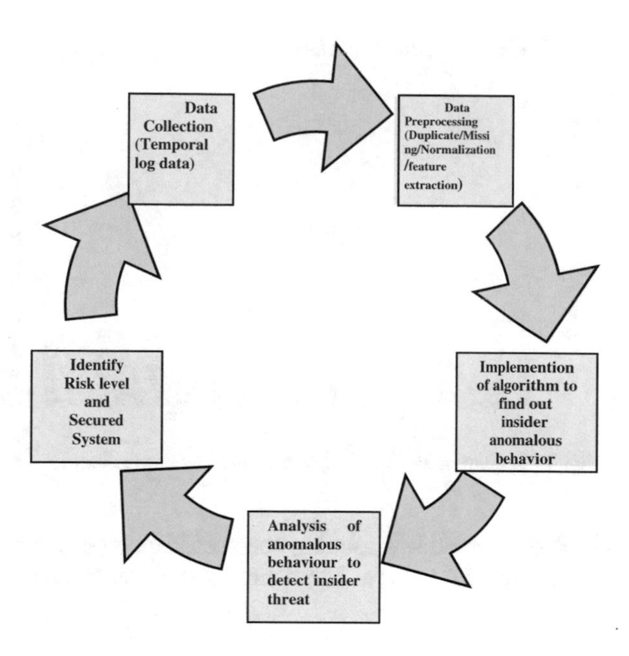

4 Data Processing and Algorithm Analysis

4.1 Data Processing

Insiders' temporal logon and logoff log data and organization user data are collected from Carnegie Mellon University. R3.2 dataset is used for this research paper. The total number of records is 8 lacs. This consists of the previous and current user data with features like userid, user_name, user email address, department, designation, date of birth, date of joining, experience, qualification, contact number, salary information, performance, achievements, physical address, id proof number, etc. All data scanned and extracted only are the required useful data for research. Remove the other data to focuse and save storage space and processing time. Therefore, only the following data features of 1000 users are considered for this paper: User_id, Role, Department, Role, Function, and Emailid (Fig. 2).

The organizational structure consists of various departments and a user-assigned role. The chart presents the total number of inside users per role. All the user data is combined and stored in one file. Check the records for missing values and remove the records as data size is large so that it will not affect the result. Insiders' daily logon and logoff temporal log data are collected. This log data consists of the following features: userid, role, date, activity (logon/logoff), machine, etc. Data are checked for missing values in each dataset. Features extracted are user, pc, and activity. The user and pc are nodes. The number of activities per pc and per users are the weighted edges. Prepare a separate logon and logoff files with the number of activity count. It

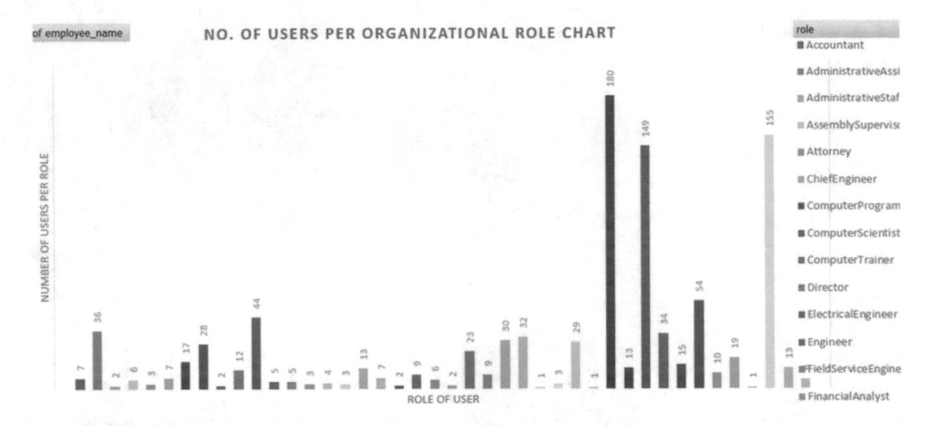

Fig. 2 Total number of users per organizational role chart

is also observed by grouping one user is logon or logoff to how many pcs and how many times? It helps to know more about user behavior in the organization.

4.2 Software Requirement

Jupyter Notebook with Python 3.7 and its required packages library are used for algorithm processing and visualization. Desktop GUI Anaconda Navigator, including distributions used to manage packages, environments, and channels is used.

4.3 Undirected Bipartite Graph Analysis

Undirected bipartite graph is constructed to show interrelationship between users and PCs. In this graph, user and pc are nodes, whereas edges represent the relationship between the users and pc. Edges' weight is the total number of logon/logoff activities. In Jupyter, it imported a bipartite graph from networks' algorithm for analysis.

G (U,V,E) is presented in user_node, pc_node, and activity_count as weighted_edges. The total number of records without missing values are 9662. This visualizes the normal and outlier data in the form of the graph.

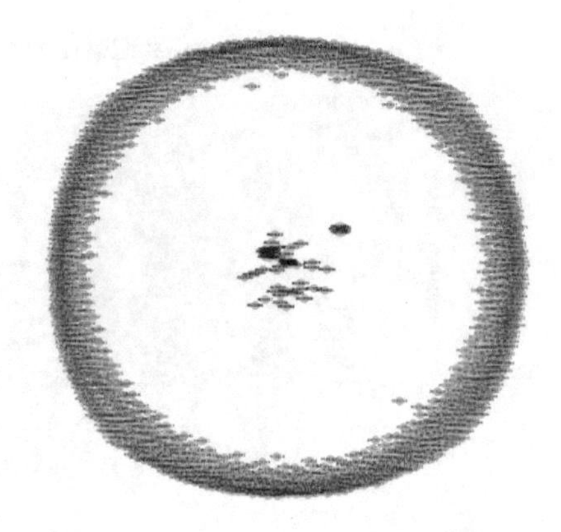

Fig. 3 Undirected bipartite graph to show a relationship between user and PC

(userid, pcid, activity_count)

```
[('AAB0754', 'PC-4470', 1),
 ('AAB0754', 'PC-5948', 345),
 ('AAB0754', 'PC-7496', 34), so on.
```

Bipartite Graph Construction

```
BG  =  nx.Graph()
BG.add_nodes_from(unodes, bipartite = 0) # users as nodes
BG.add_nodes_from(pcnodes, bipartite = 1) # PCs as nodes
BG.add_weighted_edges_from(weighted_edges)
```

The bipartite graph is shown in Fig. 3. This graph shows the behavior of data that the user connected to a pc.

4.4 Temporal Data Analysis

Insider's temporal data is processed and the minimum logon and maximum logon are found. It is further processed and the mean and mode of logon are found, and this helps to learn the machine normal behavior of the insider. Similarly, it is processed for logoff. The graphical representation of logon and logoff data are present in the following figures. The charts of temporal logon and logoff data analysis show the behavior of the insiders' logon and logoff activity, on how many pcs, and how many times (Figs. 4 and 5).

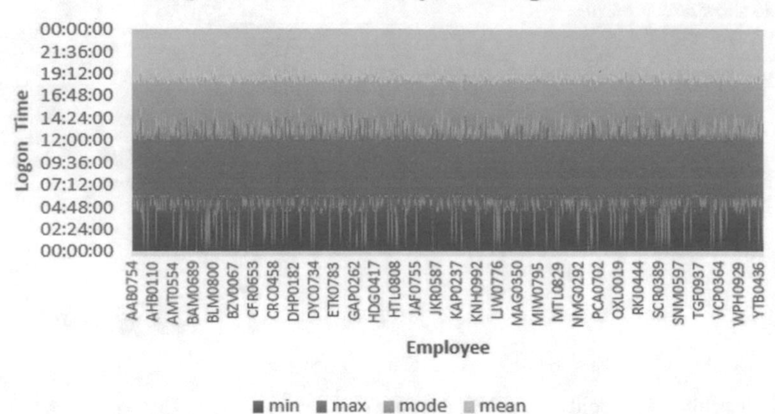

Fig. 4 Statistical data analysis for insiders' temporal logon data

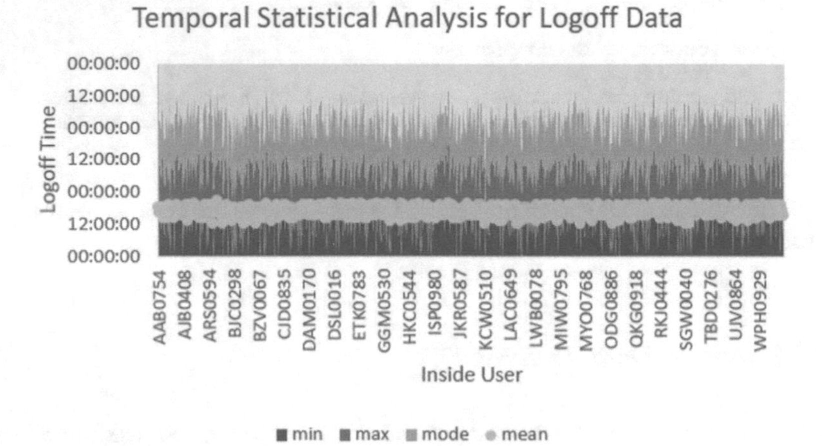

Fig. 5 Statistical data analysis for insiders' temporal logoff data

4.5 Model with Logon and Logoff Features for the Implementation of the Anomaly Detection Algorithm

In this Sklearn is used. Sklearn does not process the categorical or object parameters, it converts the time object into the numerical equivalent. A function is used to convert time to integer data. Combine the separated files of logon and logoff for the Isolation Forest input. The array is generated.

```
array([[31500, 71818, 32400, ..., 71818, 32400, 36000], [27900,
29580, 29580, ..., 29580, 29580, 25200], [30600, 31740, 31680,
..., 31740, 31680, 28800], ..., [30600, 66912, 31500, ...,
66912, 31500, 36000], [26100, 48905, 27000, ..., 48905, 27000,
28800], [ 2173, 84195, 27000, ..., 84195, 27000, 32400]],
dtype=int64)
```

Similarly, device connection and disconnection are also processed.

```
# Anomaly score
All_params_ascore =
forest.decision_function(All_params_input)
All_params_ascore
All_params_result.loc[All_params_result['ascore'] < 0]
# possible outliers
```

User	Ascore	
0	AAB0754	−0.044789
9	ABW0466	−0.010166

After processing this data, we have see 289 records of behavior is anonymous. It is also checked with K-Nearest Neighbors KNN, Histogram Base Outlier Detection HBOS, and Local Outliers Factor LOF algorithm using the PyOD library. Figure 6 shows anomaly visualization using KNN which is followed by (Fig. 7).

PyOD means Python outlier detection algorithm. It is also known as anomalous behavior detection algorithm. Algorithms are giving an outlying score. It will be compared to an internal threshold to determine the final outlier. Therefore, specific algorithms for anomaly detection are used.

Angle-based Outlier Detection algorithm shows the relationship between neighbors. The variance of cosine shows anomalous behavior. It is useful for high-dimensional data. K-Nearest Neighbor uses a mathematical classification algorithm.

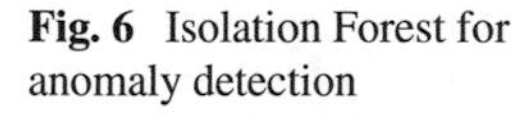

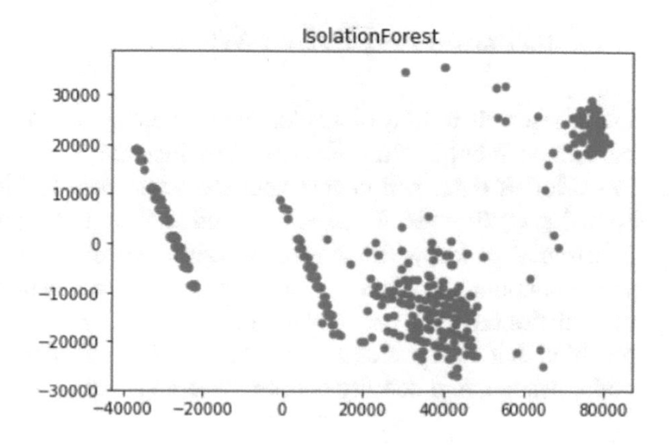

Fig. 6 Isolation Forest for anomaly detection

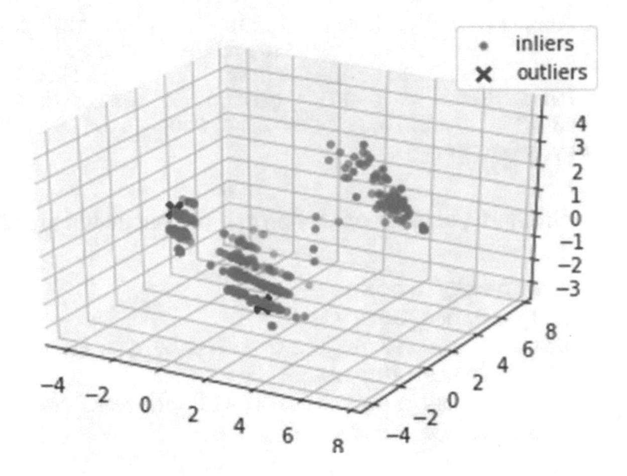

Fig. 7 All parameters for KNN model

PyOD helps KNN which includes largest, mean, and median detectors as anomaly scores.

Isolation Forest is developed on random forest and decision trees. There is a random selection of root, imagine outliers are close to the source. An outlier is identified with less number of partitions. Local Correlation Integral (LOCI) is valid for the detection of a group of an outlier.

4.6 Risk Control Model with Logon and Logoff Data

In this model, observed anomalous behavior is analyzed, and on the basis of severity of the risk, action will be taken by the organizational security executive. Detected anomaly is reported to the security department for further action. The risk control model will develop after all parameters for anomalous behavior study.

5 Conclusion and Future Work

In this research, a study of insider threat detection based on the anomalous behavior is performed. It helps to detect and control insider threats. Specifically, users, temporal logon/logoff data, and device data are processed. Insiders' logon and logoff data behavior is observed for each user and analyzed. An undirected bipartite graph is constructed to know the interrelationship between user and pc. In statistical analysis, minimum, maximum, mean, and mode are computed for given data. The values of statistics helped to understand the normal behavior of each user, and trained the machine to identify anomalous behavior of the insider. In the further part, the user node, pcnode, and activity count for weighted edge features used. In the algorithm implementation, Jupyter and Sklearn with Python are used for PyOD anomaly detection algorithms like KNN, LOF, HBOS, PCA (Principal Component Analysis). The isolation forest algorithm processes and finds out the anomalies in the data. A general model is developed on temporal data to test the anomaly by finding ascores.

The risk control architecture designed to manage insider threats is based on the anomalous behavior of the user. The risk is characterized into three levels: high, medium, and low level. Insiders' log data monitor and identify anomalous behavior.

In the future work, a different type of dataset will be used to check the accuracy and performance of the model using advanced few deep learning along with machine learning algorithms. The algorithms' implementation and analysis will be processed for future study of this research.

Acknowledgments I wish to acknowledge the Software Engineering Institute of Carnegie Mellon University and Exact Data LLC for making available data for the research study.

References

1. Almehmadi A, El-Khatib K (2017) On the Possibility of insider threat prevention using intent-based access control (IBAC). IEEE Syst J 11:373–384
2. Bao H, Lu R, Li B, Deng R (2016) BLITHE: behavior rule-based insider threat detection for smart grid. IEEE Internet Things J 3:190–205
3. Basu S, Victoria Chua YH, Wah Lee M, Lim WG, Maszczyk T, Guo Z, Dauwels J (2018) Towards a data-driven behavioral approach to prediction of insider-threat. In: 2018 IEEE international conference on big data (big data. IEEE, Seattle, WA, USA), pp 4994–5001. https://doi.org/10.1109/BigData.2018.8622529
4. Chattopadhyay P, Wang L, Tan Y-P (2018) Scenario-based insider threat detection from cyber activities. IEEE Trans Comput Soc Syst 5:660–675
5. Choi J, Bang J, Kim L, Ahn M, Kwon T (2017) Location-based key management strong against insider threats in wireless sensor networks. IEEE Syst J 11:494–502
6. Garg S, Kaur K, Kumar N, Rodrigues JJPC (2019) Hybrid deep-learning-based anomaly detection scheme for suspicious flow detection in SDN: a social multimedia perspective. IEEE Trans Multimedia 21:566–578. https://doi.org/10.1109/TMM.2019.2893549
7. Hall AJ, Pitropakis N, Buchanan WJ, Moradpoor N (2018) Predicting malicious insider threat scenarios using organizational data and a heterogeneous stack-classifier. In: 2018 IEEE international conference on big data (big data). IEEE, Seattle, WA, USA, pp 5034–5039
8. Huang X, Lu Y, Li D, Ma M (2018) A novel mechanism for fast detection of transformed data leakage. IEEE Access 6:35926–35936
9. Liu FT, Ting KM, Zhou Z-H (2008) Isolation forest. In: 2008 eighth IEEE international conference on data mining. IEEE, Pisa, Italy, pp 413–422
10. Liu L, De Vel O, Chen C, Zhang J, Xiang Y (2018) Anomaly-based insider threat detection using deep autoencoders. In: 2018 IEEE international conference on data mining workshops (ICDMW). IEEE, Singapore, Singapore, pp 39–48
11. Liu L, De Vel O, Han Q-L, Zhang J, Xiang Y (2018) Detecting and preventing cyber insider threats: a survey. IEEE Commun Surv Tutor 20:1397–1417
12. Santos E, Nguyen H, Yu F, Kim KJ, Li D, Wilkinson JT, Olson A, Russell J, Clark B (2012) Intelligence analyses and the insider threat. IEEE Trans Syst Man Cybern. - Part Syst Hum 42:331–347. https://doi.org/10.1109/TSMCA.2011.2162500

13. Saxena N, Choi BJ, Lu R (2016) Authentication and authorization scheme for various user roles and devices in smart grid. IEEE Trans Inf Forensics Secur 11:907–921
14. Smith TD (2018) Countering inside threat actors in algorithm-based media. In: 2018 IEEE international conference on big data (big data). IEEE, Seattle, WA, USA, pp 4453–4459. https://doi.org/10.1109/BigData.2018.8621940
15. Walker-Roberts S, Hammoudeh M, Dehghantanha A (2018) A systematic review of the availability and efficacy of countermeasures to internal threats in healthcare critical infrastructure. IEEE Access 6:25167–25177
16. Yaseen Q, Alabdulrazzaq A, Albalas F (2019) A framework for insider collusion threat prediction and mitigation in relational databases. In: 2019 IEEE 9th annual computing and communication workshop and conference (CCWC. IEEE, Las Vegas, NV, USA), pp 0721–0727. https://doi.org/10.1109/CCWC.2019.8666582

A Comparative Study of Text Mining Algorithms for Anomaly Detection in Online Social Networks

Sujatha Arun Kokatnoor and Balachandran Krishnan

Abstract Text mining is a process by which information and patterns are extracted from textual data. Online Social Networks, which have attracted immense attention in recent years, produces enormous text data related to the human behaviours based on their interactions with each other. This data is intrinsically unstructured and ambiguous in nature. The data involves incorrect spellings and inaccurate grammars leading to lexical, syntactic and semantic ambiguities. This causes wrong analysis and inappropriate pattern identification. Various Text Mining approaches are being used by researchers which can help in Anomaly Detection through Topic Modeling, identification of Trending Topics, Hate Speeches and evolution of the communities in Online Social Networks. In this paper, a comparative analysis of the performance of four classification algorithms, Support Vector Machine (SVM), Rocchio, Decision Trees and K-Nearest Neighbour (KNN) for a Twitter data set is presented. The experimental study revealed that SVM outperforms better than other classifiers, and also classifies the dataset into anomalous and non-anomalous user's opinions.

Keywords Anomaly detection · Social media · Rocchio algorithm · Decision trees · K-Nearest neighbour (KNN) and Support Vector Machine (SVM) · Kernel function · Gini index · Entropy and Euclidean distance

1 Introduction

Individuals express their voices, opinions and views on Online Social Media like Facebook and Twitter. This, in turn, influences their behaviours. The opinions and views expressed in online social media can be analysed and can be used in discovering patterns and future predictions. Networking through social media solves and enhances coordination problems with people residing in different geographical locations [1]. Thereby increasing the efficiency in social campaigning by publicizing the

S. A. Kokatnoor (✉) · B. Krishnan
Department of Computer Science and Engineering, CHRIST (Deemed to be University), Bangalore 560074, India
e-mail: sujatha.ak@christuniversity.in

D. S. Jat et al. (eds.), *Data Science and Security*, Lecture Notes in Networks and Systems 132, https://doi.org/10.1007/978-981-15-5309-7_4

necessary data anytime and in anyplace. On the other hand, people frequently use unstructured sentences, inaccurate grammars and incorrect spellings for searching a particular topic, or for posting their views and comments or communication through various discussion forums. This leads to ambiguities such as lexical, syntactic and semantic. Hence, it becomes a critical task to extract accurate information and discover logical patterns from such unstructured data. The solution to the problem addressed above is Text Mining.

Text Mining is the trending field that encompasses several research fields like Natural language Processing (NLP), Information Retrieval (IR), Computational Linguistics (CL), Information Extraction (IE) and Data Mining (DM). Text Mining is an expansion of Data Mining. Data Mining handles structured and organized data like databases or ERP frameworks whereas Text Mining deals with unstructured text data like user's comments, tweets and posts in social media and reviews for products or brands for online shopping websites. Both Data Mining and Text Mining use a wide variety of functions to transform presented data to valued intuitions and knowledge [2].

Text mining involves unstructured data and has to undergo several processes before proceeding to anomaly detection applications. The processes include Preprocessing, Text Transformation, Feature Selection, Pattern Identification and then Interpretation or Evaluation. Preprocessing involves tokenization, filtering and stemming process. Text mining techniques do not consider the text structure and uses Term Frequency–Inverse Document Frequency (TF-IDF) or Word2Vector model for Text Transformation. Feature selection process carefully selects the relevant attributes without affecting the performance of the model chosen and finally classification and clustering algorithms for discovery of the patterns and its predictions.

Text Mining is used to find a general opinion of the subject concerned, human behaviours and thinking patterns and community identification [3]. In most of the research works on Text Mining, the authenticated investigation and analysis of different Text Mining approaches is lacking. So this paper focuses on a few Text Mining Techniques for unstructured data from Online Social Media.

The paper is organized as follows. Section 2 features related work. Section 3 outlines the diverse Text Mining procedures. Section 4 presents comparison results and lastly the Conclusion is given in Sect. 5.

2 Related Work

The work focusses on Supervised Machine Learning Approaches employed in the classification of Social Media text data. In supervised machine learning approach, the output variable y is mapped to a function of an input variable x as given in Eq. (1). The main objective of the mapping function is to predict the value of y accurately given a new input data x.

$$y = f(x) \tag{1}$$

First a labelled training dataset is used for training the model in the supervised machine learning approaches. Later a test dataset comprising of unlabelled data is fed into the trained model for the classification into similar groups [4]. It is termed as supervised learning as a model is inferred from the labelled dataset, a dataset with input and the known output. Decision Trees, KNN, SVM and Rocchio are few examples of supervised machine learning approaches [5].

As per Nagarju Kolla et al. [6], supervised machine learning algorithms including Decision Tree, KNN, Logistic Regression, Naïve Bayes and SVM can be used for prediction of brand loyalty. Shahadat Uddin et al. [7] have identified behavioural anomalies and have predicted diseases using Logistic Regression, SVM, Decision Tree, Random Forest, Naïve Bayes, KNN and Artificial Neural Networks. Of all the algorithms compared SVM and Naïve Bayes have yielded good results. Adam Tsakalidis et al., have used Linear Regression for the Prediction of 2014 European Union elections in three different countries using Twitter and polls [8].

In this paper, an overview of SVM, Rocchio, Decision Trees and K-Nearest Neighbour algorithms are discussed.

3 Supervised Machine Learning Approaches

Learning from previous experiences is a human characteristic whereas computers are not able to do so. The main aim of this paper is to learn a target that can be used to estimate the performance of a class while learning using supervised machine learning approach. Figure 1 represents the broad categories of supervised classification algorithms as discussed in this paper.

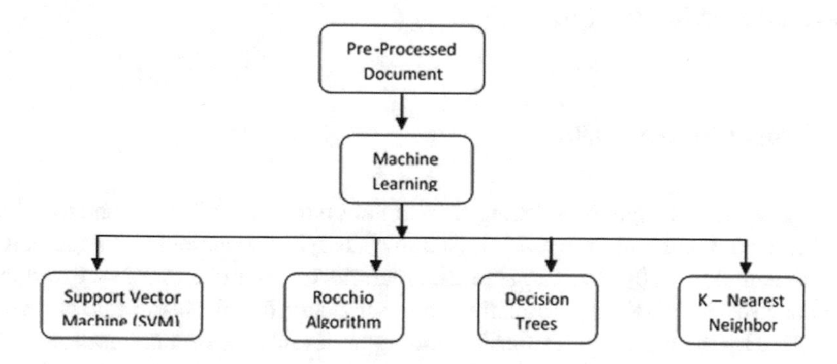

Fig. 1 Text mining using supervised machine learning algorithms

3.1 Support Vector Machine (SVM)

SVMs are supervised machine learning algorithms used for classification, regression and anomaly detection. For a dataset comprising of labels set and features set, a SVM classifier constructs a model for predicting classes for new data points. It appoints new data point/information by focusing it on one of the classes. On the off chance that there are just 2 classes, at that point it very well may be called as a Binary SVM Classifier [9]. Two variations of SVM Classifiers exists. They are Linear and Non-linear SVM Classifier. The majority of real-world issues contain non-separable data for which there are no hyperplanes that distinguish positive from negative instances in the given training dataset. One good solution is to map the data to a higher dimension space and then create a separating hyper-plane. This new high-dimensional space is also called as the transformed feature space. An appropriate kernel function is important for better classification. Some of the kernels are listed from Eqs. (2)–(5):

- Linear Kernel : $K\,(W,\ X)\ =\ W^T X$ (2)

- Polynomial Kernel : $K\,(W,\ X) = (\alpha W^T X + c)^n,\ \alpha > 0$ (3)

- Radial Basis Function (RBF) : $K\,(W, X) = \exp(-\alpha\,|W - X|^2),\ \ \alpha > 0$ (4)

- Sigmoid Kernel : $K(W,\ X) = \tanh(\alpha W^T X + c),\ \ \alpha > 0$ (5)

Here α, c and n are the kernel parameters. Where W is a training vector and is mapped into a higher dimensional space by the function ϕ and $K\,(W, X) \equiv \phi(X)$, is known as kernel function [9].

3.2 Rocchio Algorithm

It is a model that recognizes indexing, a formulation for search request and matching the requested document. It is mainly used in classifying documents into anomalous or non-anomalous [10]. Rocchio algorithm permits the user to enhance the system's efficiency by iteratively reformulating the user query based on the relevant feedback given by the user. The feedback is through relevant assessment questions. Text document comprises Synonymy—different words which have the same meanings in a natural language. This can be dealt with by re-characterizing the solicitation by utilizing the relevance feedback method. In this strategy, the user gives input that indicates important material concerning the exact subject of a specific space. The initial results are generated based on the query asked by the user. The obtained results are marked relevant or irrelevant and the same is communicated back. Hence, this technique is an incremental and iterative process based on the user's feedback and plays a dynamic part by delivering applicable outcomes that a user needs [11].

As Rocchio algorithm practices one model vector for demonstrating a class, it gives fewer prediction accurateness [12]. So according to the author, KNN (K-Nearest Neighbour) algorithm along with Rocchio algorithm would enhance the performance. A similar method to Rocchio is the Nearest Centroid Classifier. In this technique, from a given training set, the algorithm finds the centroid for each class and assigns a class to a text document that is nearest to the centroid class. Utilizing the vector space prototype and the centroid of this class, a report can be portrayed which is determined as the vector normal of its individuals. The formulation is as shown in Eq. (6):

$$\bar{\mu}(c) = \frac{1}{|\mathrm{Dc}|} \sum_{d \varepsilon \mathrm{Dc}} \vec{v}(d) \tag{6}$$

Dc signifies the total number of documents in *D*, whose class is *c* and $\vec{v}(d)$ is he normalized vector of d [13]. The calculation to find the distance between the vectors is done using Euclidean distance. The Rocchio algorithm discovers the splitting hyperplanes among the various classes based on the centroid calculated. The algorithm does classification with sphere-shaped classes which have the matching radius and this makes it hard for classification problems for 2 classes. The classification can become unsuitable when the centroid for a particular fails to signify the class structure.

3.3 Decision Trees

It is a supervised machine learning approach that is primarily used for classifying the dataset. The decision model is built based on the feature set of a given dataset. A Decision Tree classifier has a root node, internal nodes and lead nodes. The internal node indicates a test made upon an attribute resulting into branches representing the outcomes of the test. The leaf nodes are the classes or outcomes of the result of the computations from the previous levels of the decision tree. The path from root node to the leaf node represents the classification rules. As the test conditions of the given training set can be defined, we need a measure to determine the best way to split the records. The goal of the best test conditions is to determine whether it results in a uniform class distribution within the nodes, the purity of the child nodes before and after the division. The greater the pure degree, the higher the distribution of classes. We need to compare the degree of parental impurity before splitting with the degrees of impurity of the child node after splitting to assess how well a test condition performs [14]. The greater the difference, the better the test. The equations from (7)–(9) are the measurements used for the calculation of the node purity where

t represents the parent node before splitting and $p\,(i|t)$ is the proportion of instances which belongs to class i :

- Misclassification Error : $1 -_{i}^{\max} [p(i|t)]$ (7)

- Entropy Method : $-\sum_{i=0}^{c-1} p(i|t) \ln p(i|t)$ (8)

- Gini Index : $-\sum_{i=0}^{c-1} p(i|t) \times p(i|t)$ (9)

where c in Eqs 8 and 9 represents the number of classes and $0 \ln 0 = 0$ for the calculation of the entropy.

There is a ton of decision tree calculations which have been connected to content characterization including ID3, C4.5, CART, SLIQ, etc. [15]. A standout amongst the most notable decision tree calculations is ID3 and its successors C4.5 and C5.1. It is a top-down technique that recursively builds a more tasteful decision tree. At each dimension of the tree, ID3 chooses the property that has the most astounding data gain [14].

3.4 K-Nearest Neighbour (KNN)

KNN is the simplest supervised classification approach. The input dataset comprises K related training samples in a feature set and its output is associated with its class label. A new data element is classified based on the major contributions given by its neighbours and accordingly the class label is assigned to a class which is the common among its K-Nearest Neighbours [16]. To find the nearest neighbours, a number of similarity measures are used and Euclidean Distance is one of the popular methods. It is calculated by using Eq. (10):

$$D(p_i, p_j) = \sqrt{(x_j - x_i)^2 + (y_j - y_i)^2} \tag{10}$$

where p_i and p_j are the data points having (x_i, y_i) and (x_j, y_j) coordinates, respectively.

4 Results and Discussions

The supervised classification algorithms, SVM, KNN, Rocchio and Decision Trees, are implemented on Twitter Airline Sentiment Dataset [17]. The Twitter US airline

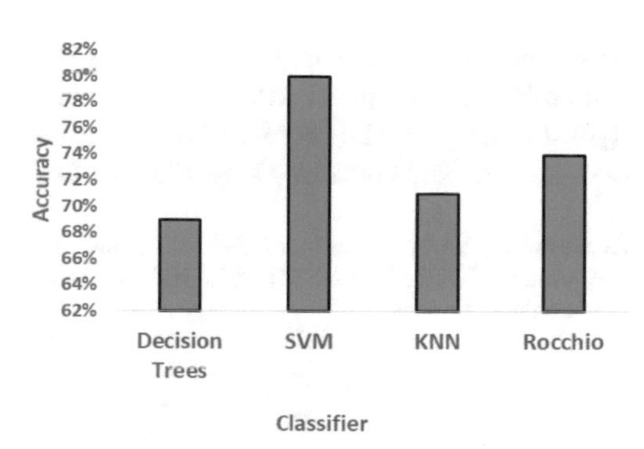

Fig. 2 Comparative results of classification algorithms

Sentiment Dataset is a collection of 15000 tweets regarding the airline companies that are categorized into positive, negative or neutral tweets. The dataset is pre-processed by the removal of Stop Words, Stemming Words and Punctuation Marks, and the text was transformed using TF-IDF model. After processing, the data has been used for the above-listed classification algorithms. The following parameters are used:

- For SVM: C=1.0, kernel='rbf' where C is the regularization parameter
- For Rocchio: α, c and n values are chosen as 1, 0.75 and 0.15, respectively. Metric='euclidean'
- For KNN: K=5, metric='euclidean'
- For Decision Tree: criterion='entropy', min_samples_split=2

Based on the analysis, it has been found that SVM Classifier yielded 80% accuracy as compared with Rocchio (74%), KNN (71%) and Decision Trees (69%) as shown in Fig. 2.

In the experimentation process, Support Vector Machine (SVM) is yielding better classification results. The classes obtained are anomalous and non-anomalous users' opinions. It indicates that Airline Sentiment Dataset, which is a high-dimensional dataset, is easily separable into classes using the Decision Plane, i.e., SVM uses an RBF kernel function and non-linear hyperplanes to separate the dataset into different classes. On the other hand, KNN produces a highly intricate resolution border based on the training dataset given. KNN has less performance because of the initial metric vector that is been chosen and that does not yield accurate discrete classes.

5 Conclusion

The Text Mining Machine Learning algorithms were surveyed, reviewed and implemented to understand their applications in Topic Modelling, Unexpected Contents detection, Discovery of Trending Topics and so on. Topic Modelling is one such popular application of Text Mining, which discovers hidden semantic structures in the text data thereby identifying the abstract topics present in a set of documents. It is observed through the survey that unsupervised machine learning algorithms are more applicable than supervised learning for Topic Modelling as the Text Dataset available is unstructured with no predefined labels. Based on the analysis done, a hybrid model could be developed which will address the limitations of the discussed classification and clustering algorithms in this paper.

Acknowledgments The authors wish to acknowledge Dr. Raju G and the faculty members of the Department of CSE of CHRIST (Deemed to be University), India for the technical and infrastructural help rendered by them.

References

1. Wang J, Li M, Chen J, Pan Y (2011) A fast hierarchical clustering algorithm for functional modules discovery in protein interaction networks. IEEE/ACM Trans Comput Biol Bioinf 8(3):607–620
2. Aggarwal CC, Wang H (2011) Text mining in social networks. Soc Networ Data Anal
3. Nie F, Zeng Z, Tsang IW, Xu D, Zhang C (2011) Spectral embedded clustering: a framework for in-sample and out-of-sample spectral clustering. IEEE Trans. Neural Network 22(11):1796–1808
4. Osisanwo FY, Akinsola JET, Awodele O, Hinmikaiye JO, Olakanmi O, Akinjobi J (2017) Supervised machine learning algorithms: classification and comparison. Int J Comput Trends Technol (IJCTT) 48(3)
5. Kaur S, Jindal S (2016) A survey on machine learning algorithms. Int J Innovative Res Adv Eng (IJIRAE) 3(11)
6. Kolla N, Giridhar Kumar M (2019) Supervised learning algorithms of machine learning: prediction of brand loyalty. Int J Innovative Technol ExplorEng (IJITEE) 8(11)
7. Uddin S, Khan A, Ekramul Hossain Md, Ali Moni M (2019) Comparing different supervised machine learning algorithms for disease prediction. BMC Med Inf Decis Mak 19, Article number: 281
8. Tsakalidis A, Papadopoulos S, Cristea AI, Kompatsiaris Y (2005) Predicting elections for multiple countries using Twitter and Polls. IEEE Intell Syst 30(2)
9. Wang Z, Qu Z (2017) Research on web text classification algorithm based on improved CNN and SVM. In: IEEE 17th international conference on communication technology (ICCT), pp 1958–1961
10. Ling HS, Bali R, Salam RA (2006) Emotion detection using keywords spotting and semantic network IEEE ICOCI 2006. In: International conference on computing & informatics. Kuala Lumpur, pp 1–5
11. Liu F, Lu X (2011) Survey on text clustering algorithm. In: Proceedings of 2nd international IEEE conference on software engineering and services science (ICSESS). China, pp 901–904
12. Sri Nikitha K, Sajeev GP (2018) An improved Rocciho algorithm for music mood classification. In: 2018 international conference on data science and engineering (ICDSE), pp 1–6
13. Vijayan VK, Bindu KR, Parameswaran L A comprehensive study of text classification algorithms. In: International conference on advances in computing, communications and informatics (ICACCI), pp 1109–1113
14. Damiran Z, Altangerel K (2014) Author identification—an experiment based on Mongolian literature using decision trees. In: 7th International conference on Ubi-media computing and workshops. Ulaanbaatar, pp 186–189
15. Patterson D, Rooney N, Galushka, M, Dobrynin V, Smirnova E (2008) SOPHIA-TCBR: a knowledge discovery framework for textual case-based reasoning, knowledge-based systems, vol 21, no 5, pp 404–414

16. Guo Gongde, Neagu Daniel, Cronin Mark (1970) Using kNN model for automatic feature selection. Lect Notes Comput Sci 3686:410–419
17. https://www.kaggle.com/crowdflower/twitter-airline-sentiment

Internet of Things (IoT) Security Threats on Physical Layer Security

Rathnakar Achary

Abstract Internet of things (IoT) interconnects things, users, and billions of tiny devices in the global network. It has been in focus due to its sensing and control capabilities with a massive deployment of machine to machine (M2M) and people to machine (P2M) devices. Due to its architectural constraints the IoT devices presents unique communication and security concerns. One of the techniques to mitigate this is by replacing the lightweight cryptographic protocols with a robust physical layer security. This research paper is to analyze the security attacks on the physical layer, and propose a security technique to achieve the security requirement of the devices connected to IoT, communicating between the legitimate users. This enables to develop a secure channel to configure the device without the application of any complex system or changes in the hardware of the device.

Keywords Internet of things · Physical layer security · Secure channel analysis · SNR · Man-in-the-middle attack · IoT security

1 Introduction

The devices connected in IoT are prevalent in a variety of applications like smart gird, intelligent transport systems, smart security, industrial automation and smart city. These diverse applications of IoT can ensure convenience to the users, but it cannot ensure the security in terms of confidentiality, integrity, and availability (CIA) [1]. An intruder gain access over the devices connected to IoT and the communication channel. It will adversely affect the security of the entire network where these devices are connected. There has been considerable interest in the development of security mechanisms based on the information-theoretic characteristics of the physical layer [13]. It is mainly focused on the symmetric key encryption system and doesn't guarantee security due to key sharing between parties. Still the possibility of providing an unbreakable security mechanism is of paramount requirement of the security service

R. Achary (✉)
Alliance College of Engineering and Design, Alliance University, Anekal, Bangalore, India
e-mail: rathnakar.achary@alliance.edu.in

D. S. Jat et al. (eds.), *Data Science and Security*, Lecture Notes in Networks and Systems 132, https://doi.org/10.1007/978-981-15-5309-7_5

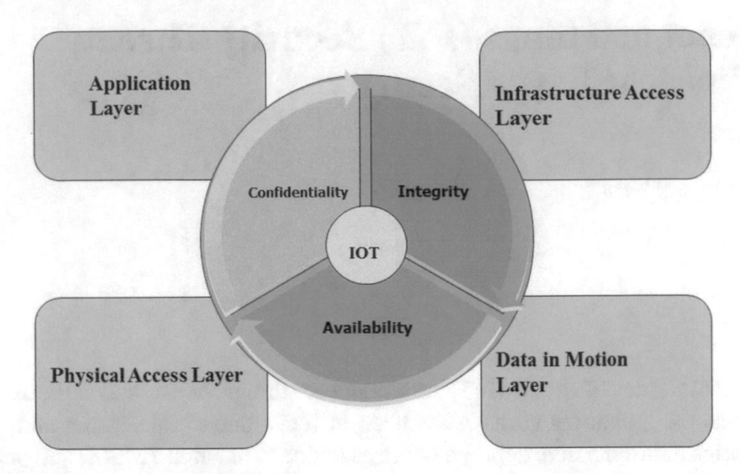

Fig. 1 CIA as the security requirements

providers, which stops eavesdropping party from unauthorized access. In this paper, a physical layer security solution to effectively provide a comprehensive mechanism to fulfill the requirements of the CIA is proposed. The CIA triad is mapped with the IoT Security requirements as shown in Fig. 1.

The other sections of the paper are structured as follows: In Sect. 2, literature review and works of others about the confidentiality and security in IoT devices are presented. Threat analysis in the physical layer of the IoT network is performed in Sect. 3. Section 4 describes the proposed mechanism for achieving better confidentiality and authentication. Section 5 provides the result analysis and the enhancement of security and the conclusion is in Sect. 6.

2 Literature Review

IoT is more vulnerable to security issues due to its physical characteristics [13]. The survey conducted by the leading security firm Kaspersky recently contended an incriminatory analysis of the security challenges of IoT [7], with and the precarious caption "*Internet of Crappy Things*". The IoT security issues are summarized as

- Security must be the essential requirement of IoT.
- Due to the resource constraints in IoT devices there are presently no guidelines on how to implement security solutions in IoT-based devices.
- IoT requires a unique security solution, which is based on the evolution of security solutions in the past 25 years.
- There is no single and comprehensive solution, it can effectively alleviate security threat on IoT devices.

According to Edith Ramirez, US Federal Trade Commission Chairwoman, the three security challenges for the future of IoT are Ubiquitous data collection, Potential for unexpected use of consumer data, and Heightened security risk [4]. She insisted corporations to improve confidentiality and to build IoT devices embedded with a security mechanism by implementing a security engrossed method, with optimized data collection. Many researchers have identified that limited computational power is one of the important challenges for the development of complex security solutions for IoT devices, and the new firmware, embedded operating systems, new software and protocols, which cause the interoperability are the other issues.

The generally used security architecture is completely centralized. This simplifies the key management and the processing task of these low-power devices [2]. In most of the wireless devices such as IoT, RFID, and wireless sensor networks [12], data and privacy protection is one of the main concerns. The general mechanism adopted to protect the integrity and confidentiality of data is by password and encryption technologies [10]. In [11], the author described that access control, identity, and authentication are the essential requirements between the parties to confirm their identity and access permissions. It also prevents the intruders entering for unauthorized access to the resources. Literature [5] identifies that the key challenges of IoT devices are the characteristics of the device itself. This is because IoT itself is an integration of multiple heterogeneous networks. This prone to the vulnerability for security challenges. In [8], security challenges affecting the transport layer security is presented. From the above literature review, it is evident that most common security challenges on IoT are man-in-the-middle attack, forgery attack, heterogeneous network attack, application risk of IPv6, wireless local area network (WLAN) application conflict, and denial of service (DoS) or distribute denial of service (DDoS).

3 Threat Analysis

IoT has a vast and varied range of domestic and industrial applications. These applications are completely different from one another and also in terms of security needs such as confidentiality, integrity, and availability. As discussed in Sect. 2, these devices have low processing power, storage capacity, and not possessing a full-fledged operating system. These limitations raise a number of problems in security challenges in the device. The security threats on IoT devices are normally different from that of the threats to other network devices. One of the main challenges the IoT devices facing are their configuration. For some of the applications the user needs to reconfigure the device security properties. This process of reconfiguration is highly vulnerable to security threats. If this reconfiguration is not performed in a secured channel, then the intruder may obtain sensitive information. This risk can be mitigated by the use of appropriate cryptographic techniques. But there will be challenges of sharing the symmetric cryptographic key between the legitimate users [9].

4 Proposed Works

The proposed work provides a solution for the two most important essentials such as initial configuration and devices that are stolen or compromised. For the former one, we have proposed the solution and for the later one the issue must be addressed through a blend of administration procedures and protocols. With the help of master configuration, we configure a device's cryptographic key in a secure and error-free manner [9]. The device has to generate a secure and fresh key on request in the form of one-time password (OTP). The message transmitter is in the form of a tiny pocket device. This device must not have any connectivity to any other network. The device works as a standalone module to securely transmit the signal to the legitimate receiver or device.

4.1 Secure Channel Analysis

The information-carrying capacity of a channel is determined by Shannon's theorem as $C = W \log_2(1 + \text{SINR})$. Let Alice ($A$) wants to transmit a message to Bob (B) in an insecure channel. An intruder Eve (E) wants to capture the packet and sniff the information. Let C_{AB} and C_{AE} be the channel capacity between A and B and A and E, respectively. The secrecy capacity C_{S} of the discrete memoryless wireless channel is given as the difference between C_{AB} and C_{AE} as

$$C_{\text{S}} = \max\{0,\ C_{AB} - C_{AE}\} \quad = [\log(1 + \text{SNR}_{AB}) - \log(1 + \text{SNR}_{AE})] \tag{1}$$

where C_{S} is the secrecy capacity, which is the maximum secrecy rate that can be attained by the legitimate users Alice and Bob, in the presence of an intruder Eve. For a better channel capacity between the legitimate sender and the receiver [6], their channel capacity C_{AB} must be always better than the channel capacity between C_{AE}. It can be achieved only if the signal-to-noise ratio (SNR) of the channel between A and B is greater than that of A and E. This approach may be considered as a possible security method. Due to the reduced throughput and low data rate of the channel, this method is greatly employed to send the keys to other communication protocols. As in Fig. 2, the communication between terminal T_1 and IoT terminal T_2 is being eavesdropped by an intruder terminal T_3. The two terminals T_2 and T_3 are connected to the same IoT, the signals at the outputs of the main channel and intruders channel are usually different [2]. The quality of the intruder channel may be significantly degraded compared to the main channel. Also it is observed that if the intruder is located further away from the transmitter, then the legitimate receiver receives an attenuated signal [3]. In the scenario explained above, Alice want to communicate a pair of elliptical cryptographic curve (ECC) keys to the receiver Bob. The rate at which the data transmission happens between the two users is relatively low. For a secured communication between A and B, the condition must be satisfied when R

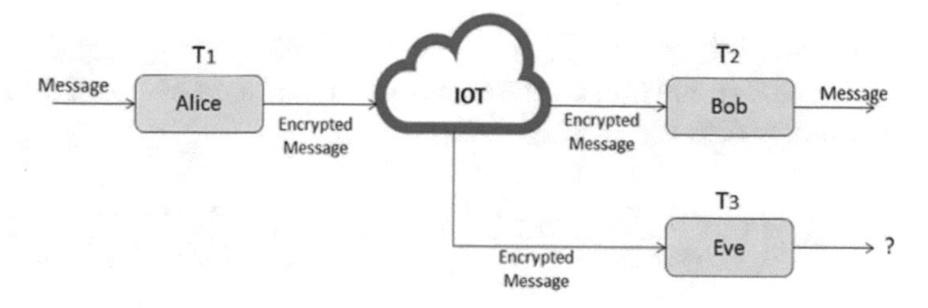

Fig. 2 Intrusion by the eavesdropper

< C_S. Where R is the date rate and Cs is the secrecy capacity of the channel [6]. To meet the required quality of service (QoS), the signal rate must be higher than the minimum rate, which is, $R_{\text{min}} = \frac{32\,\text{Kb}}{2\,\text{s}} = 16\,\text{Kbps}$

Where the numerator term is the amount of information to transmit (32 Kb) and the denominator represents the maximum time required for the configuration exchange.

$$\therefore R_{\text{min}} < R < C_S \tag{2}$$

If m and n are the two entities involved in the message transmission process, in general, the channel capacity between these two entities are

$$C_{mn} = W \log_2(1 + \text{SNR}_{mn}) \tag{3}$$

In Eq. (3), C_{mn} depends on the bandwidth W and the logarithm of signal-to-noise ratio.

$$\text{SNR}_{mn} = \frac{P_r x_{mn}}{\text{N}} \quad \therefore C_{mn} = W \log_2\left(1 + \frac{P_r x_{mn}}{\text{N}}\right) \tag{4}$$

where the noise N is the additive white Gaussian noise (AWGN), which is a function of the bandwidth $W \cdot N = W \cdot K \cdot T$, K—Boltzmann's constant and T—Temperature

$$\therefore C_{mn} = W \log_2\left(1 + \frac{P_r x_{mn}}{\text{N}}\right) = W \log_2\left(1 + \frac{P_r x_{mn}}{W \cdot K \cdot T}\right)$$

If $P_{rm} = \frac{P_{tm} \cdot G_{tm} \cdot G_{rm}}{L(d_{mn})}$ in which $L(d_{mn})$ is the free space path loss. The other losses due to the transmission channel parameter are considered as negligible.

$$L(d_{mn}) = \left(\frac{4\pi d_{mn}}{\lambda}\right)^2 \quad \therefore C_{mn} = W \log_2\left(1 + \frac{P_{t_m} \cdot G_{t_m} \cdot G_{r_m} \cdot \lambda^2}{(4\pi d_{mn})^2 \cdot W \cdot K \cdot T}\right). \tag{5}$$

By considering the parameters in Eq. (5), the channel capacity C_{AB} between Alice and Bob and C_{AE} between Alice and the intruder is indicated as

$C_{AB} = W\log_2\left(1 + \frac{P_{tm}\cdot G_{tm}\cdot G_{r_{mB}}\cdot\lambda^2}{(4\pi d_{AB})^2\cdot W\cdot K\cdot T}\right)$ $C_{AE} = W\log_2\left(1 + \frac{P_{tm}\cdot G_{tm}\cdot G_{r_{mE}}\cdot\lambda^2}{(4\pi d_{AE})^2\cdot W\cdot K\cdot T}\right)$ The difference between C_{AB} and C_{AE} are the antenna gains and the distance of the receiver from the transmitter A as $C_S = C_{AB} - C_{AE}$

$$\therefore C_S = W\log_2\left(1 + \frac{P_{tm}\cdot G_{tm}\cdot G_{r_{mB}}\cdot\lambda^2}{(4\pi d_{AB})^2\cdot W\cdot K\cdot T}\right) - W\log_2\left(1 + \frac{P_{tm}\cdot G_{tm}\cdot G_{r_{mE}}\cdot\lambda^2}{(4\pi d_{AE})^2\cdot W\cdot K\cdot T}\right)$$

To establish a link between the sender and the receiver, the receiver must be at distance of 2λ from the transmitter. Where λ is the wavelength and $\lambda = c/f$.

$$\therefore C_S = W\log_2\left(\frac{\left(1 + \frac{P_{tm}\cdot G_{tm}\cdot G_{r_{mB}}\cdot\lambda^2}{(4\pi d_{AB})^2\cdot W\cdot K\cdot T}\right)}{\left(1 + \frac{P_{tm}\cdot G_{tm}\cdot G_{r_{mE}}\cdot\lambda^2}{(4\pi d_{AE})^2\cdot W\cdot K\cdot T}\right)}\right)$$

The parameters analyzed in this paper are essentially considered to ensure that the channel between the sender and receiver is a secured channel. To prevent the intruder entering into the communication channel, jamming can be introduced. This is done by injecting noise in the intruder channel. As a proactive technique this is less feasible, because it may not be considered that there will be an intruder in the channel and also the location of the intruder is not known to the legitimate users A and B. By refereeing to the value of R_{min} in the proposed method, the key parameters considered for legitimate users are frequency, bandwidth, and distance. These parameters give back the conditions that leads to define a security area. This enables a maximum rate and the minimum distance of Eve to the legitimate users, i.e., the proposed method provides the minimum distance that an intruder should stay away from the legitimate users in order to have a minimum security rate. This creates a warning zone, i.e., the legitimate users have to verify the intruder's presence within the minimum distance if they want to communicate with a minimum target rate.

5 Result Analysis

The secrecy capacity C_S of the proposed system is analyzed by simulation using MATLAB and design parameters. Equation (2) represents the important parameters considered between A and B to find the minimum rate as R_{min}. The variations in the secret capacity C_S by changing the distance between legitimate user R_{AB} and the intruder R_{AE} is shown in Fig. 3. This gives the minimum distance between A and B with minimum rate R_{min} comparing to the transmission rate of the intruder. The intruder requires a higher rate than R_{min}, because he is located at a greater distance than R_{min} (Table 1).

Figure 4 represents the SNR value between A and B and A and E as a function of distance. To meet the secrecy requirements, the SNR of the intruder must be lower than the SNR of the legitimate receiver [3]. In Fig. 5, the curve over shifted with

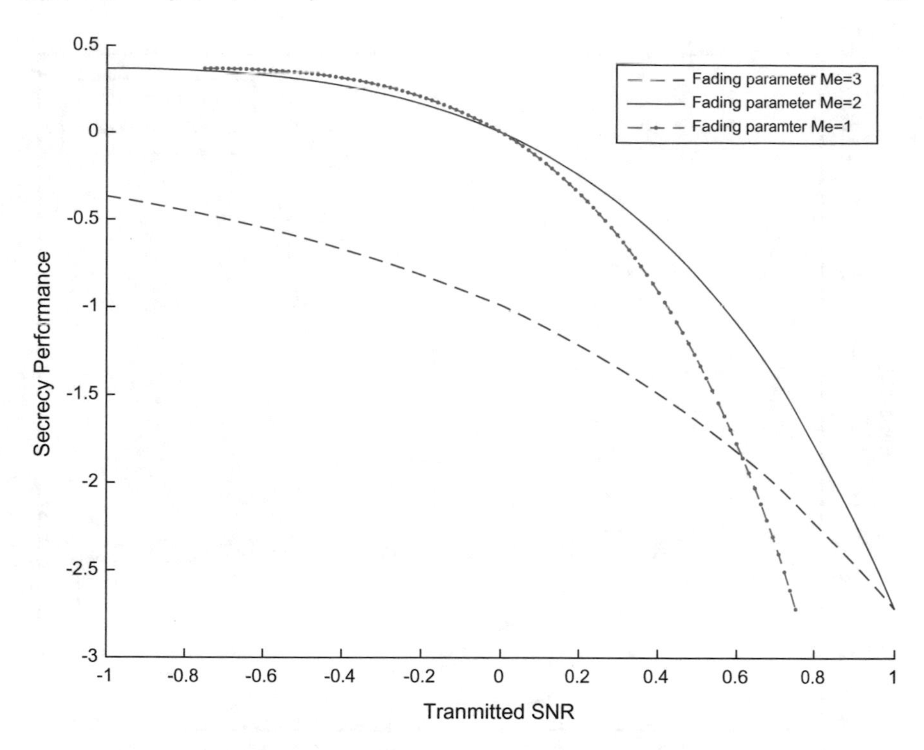

Fig. 3 Secrecy capacity as a function of users B and E distance from A

Table 1 Parameters for simulation

Parameters	Value
Pt_m	1 mW
G_{t_m}	1
$G_{r_{mB}}$	1
$G_{r_{mE}}$	10
W	1 MHz
K	$1.3807 \cdot 10^{-23}$ JK^{-1}
T	290 K
d_{AB}	1–3 Mt
d_{AE}	1–9 Mt
f	2.4 GHz
R_{min}	16 Kbps

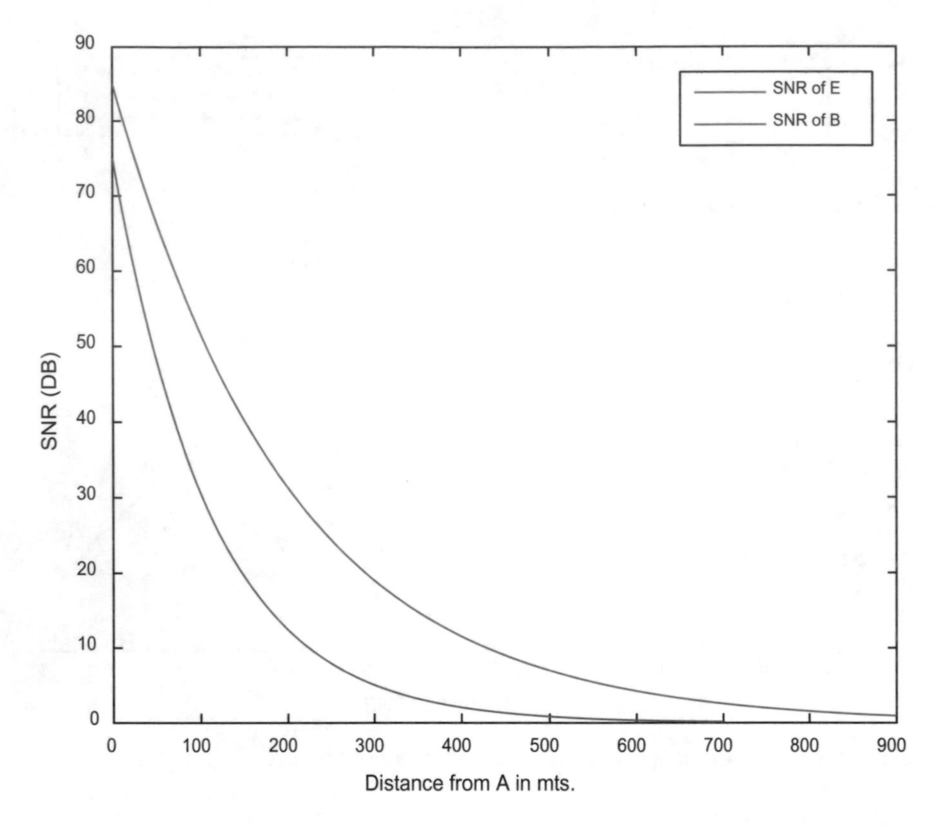

Fig. 4 SNR as a function of distance

an SNR difference of 10 dB. From the analysis results, it is clear that for secured communication between the legitimate users A and B, select a minimum rate R_{min} with a maximum distance R_{AB} and appropriate modulation and coding scheme for a more accurate propagation model.

6 Conclusion

In this research paper, a detailed survey related to the recent research in the field of IoT security challenges is conducted. By using an information-theoretic formalism it is clear that, in all of the principal channel models of wireless networking, the physical layer can in principle support reliable data transmission with perfect secrecy under realistic conditions. From the simulation results obtained from the proposed system, it is clear that the minimum rate R_{min} allotted for the legitimate parties in the communication process has to ensure that the intruder is not located inside the coverage area. This enables the distance between the legitimate users is secured.

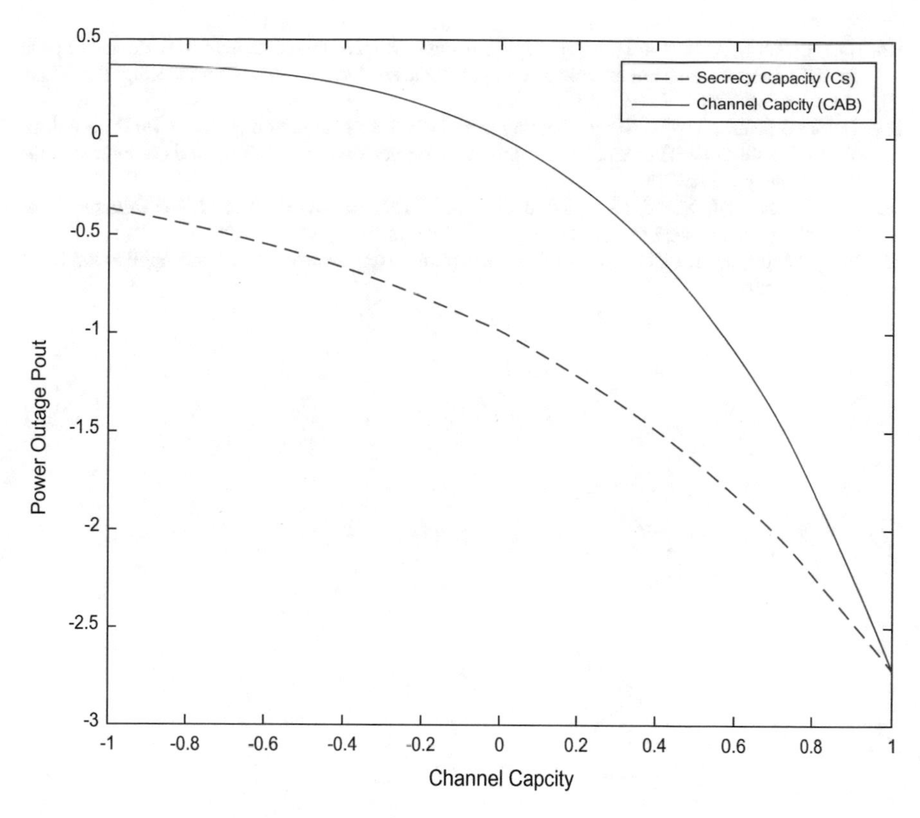

Fig. 5 Changes in power outage in dB

References

1. Akyildiz IF, Su W, Sanakarasubramaniam Y, Cayirci E (2002) Wireless sensor networks: a survey. Comput Netw 38(4):393–422
2. Alexander T, Dohler R (2013) Security threat analysis for routing protocol for low power and lossy networks (RPL) Dec 15, 2013
3. Chen M, Kwon T, Mao S, Leung V (2009) Spatial–temporal relation-based energy-efficient reliable routing protocol in wireless sensor networks. Int J Sens Netw 5(3):129–141
4. De Turck F, Vanhastel S, Volckaert B, Demeester P (2002) A generic middleware-based platform for scalable cluster computing. Future Gener Comput Syst 18(4):549–560
5. Elkhiyaoui K, Blass EO, Molva R (2012) CHECKER: on-site checking in RFID-based supply chains. In: Proceedings of the fifth ACM conference on security and privacy in wireless and mobile networks
6. Hamad F, Smalov L, James A (2009) Energy-aware security in M-Commerce and the internet of things. IETE Tech Rev 26(5):357–362
7. Internet of Crappy Things https://www.kaspersky.com/blog/internet-of-crappy-things/7667
8. Liu B, Chen H, Wang HT, Fu Y (2012) Security analysis and security model research on IoT. Comput Digital Eng 40(11):21–24
9. Mathur S, Trappe W, Mandayam N, Ye C, Reznik A (2008) Radio-telepathy: extracting a secret key from an unauthenticated wireless channel. In: Proceedings of MobiCom. pp 128–139

10. Suo H, Wan J, Zou C, Liu J (2012) Security in the internet of things: a review. In: Proceedings of the IEEE international conference on computer science and electronics engineering (ICCSEE), vol 3. pp 648–651
11. Tsudik G (2006) YA-TRAP: yet another trivial RFID authentication protocol. In: Proceedings of fourth annual IEEE international conference on pervasive computing and communications workshops. pp. 196–200
12. Wan J, Chen M, Xia F, Li D, Zhou K (2013) From machine-to-machine communications towards cyber-physical systems. Comput Sci Inf Syst 10(3):1105–1128
13. Ye T, Peng QM, Ru ZH (2012) IoT's perception layer, network layer and application layer security analysis

Data Mining Models Applied in Prediction of IVF Success Rates: An Overview

G. S. Gowramma, Shantharam Nayak, and Nagaraj Cholli

Abstract IVF success rates can be increased by the analysis of multiple factors related to the process. The automated systems which take these multiple factors as input and produce prediction as output are the need of the hour in IVF research area. The automated systems include mechanism for pre-processing of input metrics and prediction stage that is built using algorithms such as decision trees, naïve Bayes classifiers, case-based reasoning system, logistic regression, etc. In recent years, Artificial Neural Networks (ANN) and hybrid methods are also in use for IVF success rate prediction. There exists lot of literature in the area of IVF success rate prediction. Analyzing the works done so far can help the researchers to identify the research gap and work toward more efficient methods. This work is aimed to present a detailed review of existing techniques applied for the prediction of IVF success rates. Various techniques used so far have been discussed in detail. The drawback of each of the methods is identified which provides the key for future researches in this area. This area has been chosen considering the challenge and social relevance.

Keywords IVF success rates prediction · Data mining · Artificial neural network · Genetic algorithm

1 Introduction

Approximately, 70–80 million couples worldwide were suffering from the condition of infertility as per the study reported in the year 2016 by the WHO. Infertility condition and the type may vary transversely with the different parts of the world

G. S. Gowramma (✉) · S. Nayak · N. Cholli
Department of Information Science, R V College of Engineering, Bangalore, 560059, India
e-mail: gowribasu.dbit@gmail.com

S. Nayak
e-mail: Shantaram_nayak@yahoo.com

N. Cholli
e-mail: nagaraj.cholli@gmail.com

D. S. Jat et al. (eds.), *Data Science and Security*, Lecture Notes in Networks and Systems 132, https://doi.org/10.1007/978-981-15-5309-7_6

and is expected to influence 8–12% of couples worldwide. A surprising fact is that the infertility condition is found to be maximum in countries with even high rates of successful fertility cases; this significant condition was termed as "barrenness amid plenty." Among the different countries of various continents of the world, countries represent the African continent that was found to be affected heavily with the condition infertility.

Various researchers conducted different surveys across the countries of the African continent had reported that the prevalence and the occurrence of infertility was high in Nigeria representing 33% and low in the Gambia with 9% rates. An outsized very large study is carried out in the 25 countries representing the hotspot of the infertility with respect to the different influencing factors, among the world countries by the prestigious organization WHO considering the familial history and the medical status of more than 5800 couples and stated that the primary cause of the female infertility African women close to 85% of the occurred principally due to the infection, compared to only 33% of the women representing the other countries of the worldwide. Prevalence and the incident rates of primary infertility in Indian women were accessed by the WHO and reported that rates lie between 3.9 and 16.8% and were principally depend upon the geographical location. The percentage varies greatly among the different states with increasing order, Uttar Pradesh (3.7%), Himachal Pradesh, and Maharashtra, representing 3.7, 5% in Andhra Pradesh and much higher (15%) in the state of Kashmir [1].

Very less structured data was available almost none from southern India, to study the influence of different causative factors of the primary infertility. A research group collected baseline data of causative variable of primary infertility such as bacterial vaginosis and herpes simplex virus type-2 (HSV-2) infection from the 897 young (15–30 years) women volunteers of the place Mysore, Karnataka, and India between the time periods from November 2005 to March 2006. Results indicated that mean age of the women 26 years was found to have high prevalence rates of primary infertility (95%) due to the infection of HSV-2 infection. Various reproductive and clinical properties correlated with the condition of primary infertility among young, reproductive age (16–30 years) women in Mysore, India. With this report it is clear that the prevalence and the occurrence of the primary infertility rates were found to vary among the different tribes and the castes of the same region of the Indian population. The estimates performed by the different agencies' approaches were different, so the rates of measuring the infertility were not highly accurate [2].

The assessment of composite structure of IVF-related data of the multiple factors information helps in the accurate predictions of the IVF success rate and can be a true representative of the BIG data in health care. IVF-related data can be assessed applying the data mining method to predict the success rates of the IVF process more accurately their bye to take an important conclusion or suggestion which in turn help the doctor to suggest the right treatment method and the patients to take a right decision to increase their chance of becoming parents. Automated decision support to clinicians and the candidate couple can be developed by providing the more accurate success rates of IVF treatment which in turn can be made available by assessing the influence of multiple factors of IVF process applying the various data mining algorithms and tools, i.e., Logistic regression models, Case-Based Reasoning System (CBR), Decision Trees (DT), Naive Bayes classifiers, Artificial Neural Networks (ANN), and Hybrid methods with improved prediction capabilities. Scanty research was available in the literature against the applications of data mining methodologies in the IVF success rate prediction; motivates the authors to aim about presenting a review on various data mining methodologies applied in the prediction of the success rates of IVF procedure, further to understand the research gaps and to facilitate the development of novel predictive models with the enhanced predictive accuracies of the IVF process [3].

In this paper, the authors have assessed to various intrinsic and extrinsic factors which contribute toward the success rate of IVF treatment depending on various factors such as intrinsic factors, i.e., Genetic predisposition, age, body mass index, hormonal balance, embryo viability, sperm quality, endometriosis and overall patient's response level of the candidate couple, and the extrinsic factors such as medical equipment technology, treatment methods, personal experiences of doctors and embryologists, process time, stress due to the lifestyle and work pressure, etc. [4].

2 Data Mining Approaches of IVF Data Analytics

2.1 *Traditional Models (Medical Oriented)*

Malizia et al. [5] opined that the pregnancy per IVF cycle is the traditional way of measuring the outcomes of the IVF procedure. Study applied a traditional way of approach for measuring the rate of success of IVF procedure by assessing the live children's birth rates for the patients going through the fresh IVF cycle. Study collected the data related to the women undergoing the fresh IVF cycle for the first time with the fresh embryo (not frozen) from the single data source of the data between the years 2000 and 2005 depicting the data of 6164 patients undergone 14,248 IVF cycle. Two different methods applied in the data mining principally considering the maternal age were found to be the optimistic-highest probability of pregnancy, considered because after the treatment the patients never turned up to the hospital for the next consultation or not turned up for the next round of IVF cycle and conservative methods—assuming the failure of the live birth who did not report to the center. Study reported that the optimistic analysis reported the successful live birth rates as 72% and the conservative 51% among the 14,248 cycle's data of the 6164 patients. Study reported that the overall live birth rates were found to be decreased with the increased age of the female.

Velde et al. [6] compared the efficacy of the two different models practiced in measuring the success rates of the IVF process, i.e., Nelson's model developed in the year 2011 that of Templeton's model designed in the year 1990. Study identified a primary infertility data of the 5176 patients representing the population of

Dutch origin collated from the database termed human fertilization and embryology authority database (HFEA) specifying the time frame 1991–1994 for the Templeton model and the data between the years 2003 and 2007 for Nelson's model. Ongoing pregnancy and the live birth rates were the two different attributes depicted as overall success or failure after the IVF treatment cycle, and were selected for the assessment. Study reported that both the models applied were shown the similar probability of results for the Templeton model (0.597) and the for the Nelson model (0.590) in terms of curve depicting receiver operating characteristics, with very limited percent of discriminative ability on the analyzed datasets. Initially, the success present reported by both the models was dissimilar representing 21% for the Templeton model and 29% for the Nelson model. After adjusting the model with that of the change in the success rates over the time frame, both the models were much improved and shown nearly the same results, i.e., Templeton's model with 21% prediction rates versus 20%; Nelson's model 24%. Study concluded that the applied model design was good for the assessment of less number of factors and restricted to analyze only the primary infertility rates. The other conditions such as intrinsic and the various extrinsic factors were the combination, and mainly the data related to the secondary infertility was not able to analyze.

Rødgaard et al. [7] state that the IVF procedure is one among the commonly applied techniques to treat the infertility condition and was successfully practiced till today from so many long years to help the needy couples. But due to the influence of different factors, the success rates of the IVF procedure reported to be less and lost no progress in the duration of last decade. The accurate assessment of the viability condition of the fertilized embryos still remains as the challenge, and there is an urge in the advent of technology in this significant stage of the IVF cycle. Selection of viable embryos before the embryo transfer protocol certainly improves the success rates of the IVF cycle; hence, the analysis of the data representing the different attributes correlated with the embryo quality and viability shall be of great importance in measuring the success rates of the IVF cycle. Presently available techniques to identify the viability status of the embryos are based on the non-invasive approaches that work principally considering the morphology properties of the embryos such as embryos cleavage rates, formation of blastocyst, and growth-related kinetics of the transferred embryos. The method depends only on the simple attributes such as morphological properties, which may not be sufficient to predict the rates accurately. The advanced methods focus on the information of the other impactful studies such as molecular, omics based (proteomics and Metabolomics) and also most recent approach Sn RAN and mi RNA. Medical-based traditional methods still have many advantages but the biggest setback with these methods was the lack of study design to analyze the interactive efforts of the multivariable factors that influence the success rates of the IVF procedure.

2.2 Statistical and Regression-Based Models

Wilcox Lynne et al. [8] reviewed contemporary approaches such as medical-related traditional approach and epidemiologic-based statistical approaches in measuring the percentage of pregnancy rates after the treatment (IVF-ET) process, with the aim of studying the issues concern with the rates and also to come out with some useful suggestion of novel model designs to predict the IVF success rates more accurately by analyzing the data collected from the Society for Assisted Reproductive Technology sister concern of The American Fertility Society of the USA about the infertile couples undergoing in IVF-ET treatment cycle. Study reported that the probable number of live births/ET procedure is more or less similar for each IVF-ET cycle for the same couple, whereas the probabilities found to be different for single cycle accessed from different couples.

Trimarchi James et al. [9] developed a mathematical model C5.0 decision tree data mining algorithm to investigate the impact of more than 100 factors of 380 records, not able demographics of candidate couple, stimulation management, response nature, oocyte quality, embryo viability, and embryo transfer variables, on real-time influencing the IVF success rates. The results were validated by applying the statistical model developed based on the principles' logistic regression. Study reported that there is a significant increase in the predictive accuracies of the model considering more factors but not for only factors female age as a predictor of IVF outcome. Study concluded that the designed decision tree models can predict IVF outcome with 75% accuracy on known outcomes of cycles.

Baker et al. [10] applied a logistic regression design model to assess the influence of pre-implantation process factors (81) for the two main characters of the clinical pregnancy cycles such as thickness of the endometrium (mm) at down-regulation condition and concentration, and exposure time of the gonadotropin hormones. Study concluded that there is a significant need in assessing various intrinsic and extrinsic factors of the IVF process, which were needed to be assessed in order to measure the IVF success rate with accurate prediction possibilities.

LeBlanc and Kooperberg [11] developed a statistical-based model design to measure the influence of potential unmeasured variables with respect to the attributes, i.e., clinical diagnoses information of the candidate couple, number of IVF treatment cycles undergone with the outcome collected from the databases represent the information of Stanford Hospital and Clinics US. Study felt that the assessment of all the influence of unmeasurable factors is quite complex, so there are some definite concerns related to the model design in estimation of the success rates. In order to overcome the issue discussed, study proposed a novel design model constructed based on the boosting algorithms.

Kakhki et al. [12] proposed a quick decision support system for the IVF patients' designed works on the principles of Bayesian networks' statistical models. Study assessed the efficacy of four different Bayesian network classifiers, i.e., Naïve Bayesian, two constraint-based such as Bayes-N and MP-Bayes, and Greedy search and scoring-based classifier, on predicting not only the success rates but also the presence of failure by analyzing the data collected from the real-world database. Results specified that among the four classifiers applied, naïve Bayes classifier was found to be the best in measuring the success rates of the IVF procedure.

Smith et al. [13] compared two different prediction models applied in the prediction of the IVF success rates, i.e., Templeton and IVF predict considering the data of 130,960 IVF cycles undertaken in the United Kingdom between the years 2008 and 2010. Hosmer–Leme show that statistics-based calibration plots were developed for the selected datasets to measure the efficacy of both the models. Study reported that the IVF predicts model design shown noticeably superior calibration and much higher diagnostic accuracy denoting the calibration plot intercept of 0.040 (95%) and slope of 0.932 (95%) in comparison with 0.080 (95%) and 1.419 (95%) for the Templeton model. Study concluded that both models were not significant in predicting the accurate percent of the live birth rate, but this was particularly very low in the Templeton model.

Dhillon et al. [14] used multivariate logistic regression model design to measure success rates of IVF/ICSI cycles for the selected factors such as age 36 years and above, in case of male candidate: tubal factor infertility, unexplained condition of infertility, and ethnicity of the Asian population for the data collected between the years 1991 and 2007, live birth data of 9915 women who underwent IVF/ICSI treatment at any center for Assisted Reproduction clinic from 2008 to 2012, and validation of the design model was performed on data collected from 2723 women underwent treatment in the year 2013. Methodology applied was multivariable logistic regression design developed to assess the discriminatory ability using the area under receiver operating characteristic (AUROC) curve, and calibration was assessed using calibration-in-the-large and the calibration slope test. Study concluded that the model was unable to report for factors, i.e., smoking and alcoholism which invariably affects the IVF/ICSI outcome and also found to be limited in analyzing the ethnic distribution and outcomes restricted for only UK population.

McLernon et al. [15] developed a population-related cohort model design to calculate approximately the likelihood of a live birth over several inclusive cycles in IVF success rates assessed based on a couple's definite characteristics and IVF process information. Study collected data from all the licensed IVF clinics of the United Kingdom national-level data from the data source Human Fertilization and Embryology Authority register about 253 417 women candidates with IVF and ICSI treatment process representing the period between 1999 and 2008. Among the two models employed, first model utilized the information specifying before the treatment process and the later assessed additional data collected during the process of the first IVF attempt. Pre-treatment predictors of live birth influencing factors, i.e., female age and length of infertility condition, and post-treatment factors such as total number of ovules isolated, cryopreservation of quality, and period of embryos and stage at which embryo transferred were studied by the model. A total of 1,13,873 women information representing the 1,84,269 complete cycles of the IVF process was assed and concluded that among the selected 113 873 datasets, 33, 154 with the contribution of (29.1%) had a live birth after their first complete cycle and 48, 925 (43.0%) after six complete cycles.

Hafiz et al. [16] proposed a statistical-based cross-sectional study design to assess the IVF-related data of 486 patients representing the 29 variables applying a census method initially and further the parameters such as levels of mean accuracy and the point of mean area measured through curve representing the receiver operating characteristic (ROC) for identified classifiers, i.e., support vector machines classifiers, recursive partitioning (RPART) classifiers, random forest (RF) classifiers, adaptive boosting classifiers, and one-nearest neighbor classifiers. Performance of the study design was assessed by studying the properties of the selected classifier such as specificity, sensitivity of the model, both positive and negative predictive values, and probability ratios of classifiers. The results exposed the areas under the ROC curve (AUC) as 84.23 and 82.05%, respectively, and the output of RPART suggested that the rate of pregnancy is negatively influenced by the female age. Study concluded that among the different classifiers, RF and RPART were found to be more accurate in predicting IVF/ICSI success rates compared to other classifiers considered in the studied. Out of 29 variables assessed, age of female, viable embryos number, and levels of the hormone estradiol in the patient serum on the day of administration of human chorionic gonadotropin (HCG) hormone were found to have significant impact on the success rates of the IVF rates.

2.3 *Computational Algorithms Based Models*

Camilo et al. [17] proposed genetic algorithm-based IVF module (IVFm) implemented in a parallel flow fashion, to predict the success rates of the IVF cycles by analyzing the chromosomal information from the identified population and new generation in population suggested by the GA and also assess multiple combinations of selected influencing factors to identify quality individuals. Population information in the input values of the algorithm indicated the results as output values in terms of quality individuals. Results revealed that among the various IVFm operators studied, EAR-N was found to be more efficient in predicting the success rates. The performance of the operator is scaled based on the balance of the monitoring parameters such as area under search, rate of generating new information, and utilization of capable areas, suggested by the recombination process. IVFm operator applied was found to possess enhanced abilities in supporting different types of binary-coded evolutionary algorithms. Therefore, it is suggested as future work of joining the IVFm in different evolutionary algorithms for different types of applications.

Robert et al. [18] applied initially random forest and SVM classifiers' approaches, to assess the IVF treatment process data of the candidate patients and found that the statistical methods alone were found to be insignificant in analyzing the impact of the wide range of influencing parameters involved in predicting the IVF success rates. Study further applied proximity-based imputation algorithm in combination with the RF and SVM classifiers. Results explained that RF-based classifiers and the imputation methods were found to be superior in the prediction rates with the representation of 80% and 79% precision on learning levels for learning and on validation dataset, respectively. Proximity-based imputation algorithm reported 88% probability rates for the lack of pregnancy but for the success rates the prediction percent was found to be low (65.5%). Study concluded that further studies required special implementation of hybrid designs that may increase the success percentage and prediction accuracies.

Siristatidis et al. [19] proposed ANN model design based on the LVQ classifier systems to analyze the IVF success rates. Datasets representing the 300 cases underwent the IVF treatment cycle, and the noise is eliminated applying, i.e., removal of inconsistent entries and data representing the incomplete information. Model design applied posses special feature which allows the trainer to increase the number of data sets, while mining process is in progress, either at the initial learning phase of the model or at the output phase, if the results found to be unsatisfactory in terms of prediction accuracies of the IVF success rate measurement. A threshold range of the positive versus negative prediction values reported more than 75% is assumed as satisfactory results.

Durairaj and Thamilselvan [20] applied ANN computational model design to analyze the candidate couple data to predict the success rates of the IVF treatment process. The study design considered the datasets specifying the various medical procedures and diagnostic test results of the candidate couple such as thickness levels of the endometrium layer before and after the hormonal therapy; various factors explain the condition of the tubes, follicle number in the ovaries, different physiological and psychological factors including the stress-causing factors collected retrieved from the databases of the various hospitals and the infertility research centers. Collected datasets are pre-processed and used as input information to optimize ANN study design to train, test, and predict model values and further cross-validate the output information by comparing the actual values with respect to the predicted values indicating the prediction accuracies of the ANN model design. Study reported that the ANN predicted reports are found to be more accurate (close to 73%) compared to the other data mining techniques such as traditional, statistical, and other computational methods.

Durairaj and Nandha Kumar [21] described the importance of measuring the error percent of true and false pulse values, plotter, and different classifier tools along with the computational based data mining technique, i.e., multilayer perceptron network to filter the raw data to obtain the most relevant data representing different influential factors with the less noise percentage. The model design studied identifies minimum datasets applying the filter techniques which enhance the mining; the efficacy of the mining process helps in achieving the higher prediction accuracies of the IVF success rates. Study indicated that the proposed data mining technique is found to be very useful for judging the minimum set of influential factors which in turn increase the prediction accuracies of the IVF success rates.

Altay Güvenir et al. [22] applied a novel technique designed based on the ranking algorithm, termed as success estimation ranking algorithm (SERA)-type RIMARC, with the aim of not only predicting the success rates but also discovering the significant attributes and their exacting values influencing the result of IVF treatment cycles. The performance outcome of the proposed novel algorithm SERA-RIMARC is compared with the other algorithms of the practice, i.e., naive Bayes classifier and random forest data mining model designs, measuring the accuracies by studying the ROC curves, time take and the present of accuracy through tenfold stratified cross-validation analysis. Study concluded that the novel design applied (SERA-RIMARC) is found to be more accurate and potential to estimate the prediction percentages of the IVF treatment process.

2.4 *Hybrid Algorithm-Based Models*

Priyavadana et al. [23] intended to develop a hybrid prediction method for assessing IVF-related medical and clinical data applying a combined hybrid model consisting of ANN and rough set theory algorithms (RST), and neuro solution is the software tool applied. RST algorithm is not very effective as an individual, because the efficiency of the algorithm is negatively influenced by the deterministic and non-deterministic rules of the algorithm. These limitations can be overcome while in the combination with ANN, and hence improve the prediction accuracies of the hybrid system. Study reported that the hybrid system developed yields much improved results compared with the other data mining techniques. It is practical that the hybrid technique of combined use of two or more machine learning tool combinations gives much improved results than the use of single technique for mining information from the database.

Durairaj and Nandha Kumar [24] state the importance of pre-processing of selected datasets to reduce the noise information and hence to support to achieve more accurate prediction results of the IVF cycles applying data mining techniques. In order to achieve the abovementioned statement, the same research proposed a hybrid-based combinatorial algorithm designed by combining Ant Colony and Relative Reduct algorithm (ACRRA). The efficacy of the applied hybrid algorithm is compared with the existing data mining algorithms. The results indicate that the algorithm applied was effective and achieved given target by decreasing the noise data without changing the efficacy and knowledge of the model design to predict the success rate. Studies indicated that reduction algorithm component of the hybrid system helps in improving the classification accuracies of the hybrid design and also shortens the time duration of the process.

In Mahmoodi and Naderi [25], the biggest challenge was faced in the estimation of IVF success rates with more accurate prediction probabilities involving the assessment of multivariable attributes. The nonlinear relationships between the factors studied in a biggest limitation of the traditional and the statistical-based methods and limit their overall prediction accuracies. Computational methods such as ANN with Feed-Forward Neural Network (FFNN) design algorithm can address the limitation issue of nonlinear relationship of the studied attributes by comparing the actual versus with predicted values, allows the model design to validate the design outcomes, and enhances the prediction accuracies of the model design.

Gowramma et al. [26] in this paper multi-layered perceptron is used which performs well in classification as compared to linear classification algorithms with respect to TP rate, FP rate, recall, and precision. This author uses three different attribute selection algorithms in which MLP network outputs 87.7% with data cleaning using attribute selection WrapperSubsetEval with search method known as genetic search which outperforms to predict most of the influenced factors for IVF that increases the success rate of IVF treatment.

In Ghaheri et al. [27], Genetic Algorithm (GA) works based on the principle of Darwin's evaluation theory, gaining importance in assessing the fitness functions of the computation-based algorithms applied in the data mining designs. Hybrid ANN-GA-based novel designs provide the validated results and assess the genetic fitness of the applied design algorithm in data mining approaches. Implementation of the hybrid ANN-GA exhibits greater promises in achieving the higher prediction rates of the IVF cycle success rates.

3 Conclusion

The work carried out here presented the review of various data mining applications such as statistical, traditional, computational, and hybrid approaches used in the prediction of IVF success rates. The significance and drawback of each of the models are presented. The identified research gaps help in proposing future research in the area. It can be observed from the literature that there is a need for model with enhanced accuracy. IVF success rate prediction before the IVF cycles provides useful information and recommendation to decision to be taken by candidate couple planning to undergo treatment. The prediction of IVF success rate also helps the doctors to suggest the feasible method of ART to undergo. It can also be observed from the literature that there is a limitation in multivariable factors that impacts prediction. There is a limitation in establishing nonlinear relationship of selected factors. It can be addressed by applying the most advanced computation algorithm like feed-forward neural network of ANN.

Lacunae of the various methods in addressing the issues have been discussed. The issues can be addressed by using algorithms like ANN and GA. ANN model trains, tests, and validate the outcome of the model by comparing actual versus predicted values. Genetic algorithms are based on Darwin's theory of evolution, and they are used to assess the fitness values of the candidates. They take population as input and individual generation as output.

Hybrid systems are recommended to achieve better accuracy as the efficacy of most of the existing systems needs to be improved. Applying ANN and GA can serve the purpose of enhancing the accuracy of IVF success rate prediction.

References

1. Adamson PC, Krupp K, Freeman AH, Klausner JD, Reingold AL, Madhivanan P (2011) Prevalence and correlates of primary infertility among young women in Mysore, India. Indian J Med Res 134(4):440–446
2. Sabanegh ES Jr (2010) Male infertility: problems and solutions. Springer Science and Business Media, pp 82–83. ISBN 978-1-60761-193-6. https://books.google.com/books?id=YthJpK5clTMC&pg=PA82
3. Ozkaya AU (2011) Assessing and enhancing machine learning methods in IVF process: predictive modelling of implantation and blastocyst development. Submitted to the Institute for Graduate Studies in Science and Engineering in partial fulfillment of the requirements for the degree of Doctor of Philosophy. Graduate Program in Computer Engineering, Boğaziçi University
4. Gowramma GS, Nayak S, Cholli N (2019) Intrinsic and extrinsic factors predicting the cumulative outcome of IVF/ICSI treatment. IJITEE 9(2S):269–273. ISSN: 2278-3075. https://doi.org/10.35940/ijitee.b1007.1292s19
5. Malizia BA, Hacker MR, Penzias AS (2009) Cumulative live-birth rates after in vitro fertilization. New England J Med 360:236–43
6. te Velde ER, Nieboer D, Lintsen AM, Braat DDM, Eijkemans MJC, Habbema JDF, Vergouwe Y (2014) Comparison of two models predicting IVF success; the effect of time trends on model performance. Human Reprod 29(1):57–64. https://doi.org/10.1093/humrep/det393
7. Rødgaard T, Heegaard PMH, Callesen H (2015) Non-invasive assessment of in-vitro embryo quality to improve transfer success. Reprod Biomed 31(5):585–592
8. Wilcox Lynne S, Peterson HB, Haseltine FP, Martin MC (1993) Defining and interpreting pregnancy success rates for in vitro fertilization. Fertil Steril 60(1):18–25
9. Trimarchi James R, Goodside J, Passmore L, Silberstein T, Hamel L, Gonzalez L (2003) Comparing data mining and logistic regression for predicting IVF outcome. In: 59th annual meeting of the American society for reproductive medicine, Abstracts 80(Suppl 3):S100
10. Baker VL, Jones CE, Cometti B, Hoehler F, Salle B, Urbancsek J, Soules MR (2010) Factors affecting success rates in two concurrent clinical IVF trials: an examination of potential explanations for the difference in pregnancy rates between the United States and Europe. Fertil Steril 94(4). https://doi.org/10.1016/j.fertnstert.2009.07.1673
11. LeBlanc M, Kooperberg C (2010) Boosting predictions of treatment success. Public Health Sci Div Fred Hutchinson Cancer Research Center, PNAS August 3, 2010, 107(31):13559–13560
12. Kakhki SA, Malekara B, Quchani SR, Khadem N (2013) A model based on Bayesian network for prediction of IVF success rate. SASTech 2013, Iran, Bandar-Abbas, 7–8 March 2013
13. Smith ADAC, Tilling K, Lawlor DA, Nelson SM (2015) External validation and calibration of IVF predict: a national prospective cohort study of 130,960 in vitro fertilisation cycles. PLoS ONE 10(4):1–15.e0121357. https://doi.org/10.1371/journal.pone.0121357
14. Dhillon RK, McLernon DJ, Smith PP, Fishel S, Dowell K, Deeks JJ, Bhattacharya S, Coomarasamy A (2016) Predicting the chance of live birth for women undergoing IVF: a novel pre-treatment counselling tool. Hum Reprod 31(1):84–92
15. McLernon DJ, Steyerberg EW, teVelde ER, Lee AJ, Bhattacharya S (2016) Predicting the chances of a live birth after one or more complete cycles of in vitro fertilisation: population based study of linked cycle data from 113 873 women. BMJ 355:i5735. http://dx.doi.org/10.1136/bmj.i5735
16. Hafiz P, Nematollahi M, Boostani R, Namavar JB (2017) Predicting implantation outcome of in vitro fertilization and intracytoplasmic sperm injection using data mining techniques. Int J Fertil Steril 11(3):184–190. https://doi.org/10.22074/ijfs.2017.4882
17. Camilo CG Jr, Yamanaka K (2011) In vitro fertilization genetic algorithm, evolutionary algorithms, Kita E (ed), ISBN: 978-953-307-171-8, InTech
18. Robert M, Malinowski P, Milewska AJ, Ziniewicz P, Czerniecki J, Pierzyński P, Wołczynski S (2012) Classification issue in the IVF ICSI/ET data analysis. Stud Logic, Grammar Rhetoric 29(42)

19. Siristatidis C, Pouliakis A, Chrelias C, Kassanos D (2011) Artificial intelligence in IVF: a need. Syst Biol Reprod Med 57(4):179–185. https://doi.org/10.3109/19396368.2011.558607
20. Durairaj M, Thamilselvan P (2013) Applications of artificial neural network for IVF data analysis and prediction. J Eng Comput Appl Sci (JEC&AS) 2(9):11–15. ISSN No: 2319-5606
21. Durairaj M, Nandha Kumar R (2013) Data mining application on IVF data for the selection of influential parameters on fertility. Int J Eng Adv Technol (IJEAT) 2(6):262–266. ISSN: 2249-8958
22. Altay Güvenir H, Misirli G, Dilbaz S, Ozdegirmenci O, Demir B, Dilbaz B (2015) Estimating the chance of success in IVF treatment using a ranking algorithm. Med Biol Eng Comput 53:911–920. https://doi.org/10.1007/s11517-015-1299-2
23. Priyavadana V, Sivashankari A, Senthil Kumar R (2017) A comparative study of data mining applications in diagnosing diseases. IRJET 02(07):1046–1053. e-ISSN: 2395-0056
24. Durairaj M, Nandha Kumar R (2017) Feature reduction by improvised hybrid algorithm for predicting the IVF success rate. Int J Adv Res Comput Sci 37–39
25. Mahmoodi M, Naderi A (2016) Applicability of artificial neural network and nonlinear regression to predict mechanical properties of equal channel angular rolled Al5083 sheets. Latin Am J Solids Struct 13(8):1515–1525
26. Gowramma GS, Mahesh TR, Gowda G (2017) An automatic system for IVF data classification by utilizing multilayer perceptron algorithm. In: ICCTEST-2017. ISBN 978-81-931119-5-6, vol 2, pp 667–672. https://doi.org/10.21647/icctest/2017/49043
27. Ghaheri A, Shoar S, Naderan M, Hoseini SS (2015) The applications of genetic algorithms in medicine. Oman Med J 30(6):406–416

TreeXP—An Instantiation of XPattern Framework

Thulasi Accottillam, K. T. V. Remya, and G. Raju

Abstract Most of the data generated from social media, Internet of Things, etc. are semi-structured or unstructured. XML is a leading semi-structured data commonly used over cross-platforms. XML clustering is an active research area. Because of the complexity of XML clustering, it remains a challenging area in data analytics, especially when Big Data is considered. In this paper, we focus on clustering of XML based on structure. A novel method for representing XML documents, Compressed Representation of XML Tree, is proposed following the concept of frequent pattern tree structure. From the proposed structure, clustering is carried out with a new algorithm, TreeXP, which follows the XPattern framework. The performances of the proposed representation and clustering algorithm are compared with a well-established PathXP algorithm and found to give the same performance, but require very less time.

Keywords XML · XML clustering · Structure-based XML clustering · XPattern · PathXP · TreeXP

1 Introduction

XML clustering and classification are the two research areas which pose several challenges. The most influential factors of the data mining algorithms are the type of information they use called local information and global information. While performing XML clustering, analyzing the merits and demerits of local and global information

T. Accottillam (✉) · K. T. V. Remya
Department of Information Technology, Kannur University, Kannur 670567, Kerala, India
e-mail: thulasi.accot@gmail.com

K. T. V. Remya
e-mail: remyaelib@gmail.com

G. Raju
Department of Computer Science and Engineering, School of Engineering and Technology, CHRIST (Deemed to be University), Kanminike, Kumbalagod, Bengaluru 560074, India
e-mail: kurupgraju@gmail.com

D. S. Jat et al. (eds.), *Data Science and Security*, Lecture Notes in Networks and Systems 132, https://doi.org/10.1007/978-981-15-5309-7_7

would be highly desirable [1]. Data mining algorithms analyze the underlying structure of the data and provide solutions to the problem, but they are inadequate to depict the obtained results. So, providing a self-descriptive result is beneficial. An influential factor is the identity of the dataset to be analyzed specifically when selecting a representation for XML documents.

XML document clustering can be performed in three different ways—based on only the structure of the document, only the content of the document, or using both content and structure. This paper focuses on the structure-based approaches for XML clustering. The general algorithms for XML clustering mainly have three phases—document representation, similarity computation, and document clustering based on the similarity. Among many approaches for representing XML structure, the most commonly adopted ones are trees, vectors, paths, and graphs [2]. There is a wide variety of similarity measures available for comparing XML documents like vector-based similarity, tree-based similarity, path-based similarity, etc. These measures are eminently relying upon the representation used. In the final phase of clustering methodology, the documents are organized into number of collections based on their resemblance. Many algorithms are available for this approach, which can be categorized into two, titled hierarchical clustering and flat clustering. The flat clustering algorithms create an individual set of clusters but the hierarchical clustering algorithms provide hierarchies of clusters.

An extensive study regarding the structure-based XML clustering and their applications are discussed in [3] and according to the paper, while dealing with homogeneous datasets complex representation like tree representation and similarity measures like edit distance measure need to be applied. But while considering a huge number of documents, lighter representations such as paths and frequency-based similarity measures are the better choice. Finally, for heterogeneous datasets, simple tag or metadata-based approaches may be sufficient.

Based on the study of different methods for structure-based XML clustering, a novel called TreeXP is presented. The new algorithm is designed based on a novel representation of XML documents named compressed representation of XML tree—CRXTree, which is an extension of classical FPTree representation. The proposed approach is found to be much faster than the well-established PathXP algorithm.

2 XPattern—A Pattern-Based Clustering Framework

The XPattern is a framework used for clustering XML documents using patterns. The patterns are the distinguishable information extracted from the documents like frequent subtrees, paths, words or expressions, height or width of the tree, number of distinct edges, etc. Patterns can also be any combination of these features. The XPattern framework uses a different methodology for clustering XML documents, which allows to group documents according to their characteristic features rather than their direct similarity [4] (Fig. 1).

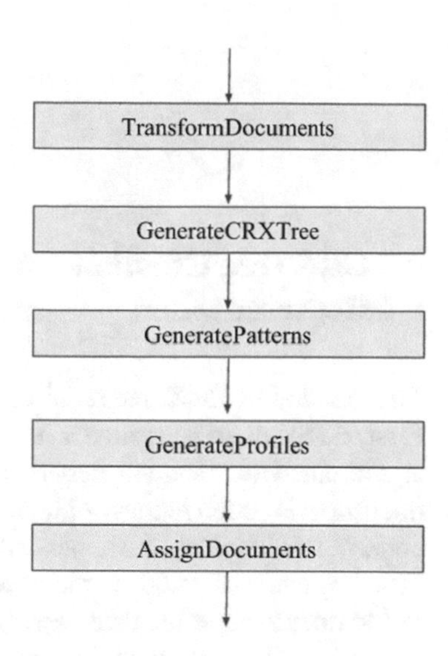

Fig. 1 XPattern framework

3 The PathXP Algorithm—An Instance of XPattern Framework

PathXP is an instantiation of the XPattern framework [4]. In PathXP, a feature f will be called a pattern if it is a maximal frequent path, i.e., if it is contained in at least minsup percent of documents in *D*, where minsup is a user-defined minimum support parameter and *f* is not a sub-path of any other frequent path:

$$\text{frequent}(f) \Leftrightarrow \exists_{D' \subseteq D} \forall_{d' \in D'} f \in d' \bigwedge \frac{|D'|}{|D|} \geq \text{minsup} \tag{1}$$

$$\text{pattern}(f) \Leftrightarrow \text{frequent}(f) \bigwedge \neg \exists_{f' \in F} (\text{frequent}(f') \bigwedge f \subset f') \tag{2}$$

The number of clusters is represented by *k,* and the initial value for the minsup is assigned as 1/*k*. Then, the maximal frequent paths are mined from input dataset *D*. The minsup is divided into two until the total count of identified paths is higher than *k* and the process is repeated. The set of mined patterns *P* is clustered together by complete link AHC algorithm using the following similarity measure:

$$sim(p_1, p_2) = \frac{\sum_{d \in D} \min\{m(p_1, d), m(p_2, d)\}}{\sum_{d \in D} m(p_1, d) + m(p_2, d) - \min\{m(p_1, d), m(p_2, d)\}} \tag{3}$$

Conclusively, all the documents are associated with concerned profiles based on the connection strength. For PathXP algorithm, connection strength is calculated by dividing the count of patterns inside document *d* as well as profile $\pi_{i,}$ by the size of $\pi_{i.}$

$$str(d, \pi_i) = \frac{\sum_{p \int \pi_i} m(p, d)}{|\pi_i|} \quad (4)$$

4 CRXTree Creation—A Novel XML Structure Representation

The creation of CRXTree requires complete scan of the entire documents only once. First, a root node is created with a user-chosen name, say *root*. The first document is examined for element details and added to the root node. The root element of document is added to the *root* node of the CRXTree along with frequency 1. Then the consecutive elements are appended to the node with their corresponding frequencies. The frequency of a node is increased when another node with the same name occurs in the document at the same level, else new node with frequency 1 is created. Like this, all the documents in the dataset are added to the CRXTree.

Along with the creation of CRXTree, the documents associated with each element are also can be noted. Hence, there is no need for subsequent scan(s).

The CRXTree differs from the conventional FPTree in some cases. In FPTree, there is a connection between elements having the same name, but in CRXTree, there is no connection between the elements having the same name. Another important difference is the CRXTree that contains all the elements of the dataset, but in FPTree only frequent items are stored. So, it is possible to recreate the entire document structure from the CRXTree.

5 Proposed Method—TreeXP

The new instantiation of XPattern framework uses the CRXTree for representing the XML documents. The frequent patterns are generated from the CRXTree, and the document pattern co-occurrences are maintained in a matrix. Then, these patterns are grouped into profiles, and each document is assigned to the profiles having the maximum connection strength, similar to the PathXP algorithm [4]. The presented method differs from the PathXP algorithm in generation of patterns, but similar in creating profiles and document assignment.

Tree XP Algorithm:

Input: set of XML documents *D*, number of clusters *k*
Output: set of k clusters C

```
xml_tree ← TransformDocuments(D)
crx_tree ← GenerateCRXTree(xml_tree)
minsup ← |D| / k;
P ← GeneratePatterns(crx_tree, minsup);
while |P| < k do
        min_sup ← min_sup/2;
        P ← GeneratePatterns(crx_tree, minsup);
end while
doc_pattern_matrix← GeneratePatternMatrix(xml_tree,P)
doc_sim_matrix← GenerateProfiles(doc_pattern_matrix,k)
C ← assignDocuments(doc_sim_matrix)
```

6 Example

Consider the dataset (*D*) which contains documents $doc_1, doc_2, \ldots, doc_8$, where doc_1, doc_2, doc_3, and doc_4 are the details of books and doc_5, doc_6, doc_7, and doc_8 are the details of journal papers [4].

First, the documents are represented as a rooted labeled ordered tree. The CRXTree used to represent the documents is portrayed in Fig. 2. This CRXTree is mined for frequent patterns with the *min_sup* = 4. The patterns generated are depicted in Table 1.

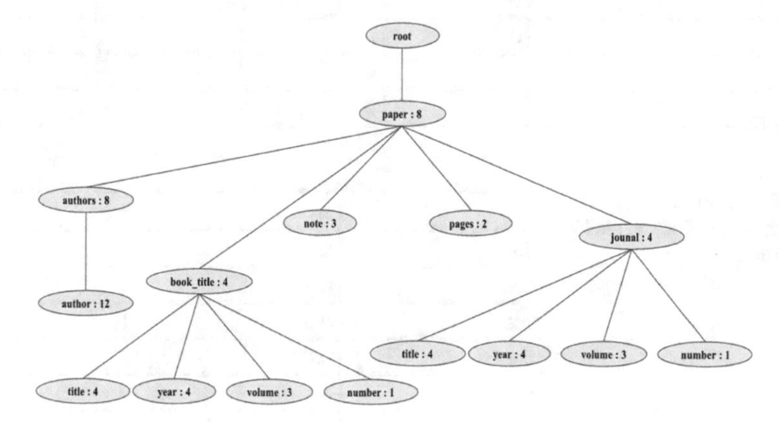

Fig. 2 The CRXTree representation for the dataset *D*

Table 1 Patterns extracted from the documents in D

Id	Pattern
p_1	Paper/authors/author
p_2	Paper/book title/title
p_3	Paper/book title/year
p_4	Paper/journal/title
p_5	Paper/journal/year

Here, the number of obtained patterns is greater than the number of expected clusters, so the minimum support is not decreased. Therefore, the algorithm continues with grouping patterns into profiles according to Eq. 3. The complete link AHC algorithm creates two profiles: the first one (π_1) contains patterns p_1, p_4, and p_5, while the second one (π_2) contains patterns p_2 and p_3. Finally, each document is assigned to the profile having the highest connection strength, using Eq. 4. The pattern document co-occurrences are presented in Table 2.

According to the connection strength presented in Table 3, documents doc_1, doc_2, doc_3, and doc_4 are assigned to the cluster represented by profile π_2, while documents doc_5, doc_6, doc_7, and doc_8 are assigned to the cluster represented by profile π_1. This concludes the algorithm.

Table 2 Document pattern co-occurrences

	p_1	p_2	p_3	p_4	p_5
doc_1	1	1	1	0	0
doc_2	1	1	1	0	0
doc_3	2	2	1	0	0
doc_4	2	2	1	0	0
doc_5	1	0	0	1	1
doc_6	1	0	0	1	1
doc_7	2	0	0	1	1
doc_8	2	0	0	1	1

Table 3 The connection strength for the documents to both profiles

	π_1	π_2
doc_1	0.33	1
doc_2	0.33	1
doc_3	0.67	1
doc_4	0.67	1
doc_5	1	0
doc_6	1	0
doc_7	1.33	0

Table 4 Details of the datasets

Dataset	Size	Number of documents	Classes	Distinct labels	Average height[a]
Eg_Docs	1.6 KB	8	2	11	4
Sigmod	1.3 MB	140	2	39	9
Homogeneous	1.5 MB	300	3	12	5
Heterogeneous	16.3 MB	1000	10	74	7

[a]Average height of the XML document tree

7 Empirical Evaluation

For the empirical analysis, four different algorithms including the classic k-means algorithm and agglomerative hierarchical algorithm also implemented along with the PathXP and TreeXP algorithms. The algorithms are implemented using R programming language commonly used among statisticians and data miners for data analysis.

7.1 Dataset

The algorithms are tested against both homogeneous and heterogeneous datasets which are both real and synthetic. They are

- Sample documents given in the example, named ***Eg_Docs.***
- SIGMOD record articles database [5] describing Sigmod articles, named ***Sigmod***.
- Synthetic homogeneous datasets created using similar DTDs containing book details, named ***Homogeneous (Homo)***.
- XML datasets available at XML Data Repository provided by University of Washington [6], named ***Heterogeneous (Het)***.

The details of the data sets are given in Table 4.

7.2 Precision

When the algorithm returns substantially more relevant results, the precision value will be high. Here, k-means and AHC algorithms produce the lowest value for precision since they fail to identify the correct clusters. Clustering quality for both PathXP and TreeXP is exactly the same having the higher value of precision. Hence, algorithms PathXP and TreeXP produce good clustering results compared to the other two clustering algorithms (Table 5).

Table 5 Precision for different clustering algorithms

	Eg_Docs	Sigmod	Homo	Het
K-means	0.75	0.17	0.14	0.05
AHC	0.85	0.15	0.13	0.11
PathXP	1	1	0.92	1
TreeXP	1	1	0.92	1

Table 6 Execution time (in minutes) for different clustering algorithms

	Eg_Docs	Sigmod	Homo	Het
K-means	0.20446418	37.2786355	6.049525976	30.42085206
AHC	0.21606114	34.29453096	5.989618683	21.05051032
PathXP	0.6650707	9.78498708	20.49168396	21.44562043
TreeXP	0.2345327854	3.81485796	7.710786581	7.04020172

The tag-only approaches are the fastest method when document size is small, but time required is very high when documents have larger size. From Table 6, we can conclude that TreeXP algorithm is ***three*** times faster than PathXP algorithm in all the situations.

The experiments are carried out with tag-only approach and PathXP algorithm along with the modified version TreeXP. To establish credibility as well as effectiveness of all these methods, we need to consider similar algorithms and carry out an extensive comparative analysis using more datasets. Implementation and results are promising. Major outcome of the study is the identification of several possibilities for the development of new approaches for structure-based XML clustering. Considering the results of PathXP and TreeXP, TreeXP maintains the same clustering quality of PathXP with much lesser time.

8 Conclusion and Future Works

The proposed algorithm TreeXP addresses some of the challenges as follows. The clusters are represented by a set of patterns, which makes identification and interpretation easier. Since the documents are not compared with each other explicitly, it is capable of including global information in the form of profiles.

A major research area that can be explored in relation to the study is the problem of clustering of large volume of XML datasets. In other words, the clustering of XML documents using the principles of Big Data. Also, the study points to the possibility of exploring collection of frequent patterns rather than maximal, by including additional, characteristic, non-maximal patterns.

References

1. Aggarwal C, Ta N, Wang J, Feng J, Zaki M (2007) Xproj. In: Proceedings of the 13th ACM SIGKDD international conference on knowledge discovery and data mining, KDD'07
2. Piernik M, Brzezinski D, Morzy T, Lesniewska A (2014) XML clustering: a review of structural approaches. Knowl Eng Rev 30(03):297–323
3. Thulasi A, Remya KTV, Raju G (2017) Structure based XML document clustering: a review. In: 2017 international conference on Infocom technologies and unmanned systems (trends and future directions) (ICTUS)
4. Piernik M, Brzezinski D, Morzy T (2015) Clustering XML documents by patterns. Knowl Inform Syst 46(1):185–212
5. Sigmodrecord.org. (n.d.) SIGMOD Record – SIGMOD Record Site. https://sigmodrecord.org
6. Aiweb.cs.washington.edu. (n.d.) UW XML Repository. http://aiweb.cs.washington.edu/research/projects/xmltk/xmldata/www/repository.html

Recommendation of Food Items for Thyroid Patients Using Content-Based KNN Method

Vaishali S. Vairale and Samiksha Shukla

Abstract Food recommendation system has become a recent topic of research due to increase use of web services. A balanced food intake is significant to maintain individual's physical health. Due to unhealthy eating patterns, it results in various diseases like diabetes, thyroid disorder, and even cancer. The choice of food items with proper nutritional values depends on individual's health conditions and food preferences. Therefore, personalized food recommendations are provided based on personal requirements. People can easily access a huge amount of food details from online sources like healthcare forums, dietitian blogs, and social media websites. Personal food preferences, health conditions, and reviews or ratings of food items are required to recommend diet for thyroid patients. We propose a unified food recommendation framework to identify food items by incorporating various content-based features. The framework uses the domain knowledge to build the private model to analyze unique food characteristics. The proposed recommender model generates diet recommendation list for thyroid patients using food items rating patterns and similarity scores. The experimental setup validated the proposed food recommender system with various evaluation criteria, and the proposed framework provides better results than conventional food recommender systems.

Keywords Content-based · K-Nearest neighbor · Term frequency · Inverse document frequency · Food preference matrix

V. S. Vairale (✉) · S. Shukla
CHRIST (Deemed to be University), Bengaluru, India
e-mail: vaishali.sheshrao@christuniveristy.in

S. Shukla
e-mail: samiksha.shukla@christuniversity.in

D. S. Jat et al. (eds.), *Data Science and Security*, Lecture Notes in Networks and Systems 132, https://doi.org/10.1007/978-981-15-5309-7_8

1 Introduction

Recommendation frameworks are gaining popularity for e-commerce websites and for online selection services like music, movies, books, and product sales. Still, recommendation of food is relatively a new research domain based on not only user's preferences but also fulfilment of the nutritional needs. The unhealthy dietary choices of individuals are mainly the reason for rapid growth in the cases of diabetes and obesity [1, 2]. The study [3] indicated that how the dietary factors affect the health issues like malnutrition, thyroid disorder, obesity and overweight, and diabetes. The important task of food computing is to find suitable food list for users and meet their preferences for food and nutritional needs [4]. The diet recommendations for users are relevant to individual's health issues. So, the diet recommender system builds a trade-off among health conditions and personalized food preferences.

In this work, the content-based techniques are explored for making personalized food recommendations. To achieve this, we developed a recommender system and set up the evaluation benchmark. The study adopted the new variations in content-based techniques to generate food items. The work is validated using performance metrics by capturing the accuracy level for the predicted ratings. Content-based methods and K-Nearest Neighbor (KNN) algorithm are evaluated together to validate the results.

This article is presented in the following sections. In Sect. 2, previously proposed recommender systems and approaches are discussed. Section 3 presents a proposed framework and describes a methodology applied in this work. Section 4 analyzes the experiments performed and discusses the results. Lastly, in Sect. 5, the conclusion is given by providing the main aspects of this research work.

2 Related Work

Many individuals have faced the issues while making healthy food choices in order to lessen the risk of diseases like thyroid, obesity, and diabetes. The chronic diseases are very much relevant to what type of food we eat. So, the food recommendations are not only based on user's food preferences but also need to consider user's health condition. Therefore, building a recommender model to incorporate these two components is becoming the core issue for healthy and nutrition-related food recommender system.

Many methods have been tried to consider health issues while recommending food by substituting ingredients [5, 6], considering calorie counts [7], suggesting food plans [8], and adding nutritional facts [9–11]. Ge et al. [7] evaluated the weighted mean for the preference and health component. The work updated the weights which were manually adjusted by the individual user. Nitish et al. [9] applied the food recognition techniques to get food-relevant data and obtain the nutritional values for recommendations. Ng and Jin [11] presented a personalized recipe recommender system for toddlers. The system incorporated the standard nutrition guidelines provided by the government of the US. Ribeiro et al. [12] considered more factors which included nutrition, preferences, and the cost for meal recommendation system for adults.

Forbes and Zhu [13] considered food ingredient data and extended the matrix factorization method for recipe recommender system. Bianchini et al. [14] generated the relevant recipes by comparing food features from both recipes' data and users' profiles. Min et al. [15] applied content-based model and considered food items as documents and food preferences as keywords. They proposed a recipe recommender system by using probabilistic topic model method. Lin et al. [16] incorporated many features of food items like ingredients, courses, dietary, cuisines, and preparation. The existing methods are trying to make a balance between user's health and food preferences using fusion methods like weighted summation [7].

3 Proposed Framework

The proposed model for food recommendation is presented in Fig. 1. The content-based learning is used to create a vector space model to generate a food preference matrix. The similarity score is embedded with the preference vector. The rating prediction is computed and generates a recommended food list.

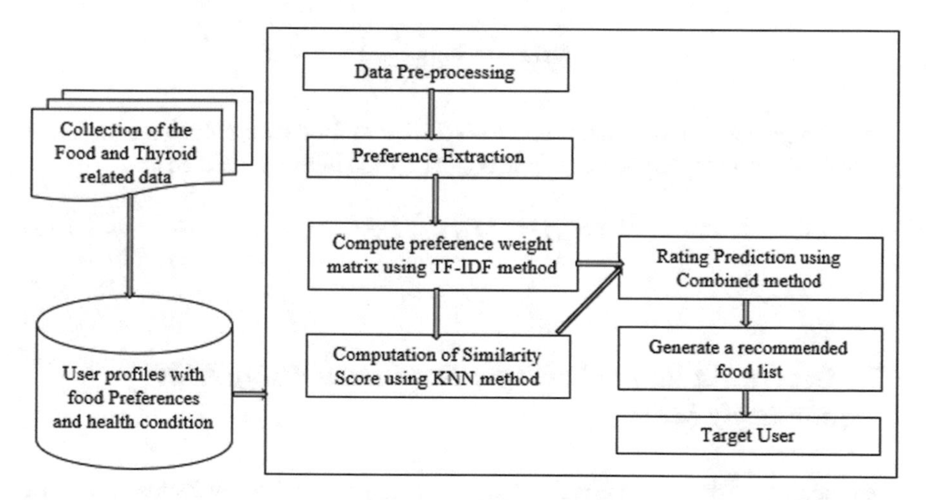

Fig. 1 Proposed framework

3.1 Generate a Food Preference Weight Vector Using TF–IDF Method

Content-based recommendation techniques basically match the features of an item with the user's profile. Then, the methods recommend the items with the highest matching score. Vector space model is used to present the vectors of weights with specific food data. Each food preference is a feature of it. Their weights are related to their relevance in the food document. Among various term weighting methods, the TF—Term Frequency and IDF—Inverse Document Frequency metric are commonly used in content-based recommendation models. TF–IDF consists of two terms. Term frequency is estimated as follows:

$$TF_{p,f} = \frac{N_{p,f}}{\sum aN_{a,f}} \tag{1}$$

where, for a food item f and a preference p, $N_{p,f}$ is the p times that food item f has preferred by the user.

However, the food items which are available in different documents may not be useful to distinguish various relevance levels among them. The relevance of rare keywords than frequent keywords can be found using IDF metric.

$$TDF_f = \log\left(\frac{F}{F_f}\right) \tag{2}$$

where F represents all food items and F_f is the list of food items which are preferred f times. The food preference's weight $W_{p,f}$ of a food item f is defined as

$$W_{p,f} = TF_{p,f} \times IDF_f \tag{3}$$

3.2 Generate a Rating Value for Food Items Using KNN Similarity Score

The proposed work crawls the various food- and nutrition-related websites. Basically, we have collected the food database which consists of food type, ingredients, quantity, and nutritional values. The proposed model is trained with user profile and food database along with food preferences of users. KNN method estimates a similarity score between food item vector and user profile vector. The proposed method uses the similarity score to calculate the predicted rating. The proposed algorithm combines the content-based and KNN methods together which is presented below.

Step 1: Use the content-based weighted preference vector $W_{p,f}$ to calculate a rating vector 'v' for each user 'u' in the food dataset.
$r_{u,f} = 1$ User u preferred the food item f by providing rating.
$r_{u,f} = 0$, user u not preferred the food item f.
Step 2: Estimate the similarity $S_{a,u}$ among the other users for an active user using Pearson correlation among their preference rating vectors.

$$S_{a,u} = \frac{\sum_{f=1}^{n} (r_{a,f} - \bar{r}_a) \times (r_{u,f} - \bar{r}_u)}{\sqrt{\sum_{f=1}^{n} (r_{a,f} - \bar{r}_a)^2 \times \sum_{f=1}^{n} (r_{u,f} - \bar{r}_u)^2}}$$

where $r_{a,f}$ and $r_{u,f}$ are the ratings given by two users a and u for a food item f; $\bar{r}_u$ and $\bar{r}_a$ are the mean ratings, and n is the total list of food items.
Step 3: Select k neighbors from the neighborhood with the decreasing values of similarity scores obtained from other neighbors for the user a.
Step 4: Predict the rating score using weighted averages of selected neighbor's ratings.

$$Pred_{a,f} = \bar{r}_a \frac{\sum_{u=1}^{k} (r_{u,f} - \bar{r}_u) \times S_{a,u}}{\sum_{u=1}^{k} S_{a,u}}$$

where $Pred_{a,f}$ is the rating prediction of the user a for food item f; k is the number of neighbors for user a.

4 Experimental Setup

The proposed recommender model focuses on obtaining the food list suitable for thyroid patients. We have collected the food list required for thyroid patients with nutritional values from various Indian food websites. We have crawled the users' profiles from the various food-related blogs and forums. The user profiles contain the preferences and their thyroid issues. The work develops the user–food item rating using user and food datasets together. The model is trained with food dataset along with thyroid user profiles. The user profiles are categorized into thyroid disorder types (Hyperthyroid, hypothyroid, normal) and among food preferences (Veg and nonveg). Each user profile has the preferred and disliked food items with their nutritional values. The preference vector is generated using Eq. 3 with TF-IDF content-based.

Cross-validation is used for validation of recommendation process. The method is used widely in recommendation systems to estimate how a predictive model performs accurately [17]. Root Mean Square Error (RMSE) and Mean Absolute Error (MAE) metrics computed the deviation among predicted and actual ratings. These metrics are applied for validation of recommendation error. The baseline algorithms, content-based method and KNN method used in this work, are computed using our specific food dataset. The major aim is to determine which recommendation method alone

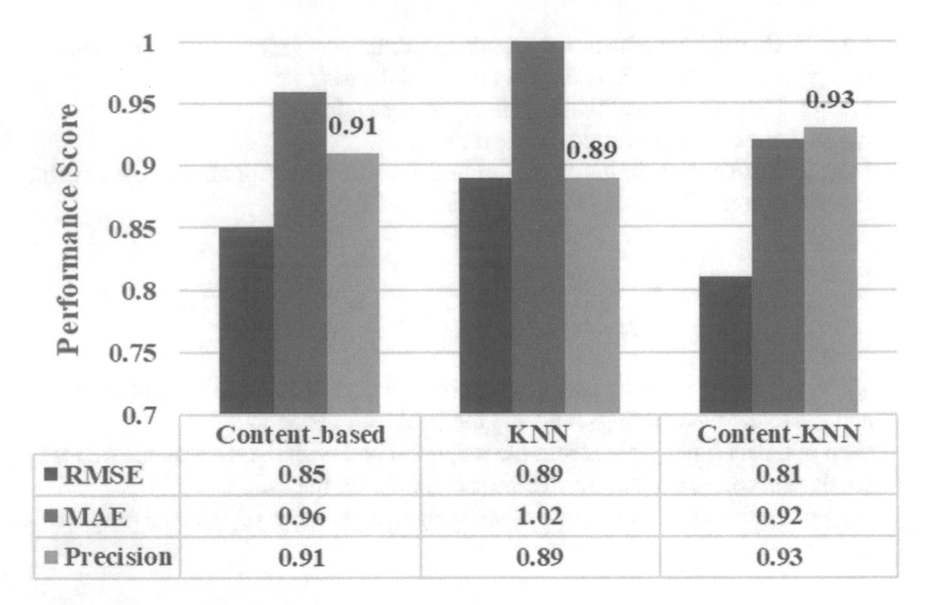

	Content-based	KNN	Content-KNN
RMSE	0.85	0.89	0.81
MAE	0.96	1.02	0.92
Precision	0.91	0.89	0.93

Fig. 2 Performance score of recommendation process

or combination of both has provided the best performance. Test results are provided in Fig. 2.

The obtained results received from experiments show that the recommender model performs well with the combination of content-based and KNN method. The proposed recommender model achieves 93% of precision for all users' profiles.

5 Conclusion

In this article, the proposed recommender model recommends the food list based on user preferences and health conditions. The work develops a classification model to generate a recommended food list based on user's profile. The model then combined the content features by combining preference weight matrix obtained from content-based method and similarity score obtained from KNN method. The proposed approach builds the users' food preference vector and then transforms the similarity score to get the predicted rating value. The prediction scores are then used to compute the performance of the recommendation framework. The combined method evaluated on dataset and tested the results to obtain personalized food recommendations. The future work may explore other recommendation algorithms and may include more features like meal cost, season of the year, and others.

References

1. Tirosh A, Calay ES, Tuncman G, Claiborn KC, Inouye KE, Eguchi K, Alcala M (2019) The shortchain fatty acid propionate increases glucagon and FABP4 production, impairing insulin action in mice and humans. Sci Trans Med 11(489)
2. I.D. Federation (2015) IDF Diabetes Atlas. International Diabetes Federation
3. G. R. F. Collaborators (2018) A systematic analysis for the global burden of disease study 2017. The Lancet 392(10159):1789–1858
4. Min W, Jiang S, Liu L, Rui Y, Jain R (2019) A survey on food computing. ACM Comput Surv 52(5):92:1–92:36
5. Elsweiler D, Trattner C, Harvey M (2017) Exploiting food choice biases for healthier recipe recommendation. In: International ACM SIGIR conference on research and development in information retrieval, pp 575–584
6. Teng C-Y, Lin Y-R, Adamic LA (2012) Recipe recommendation using ingredient networks. In: Proceedings of the ACM web science conference, pp 298–307
7. Ge M, Ricci F, Massimo D (2015) Health-aware food recommender system. In: Proceedings of the conference on recommender systems, pp 333–334
8. Elsweiler D, Harvey M (2015) Towards automatic meal plan recommendations for balanced nutrition. In: Proceedings of the 9th ACM conference on recommender systems, pp 313–316
9. Nitish N, Vaibhav P, Ramesh J (2017) Health multimedia: lifestyle recommendations based on diverse observations. In: Proceedings of the ACM on international conference on multimedia retrieval, pp 99–106
10. Nitish N, Vaibhav P, Abhisaar S, Jonathan L, Runyi W, Ramesh J (2017) Pocket dietitian: automated healthy dish recommendations by location. In: International conference on image analysis and processing, pp 444–452
11. Ng Y-K, Jin M (2017) Personalized recipe recommendations for toddlers based on nutrient intake and food preferences. In: Proceedings of the 9th international conference on management of digital ecosystems, pp 243–250
12. Ribeiro D, Ribeiro J, Vasconcelos MJM, Vieira EF, de Barros AC (2018) Souschef: improved meal recommender system for portuguese older adults. In: Information and communication technologies for ageing well and e-health. Springer International Publishing, pp 107–126
13. Forbes P, Zhu M (2011) Content-boosted matrix factorization for recommender systems: experiments with recipe recommendation. In: Proceedings of the fifth ACM conference on recommender systems, pp 261–264
14. Bianchini D, De Antonellis V, Melchiori (2015) A web-based application for semantic-driven food recommendation with reference prescriptions. In: Web information systems engineering. Springer International Publishing, Cham, pp 32–46
15. Min W, Bao B-K, Mei S, Zhu Y, Rui Y, Jiang S (2018) You are what you eat: exploring rich recipe information for cross-region food analysis. IEEE Trans Multimedia 20(4):950–964
16. Lin C-J, Kuo T-T, Lin S-D (2014) A content-based matrix factorization model for recipe recommendation. In: Advances in knowledge discovery and data mining. Springer International Publishing, Cham, pp 560–571
17. Kohavi Ron (1995) A Study of cross-validation and bootstrap for accuracy estimation and model selection. Int Joint Conf Artif Intell 14(12):1137–1143

Survey on DNS-Specific Security Issues and Solution Approaches

N. Usman Aijaz, Mohammed Misbahuddin, and Syed Raziuddin

Abstract The domain name system (DNS) is the most essential part of primary Internet infrastructure that translates hostnames to IP addresses. DNS server is available to all users to resolve their hostnames. They are not configured with any administrative control. DNS is an open resolver so that it is a hot cake for attackers to perform various attacks such as cache poisonings, DOS, and DDOS attacks. DNS only converts domain names to their matching IP addresses without providing any protection when sending messages. The man-in-the-middle can easily snoop into the insecure channel and can get the user's sensitive information. Due to this drawback of DNS, we came across another more secure layer of DNSSEC. DNSSEC uses certificates to authenticate the servers, and hence, more secure than DNS. It protects the DNS from other vulnerabilities as well. However, DNSSEC has its security concerns like confidentiality, packet fragmentation, amplification attack, and mainly it is a strong source of a DDOS attack. In this paper, we discuss how to increase the security of DNSSEC protocol by adding an extra layer of security using the EV-SSL certificate to overcome DNS security threats.

Keywords DNS · DOS · DDOS · DNSSEC · EV-SSL

N. Usman Aijaz (✉)
VTU RRC, Brindavan College of Engineering, Bangalore, India
e-mail: uaijaz9@gmail.com

M. Misbahuddin
Center for Development of Advanced Computing, Bangalore, India
e-mail: misbah@cdac.in

S. Raziuddin
Deccan College of Engineering, Hyderabad, India
e-mail: informraziuddin@gmail.com

D. S. Jat et al. (eds.), *Data Science and Security*, Lecture Notes in Networks and Systems 132, https://doi.org/10.1007/978-981-15-5309-7_9

1 Introduction

On the Internet, every computer that is connected to it is uniquely identified by the address known as Internet Protocol address or IP address, to recognize and get to it. These IP addresses are 32-bit numbers which are extremely hard to remember. As a human being, we can easily remember names compared to numerical. To overcome the difficulty and to access any system on the Internet through name instead of its IP address, a service is provided by the Internet and it is called a Domain Name System (DNS). DNS provides a means to map IP addresses to names. Without DNS, we would have to remember that 'www.amazon.com' is the IP address 72.21.207.65, which is very difficult. DNS is the most successful and largest distributed hierarchical database as shown in Fig. 1.

If a client wants to know the IP address for the hostname www.example.com, the client first accesses one of the root servers that does not have the IP address of www.example.com so it gives the IP address of the TLD servers; then the client contacts one of the TLD servers. The TLD server yields the IP address of an authoritative server; for example.com, the client contacts one of the authoritative servers, which returns the IP address of the hostname www.example.com. This may happen either iteratively or recursively. If it is recursive, then the burden of finding the IP address of a domain name is on root DNS, and if it is iterative then local DNS performs the relevant task as shown in Fig. 2.

In recent years, a large number of DNS attacks have been discovered such as Man-in-the-Middle, NXDomain, DNS ID Spoofing attacks, Malware, phishing, DOS and DDOS attacks, cache poisoning attack, information leakage attack, etc. These attacks lead to very serious consequences. Some are simple like making a service unapproachable, for example, a site isn't available, and some are risky like diverting to an inappropriate site through which attackers can pick up and get to our confidential information such as passwords, credit/debit PINs, and bank account details that may prompt tremendous money-related misfortune.

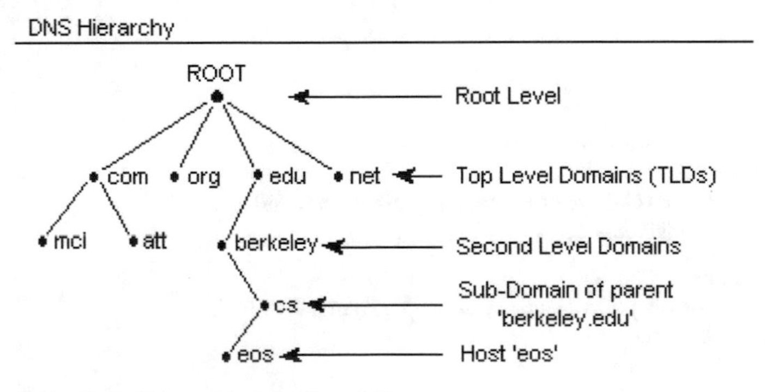

Fig. 1 DNS distributed hierarchical database [1]

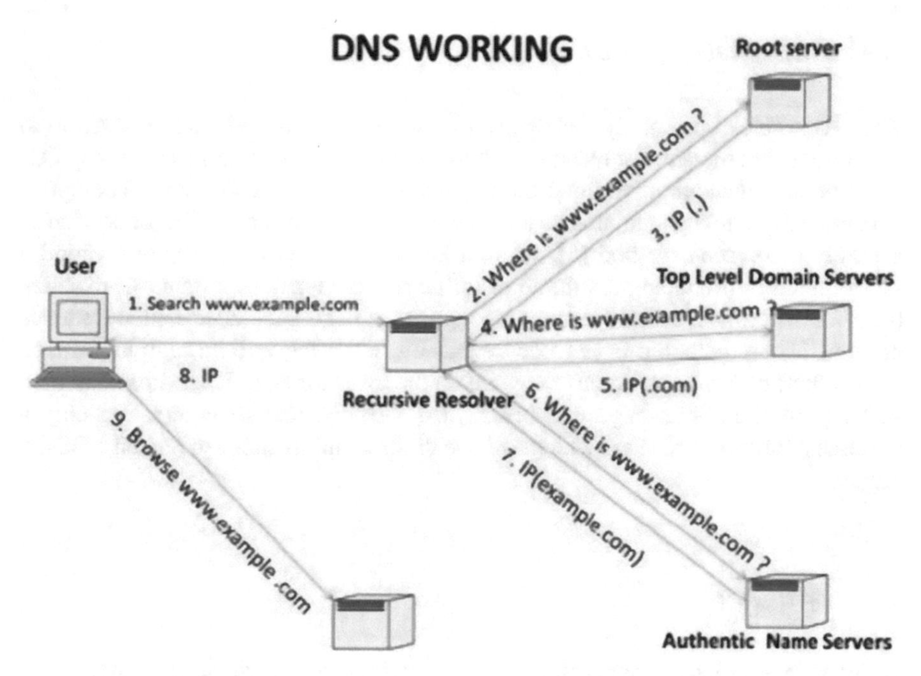

Fig. 2 Working of DNS [2]

2 DNS and Security Threats Related to DNS

DNS is an open resolver that is available to provide name resolution to all clients regardless of their domain. They are not designed with any managerial control. Open resolver is on the hotcake for attackers and may be susceptible to mischievous activities and DNS cache poisoning attacks, DOS/DDoS attacks, malware (malicious software), phishing, resource utilization attacks, DNS ID Hacking, etc. against DNS server [3]. Some of these attacks are discussed in this section.

2.1 DNS Cache Poisoning Attack

Here, DNS is allowed to store malicious data by the hackers, which results in the wrong mapping of URL to an IP address of the authoritative server which is under the control of hackers [4]. For instance, bogus addresses for some bank.com or any agency's site harmed with the IP address of web server controlled by hackers, where hackers facilitated a digitally indistinguishable replica of the original site, and all clients will be a phony victim to cache poisoning attacks. The end clients are defrauded for login, password, account information credit card numbers, etc. This information will be recorded and utilized later on by attackers.

2.2 DOS/DDOS Attack

The DOS attack commonly makes use of one PC and one Internet connection to flood a targeted machine or resource, whereas DDOS attack makes use of many PCs and Internet connections to flood the targeted resource [4]. DOS attacks comprise exceptionally damageable attacks to downfall or degrade the quality of service in a barely unforeseen method [5], while a DDOS attack is an endeavor to flood a victim through producing a volume of traffic by an enormous number of machine [6, 7]. DDOS attack utilizes a countless number of negotiated hosts called Zombies or Bots that are gathered from insecure PCs that enter the web through high-speed connections by using malicious software on the insecure PCs. These hosts are then gathered to shape one huge network named a Botnet that is anticipating only a command from the attackers who may have different motivations to launch a DDOS attack.

2.3 Spoofing

A spoofing attack may be a condition during which an individual or program can effectively pretend as another by misrepresenting information, to realize an illegitimate advantage. Many communication channels do not have mechanisms for authenticating the message to know whether it is from source or destination, and thus are susceptible to spoofing attacks when additional precautionary measures are not taken by applications to authenticate the identity of the sending or receiving host.

An example of spoofing is where the man-in-the-middle pretends to be the sender and then sniffs through the packets, modifies them, and ultimately sends the modified messages to the receiver. Spoofing can also be the case when the attacker sniffs the packets and reads/modifies the sensitive personal information.

2.4 Phishing

Phishing is a fraudulent attack used to steal user confidential information like login credentials and credit/debit card numbers. It is often used for malicious reasons, by impersonating as a reliable entity in an electronic communication. It is carried out by email spoofing or instant messaging. This kind of attack usually redirects users to enter their confidential information at a bogus site controlled by the attackers. These sites look identical to a legitimate website due to which it is very difficult for the user to identify them as fake.

An example of phishing is where the victim opens a website which is a fake one but looks exactly like a legitimate one. In this case, the victim enters his/her personal sensitive information which is obtained by the attacker, hence, using those details to hack or modify other accounts.

2.5 *Identity Theft*

Identity theft is the deliberate use of someone else's identity. It is a crime in which an imposter obtains key pieces of information such as Name, Address, Bank Account numbers, and credit/debit card numbers without their consent, and uses it for their own personal gain. Identity theft can occur when the attacker uses the victim's personal information to access financial benefits or for some other offense. In this case, the victim is not known whether her identity has been stolen. The attacker very cleverly steals the identity of the users.

2.6 *Forgery*

It is the process of creating, molding, or imitating objects, figures, or documents with the intention of cheating or changing the public perception to make a profit by selling a counterfeit item [8].

2.7 *Eavesdropping*

Eavesdropping is listening to others privately or secretly without their consent. Network eavesdropping is a network layer attack, which emphases capturing small packets from networks transferred by other computers and reading data content in pursuit of certain types of information. This type of network attack is one of the most effective due to the lack of commonly used encryption services. These types of attacks are usually performed by Black Hat hackers. On the other hand, even government agencies, such as the National Security Agency, are also involved [8].

2.8 *Packet Sniffing*

Sniffing is a function through which a person can capture packets of data flowing over a computer network. The software or device that performs this task is called a packet sniffer. By using a packet sniffer, it is possible to capture passwords, IP addresses, protocols, and other information on the network which will help the attacker to intrude the network. All network data travels across the Internet and then into and out of PCs as an individual, variable-size data packets. Since the typical PC user never sees this raw data, many spyware systems secretly send superficial information from the user's computer without their realization [8].

2.9 Tampering

Packet tampering involves deliberately modifying the contents of packets through the unauthorized channel as they travel over the Internet or exchange data on computer disks after entering a network. For example, an attacker might place a tap on the network line to capture packets as they leave the computer. Modern networks normally drop packets when the load temporarily surpasses their buffering limits [8].

2.10 Man-in-the-Middle Attack

MITM is an attack where an attacker intercepts communications between two parties either to secretly eavesdrop or modify traffic traveling between them, and they believe that they are communicating directly to each other. An example of intermittent attacks is active eavesdropping, in which the attacker forms an independent connection with the victim machine and sends a message between them to convince that they are talking directly to each other in a private connection [8].

3 Solution to DNS Security Threats

3.1 DNSSEC (DNS Security Extension)

DNSSEC (DNS Security Extension) as defined in [RFC4033-4035], proposed and standardized in 1997, resolves some security issues associated with DNS protocols. DNSSEC protects data sent by DNS servers against fraud. This ensures two security objectives, such as authentication and integrity of the data source. DNSSEC makes stronger authentication in DNS using digital signatures based on public-key cryptography [9].

DNSSEC secures zone files using cryptography. Therefore, each zone has at least one pair of the key. The child zone's public key is signed by the private key of the parent zone, parting only the root that is signed itself. This method creates a trust chain (chain of trust). According to Shulman and Wadner, Domain Name System Security Extensions (DNSSEC) standards such as RFC4033, RFC4034, and RFC4035 are intended to address the issue of cache poisoning in DNS, through data integrity and original authenticity via a cryptographic digital signature over DNS resource records [10].

3.2 SSL Certificates

These certificates are small-sized files that digitally bind a cryptographic key to an organization's information like hostname, server name or domain name, and organizational uniqueness and geographical locality of the organization. SSL is used to secure logins, data transfer, and credit card transactions. An organization wants to install the SSL certificate onto its web server to initiate a secure reliable session with web browsers. Once a secure session is made, all web traffic links between the web server and the web browser will become secure and reliable. When the SSL certificate is properly connected to the server, the application protocol known as HTTP will change to HTTPs, where 'S' means 'secure'. Depending on the type of certificate purchased for the website and which browser is being used to surf the Internet, a browser will show a padlock or green bar in the browser search bar before the website link, when someone visits a website that has an SSL certificate installed as shown in Fig. 4.

A green-locked padlock indicates that the website is secure under HTTPS, and if there is a red unlocked padlock symbol it shows that website is not secured under HTTPS. SSL certificates are also known as public-key cryptography certificates PKC. This type of cryptography has the power of two keys that are long strings of randomly generated numbers. One is called a private key; the other is the public key. The server knows the public key and is available in the public domain, which can be used to encrypt the messages. It uses asymmetric public-key encryption and decryption process, for example, if Alice sends a message to Bob, she locks it with Bob's public key, but the only way to decrypt it is to unlock it with Bob's private key. Bob is the only person with his private key, so Bob can only use it to unlock Alice's message. If a hacker blocks before Bob unlocks it, they get a cryptographic code, which they can't break without Bob's private key, which is very secure and hidden [11].

Advantages of an SSL certificate are as follows [12]:

- SSL increases customer trust and revenue;
- It increases your Google Rankings;
- It protects from Google Warning;
- SSL protects from phishing and other attacks.

3.3 Extended Validation Certificate

The EV-SSL (Extended Validation SSL) certificate is the highest form of SSL certificate available on the market so far as shown in Fig. 3 [11]. Websites phishing, or fraudulent or identity theft, continue to pose a foremost threat to legitimate/authentic websites and online services' phishes began using certificates to trick their victims into viewing their sites more faithfully and submitting their financial or personal data. EV-SSL can help solve this problem by bringing the site operator's verified

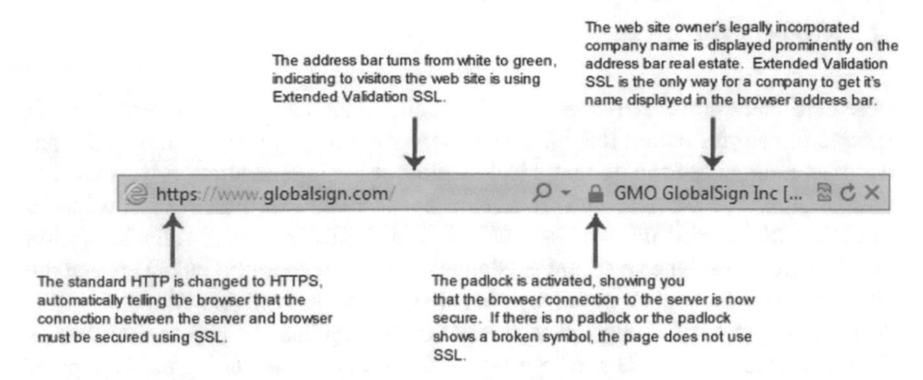

Fig. 3 EV-SSL certificate. https://www.globalsign.com/en-in/ssl-information-center/what-is-an-extended-validation-certificate

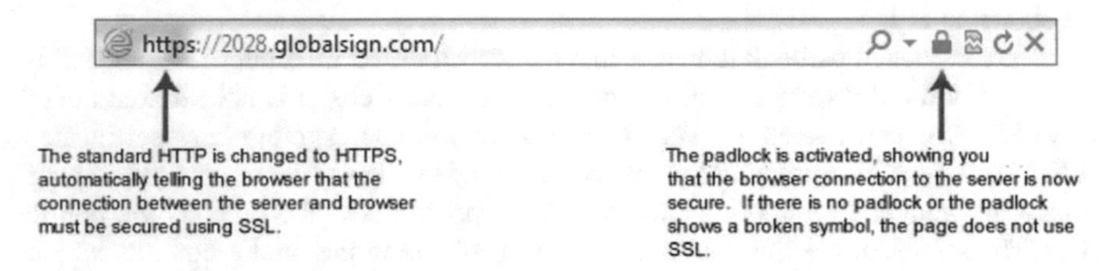

Fig. 4 SSL certificate. https://www.globalsign.com/en-au/ssl-information-center/what-is-an-ssl-certificate/

identity to the front and center of the website—displaying it directly in the address bar making verification user-friendly and easy on the client-side (Fig. 4).

4 Benefits of DNSSEC and SSL

Complete use of DNSEC will ensure that the end user is connected to the true web site or other service corresponding to a particular domain name. Although this will not cater to all the security issues of the DNS, it does protect a critical part of it, the directory lookup complementing it with other technologies such as SSL that protects the conversation, and provide a platform for yet to be developed security improvements.

4.1 Forgery

In the DNS lookup, the user computer is not always waiting for the address of www.example.com. Hence, the hacker has to schedule his forgeries such that they arrive when a user computer is waiting for the reply. There are several definite ways for the hacker to do forgery of data supplied over the Internet. Some of them are

- Attack repeatedly by sending a forged packet one after another. As a result, the probability of arriving one of the forgeries at the right time increases.
- Hacker can poke user PC to trigger/start a known DNS lookup at a known time.
- User computer stores incoming DNS results for queries in a cache and reuses those results for a while for faster navigation using this as a plus point a hacker caches the DNS records in advance and redirects the user to malicious websites.

In all of this, the hacker must win the race against the legitimate/standard DNS response from the example.com administrator. If the user receives a standard DNS response before the duplicate response, it stops listening and then ignores the duplicate response received.

- HTTPS says it protects the user from fake websites. Suppose a hacker duplicates a DNS response. The user's computer connects to the hacker's web server. But HTTPS warns the user about the "invalid certificate" and avoids displaying fake web pages. But unfortunately, HTTPS still hasn't shown him the right web page.
- HTTPS protects the integrity of web pages, not availability. It also has usage issues that prevent it from being used on most of the web pages.
- Using DNS in conjunction with HTTP or HTTPS is not protected from duplicate web pages. But using DNSEC with HTTP or HTTPS can protect against duplicate web pages but at the cost of rejecting the service.

4.2 Tampering

Tampering of packets comprises changing the contents of packets as they move from one point to another over the Internet or modifying data on computer disks after penetrating a network as they travel the communication links over the network [8]. A hacker can tamper the DNS records or the data being transferred over the channel to redirect the user to the malicious website. But to tamper the DNS records stored a hacker must introduce false records in the DNS resource records but they won't be signed by the zone, and hence will get filtered out. Also, the data traveling through the communication links cannot be changed as the SSL connection is running on the server.

4.3 Cache Poisoning Attack

An example of DNS cache poisoning is where hackers swap a genuine IP address in the DNS cache with an IP address they control. Users fetching information from this DNS cache are forwarded to the hacker's compromised site. The redirected site may be a malware download site or one designed to collect confidential information.

Another way of preventing cache poisoning is by adding security to the DNS using tools such as DNSSEC. Secure DNS (DNSSEC) uses cryptographic digital signatures signed with a trusted pair of the public key(s)/private key(s), and these certificates are stored in the zone files to validate the legitimacy and integrity of the information. The attacker cannot forge the digitally signed certificates and hence, DNS cache poisoning is prevented.

4.4 Eavesdropping

This attack is cured using SSL as encrypted data is transferred over network links which cannot be decrypted by the hacker; hence, they don't know what messages are exchanged between the end entities and therefore cannot alter them.

5 Conclusion

Denial of service attack and especially Distributed DOS attacks are dangerous for the Internet and web services. According to the surveys, the percentage of attacks is at an exponential rise with new and sophisticated techniques. In this paper, we discussed DNS and its security threats like cache poisoning, DOS and DDOS attacks, spoofing, phishing, identity theft, forgery, etc. To secure DNS, IFTF added extensions to it known as DNSSEC. DNS together with HTTP or HTTPS cannot protect against fake web pages. But using DNSSEC with either HTTP or HTTPS can protect against fake web pages but with a cost of denial of service. DNSSEC uses cryptographic digital signatures signed with a trusted pair of the public key(s)/private key(s), and these certificates are stored in the zone files to authenticate the authenticity and integrity of the information. The attacker cannot forge the digitally signed certificates, and hence, DNS Cache poisoning is prevented. Although using DNSSEC over DNS cures a lot of DNS security problems, they are still some security issues left that cannot be cured using DNSEC. DNSSEC does not do any data encryption on the DNS and isn't involved at all once the web interaction has started. This means that anyone watching the network could easily steal information. DNSSEC and SSL are two vital components of a layered approach to Internet security. When woven together, DNSSEC and SSL provide users the confidence to reliably trust they're on the right websites and talking to the right people in a verifiably secure way.

References

1. https://www.inetdaemon.com/tutorials/internet/dns/operation/hierarchy.shtml
2. https://aws.amazon.com/route53/what-is-dns/
3. Jalalzai MH, Shahid, WB, Iqbal MMW (2015) DNS security challenges and best practices to deploy secure DNS with digital signatures. In: Proceedings of 2015 12th international Bhurban conference on applied sciences and technology (IBCAST) Islamabad, Pakistan, 13–17 January 2015
4. Zou F, Zhang S, Pang L, Li L, Li J, Pei B (2016) Survey on DNS security. In: 2016 IEEE icon data science in cyberspace
5. Todd B (2000) Distributed Denial of service attacks, Feb. 18, 2000.http://www.linuxsecurity.com/resource/files/intrusion/detection/ddos–whitepaper.html
6. Peng T, Leckie C, Ramamohanarao K (2007) Survey of network-based defense mechanisms countering the DoS and DDoS problems. ACM Comput Surv 39(1), Article 3
7. Tariq U, Hong M, Lhee K (2006) A comprehensive categorization of DDoS attack and DDoS defense techniques. ADMA LNAI 4093:1025–1036
8. Wikipedia. https://en.wikipedia.org
9. https://www.icann.org/resources/pages/dnssec-what-is-it-why-important-2019-03-05-en
10. Shulman H, Waidner M (2014) Towards forensic analysis of attacks with DNSSEC. In: Proceedings of the 2014 IEEE security and privacy workshops, SPW'14, May 17–18, 2014, pp 69–76
11. GlobalSign GMO INTERNET GROUP, "SSL Information Center". https://www.globalsign.com/en/ssl-informationcenter/what-is-an-ssl-certificate/
12. https://www.globalsign.com/en-au/ssl-information-center/what-is-an-ssl-certificate/

Algorithms for the Metric Dimension of a Simple Graph

Xavier Chelladurai and Joseph Varghese Kureethara

Abstract Let $G = (V, E)$ be a connected, simple graph with n vertices and m edges. Let $v_1, v_2 \in V$, $d(v_1, v_2)$ is the number of edges in the shortest path from v_1 to v_2. A vertex v is said to distinguish two vertices x and y if $d(v, x)$ and $d(v, y)$ are different. $D(v)$ as the set of all vertex pairs which are distinguished by v. A subset of V, S is a metric generator of the graph G if every pair of vertices from V is distinguished by some element of S. Trivially, the whole vertex set V is a metric generator of G. A metric generator with minimum cardinality is called a metric basis of the graph G. The cardinality of metric basis is called the metric dimension of G. In this paper, we develop algorithms to find the metric dimension and a metric basis of a simple graph. These algorithms have the worst-case complexity of $O(nm)$.

Keyword Metric dimension · Resolving set · Locating set · Metric basis · Algorithms · BFS · Spanning tree

1 Introduction

Let $G = (V, E)$ be a connected, simple graph with n vertices and m edges. Let, for v_1, v_2 of V, $d(v_1, v_2)$ be the distance between vertices v_1 and v_2, which is the number of edges in the shortest path from v_1 to v_2. A vertex v is said to distinguish two vertices x and y if $d(v, x)$ and $d(v, y)$ are unequal. The concept of metric generator of a graph was introduced by Alejandro Estrada-Moreno and others [3] as a set of vertices which distinguish every pair of vertices of the graph. That is, a set of vertices S is a metric generator of G, if for every pair of vertices u, v of G, there is at least one element of S which distinguishes u and v. Trivially, the whole vertex set V is a metric generator of G because, for any pair of vertices u, v, the vertex u distinguishes u and v. Also, v

X. Chelladurai · J. V. Kureethara (✉)
CHRIST (Deemed to be University), Bangalore 560029, India
e-mail: frjoseph@christuniversity.in

X. Chelladurai
e-mail: xavier.c@christuniversity.in

D. S. Jat et al. (eds.), *Data Science and Security*, Lecture Notes in Networks and Systems 132, https://doi.org/10.1007/978-981-15-5309-7_10

distinguishes u and v. A metric generator with minimum cardinality is called a metric basis of the graph G. The cardinality of metric basis is called the metric dimension of G. The metric dimension of the Cartesian Product [2], of the corona product [21], its relation with partition dimension [20] and resolvability [4] are of great academic interest. The local metric dimension was studied in [15] whereas strong metric dimension evoked much more interest [13, 14]. The related concept of partition dimension [6, 7, 20, 22] has also been studied extensively. The concept of metric generators was also studied by Slater [16, 18] where he called them locating sets [9, 19] as he designed the concept from the applications in identifying uniquely the intruders of networks. Harary and Melter in [8] also introduced the same concept with the name resolving sets. Studies on resolving domination [1], resolvability and the upper dimension [5] etc. evoked much interest among the researchers. Applications of resolving sets to the navigation of robots in networks are discussed in [12], and the applications to chemistry were discussed in [10, 11]. Alejandro Estrada-Moreno and others [3] have also generalized the concept of metric generators and defined k-metric generators. Let k be an integer less than n. A set of vertices S is called a k-metric generator if every pair of vertices of G is separated by at least k vertices of S. A graph G is said to be k-metric dimensional if k is the largest integer such that G has a k-metric generator [17].

In this research, we introduce a few data structures to efficiently represent the graph and develop the algorithm to find the metric dimension and metric basis of a simple graph. We have also developed algorithms for k-metric generators and k-metric dimension of graphs.

2 Metric Dimension and k-Metric Dimension

In the graph G in Fig. 1, $d(v_1, x) = 1$, and $d(v_1, y) = 2$, So, v_1 distinguishes x and y. $d(u, x) = 1$ and $d(u, y) = 1$, So, u does not distinguish x and y.

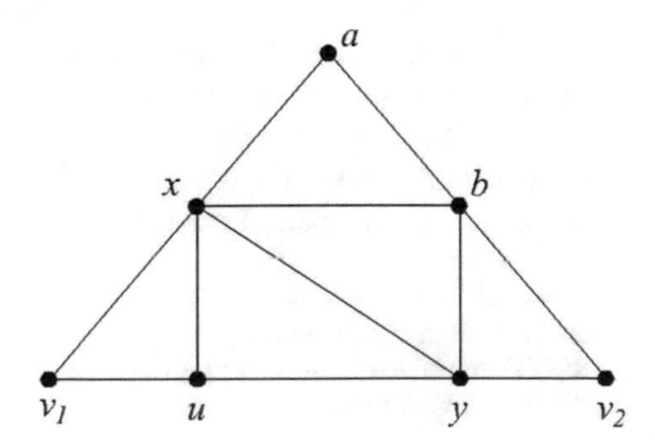

Fig. 1 Graph $G = (V, E)$

2.1 *Metric Dimension of a Path and Cycle*

For a path, there are two pendant vertices. From every vertex of the graph to any of the pendant vertices, the distance is unique. Hence, the metric dimension of a path is 1. In fact, each of the pendant vertices independently qualifies to be a metric generator. It is a well-known fact that a graph has a metric dimension 1 if and only if it is a path. A closed path is a cycle. A single vertex cannot be a metric generator. However, any adjacent vertices form a metric basis. Hence, the metric dimension of a cycle is 2.

To find the metric dimension of a graph, we can begin with constructing the distance matrix of the graph. Distance matrix of a graph G is defined as follows.

$D(G) = d_{ij}$, where d_{ij} is the distance between the vertices v_i and v_j.

Given a graph with n vertices, formulating the distance matrix and then finding metric dimension is not that simple.

2.2 *Metric Basis of a General Graph*

We now illustrate the finding of the metric dimension of a general graph.

The metric dimension is neither 1 nor 8. In Table 1, we present the distance matrix of graph in Fig. 2.

Table 2 shows the list of all vertices that distinguishes every pair of vertices of the graph in Fig. 2.

Table 1 Distance matrix of the graph in Fig. 2

	v_1	v_2	v_3	v_4	v_5	v_6	v_7
v_1	0	1	2	3	2	2	1
v_2	1	0	1	2	1	2	1
v_3	2	1	0	1	2	3	2
v_4	3	2	1	0	1	2	2
v_5	2	1	2	1	0	1	1
v_6	2	2	3	2	1	0	1
v_7	*1*	1	2	2	1	1	0

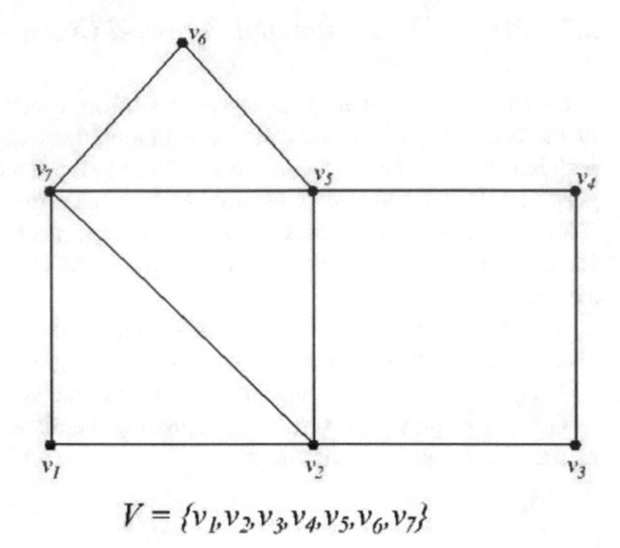

Fig. 2 A non-acyclic graph with seven vertices and nine edges

Table 2 Distance matrix of the graph in Fig. 2

Pair	v_1	v_2	v_3	v_4	v_5	v_6	v_7	$\|D(G)\|$
v_1v_2	*	*	*	*	*			5
v_1v_3	*		*	*		*	*	5
v_1v_4	*	*	*	*	*		*	6
v_1v_5	*		*	*	*	*		5
v_1v_6	*	*	*	*	*	*		6
v_1v_7	*			*	*	*	*	5
v_2v_3	*	*	*	*	*	*	*	7
v_2v_4	*	*		*			*	4
v_2v_5	*	*	*	*	*	*		6
v_2v_6	*	*	*			*		4
v_2v_7		*	*			*	*	4
v_3v_4	*	*	*	*	*	*		6
v_3v_5			*		*	*	*	4
v_3v_6		*	*	*	*	*	*	6
v_3v_7	*		*	*	*	*	*	6
v_4v_5	*	*	*	*	*	*	*	7
v_4v_6	*		*	*		*	*	5
v_4v_7	*	*	*	*		*	*	6
v_5v_6		*	*	*	*	*		5
v_5v_7	*			*	*		*	4
v_6v_7	*	*	*			*	*	5

2.3 *Algorithm 1: To Find* k *Such that the Graph* G *Is* k-*Metric Dimensional*

Input: A simple connected graph $G = (V, E)$ with n vertices $\{v_1, v_2, v_3, \ldots, v_n\}$ and m edges represented by the Boolean adjacency matrix $A(n, n)$ which has only $2m$ true values.

Step1:Find the Distance matrix, *distance(n,n)* where *distance* $_{ij}$ is the distance between v_i and v_j where i and j values range from *1* to *n*.

Step2 ***For*** *every pair of vertices* (v_i, v_j), $i \neq j$, *do the following*:

Step2.1 For k=1 to n do the following

$d_k(v_i, v_j) = 1$ if v_k distinguishes v_i and v_j and 0 otherwise.

End For

Step2.2 Calculate $d(v_i, v_j) = \sum_{k=1}^{k=n} d_k(v_i, v_j)$

End For

Step3

$k = \min\{d(v_i, v_j)$, for i and j ranging from 1 to n.

Output The result is k. This means that metric dimension of graph G is k.

Correctness of Algorithm 1.

In the above algorithm, in Step **3**, assume that the value k is as follows:

$$k = d(v_i, v_j), \text{ for specific values of } i \text{ and } j.$$

This means that v_i and v_j are distinguished by exactly k vertices of G. This shows that there is no $(k + 1)$-metric generator.

Also, since k is chosen as the minimum among all vertex pairs, for any other vertex pair (v_s, v_t), we have $k \leq d(v_s, v_t)$.

This means that v_s and v_t are distinguished by more than k vertices.

So, we have established that every pair of vertices is distinguished by at least k vertices. By choosing $S = V$, this shows that S is a k-metric generator.

2.4 Illustration of the Algorithm

Consider $V = \{v_1, v_2, v_3, v_4, v_5, v_6, v_7\}$ *as* shown in Fig. 2.

Step 1: Table 1 gives the distance matrix of the graph.

Step 2: For $i = 1, j = 2$, let us calculate

Step 2.1

For $k = 1$ to n do the following

$d_k(v_1, v_2) = 1$ if v_k distinguishes v_1 and v_2 and 0 otherwise.

End For

This is derived as (1 1 1 1 1 0 0).

In a similar manner, we can calculate the column matrix for every pair of vertices v_i and v_j such that $i < j$. The full matrix is derived from Table 2 and is shown below:

Step 2.2 For every row in the above matrix, calculate the row sum. This is shown in Table 2 as the last column.

Step 3: In the last column of Table 2, choose the minimum value as k.
We get $k = 4$.
This means that the graph is 4-metric dimensional.

2.5 Algorithm 2: Greedy Algorithm to Find the Metric Basis

Input: A graph G with n vertices, m vertices and given that it is k-metric dimensional. The notations and data structures given in Algorithm 1 are assumed here to hold true.

Step 0 : Initialization

Let S = Empty set. S is the metric generator. Initially it is empty and we shall add vertices one by one in such a way that each vertex distinguishes maximum number of vertex pairs.

Let $W = \{(v_i, v_j)$ such that $i \neq j$ *and* i, j *from* 1 *to* $n\}$. W is the set of all vertex pairs not distinguished by any vertex in S. Since S is empty, W has all the vertex pairs.

Step 1

Do While W is non-empty

Step 1.1: For every *vertex* v_i *of G which is not a member of S* do

For every member (v_s, v_t) of W, do the following

$d(i) = \sum_{s,t=1}^{n} d_i(v_s, v_t)$

$d(i)$is the number of vertex pairs distinguished by v_i

End For

End For

Step 1.2:

Fix the value k such that $d(k) = Max\{ d(i) where\ v_i \notin S\}$

We choose v_k not in S, in such a way that v_k distinguishes maximum number of vertex pairs of W.

Let $S = S \cup \{v_k\}$. We add v_k to the metric generator S.

Step 1.3

From W, Remove all (v_i, v_j) distinguished by v_k

Continue Do-loop

Output

S is the metric generator with minimum cardinality and hence the metric basis.

2.6 *Illustration of Algorithm 2*

Consider $V = \{v_1, v_2, v_3, v_4, v_5, v_6, v_7\}$ as shown in Fig. 2.

The vertex set V is trivially a metric generator. Now we try to make small matrices to build a smaller metric generator with greedy approach.

Step 1: In Table 2, identify the column with maximum number of true values. There are totally 21 rows. The column v_1 has maximum true values, 18. This means that v_1 distinguishes all 17 pairs except four (v_2, v_7), (v_3, v_5), (v_3, v_6), and (v_5, v_6).

In this iteration, the metric Generator $MG = \{v_1\}$.

Step 2: Delete the 17 rows from the matrix and retain only the four rows which are not distinguished by v_1. We get the matrix as shown in Table 3.

Step 3: In Table 3, identify the column with maximum number of true values. There are totally 4 rows. There are two columns v_3 and v_6 having maximum true

Table 3 Distinguisher matrix after the removal of v_1

Pair	v_1	v_2	v_3	v_4	v_5	v_6	v_7	|D(G)|
v_2v_7		*	*			*	*	4
v_3v_5			*		*	*	*	4
v_3v_6		*	*	*	*	*	*	6
v_5v_6		*	*	*	*	*		5

values, 4. This means that v_3 distinguishes all four pairs. So, we add v_3 to the metric generator.

In this iteration, the metric generator $MG = \{v_1, v_3\}$.

Also, v_6 distinguishes all the four pairs. So, $\{v_1, v_6\}$ is also another metric generator.

In this case, we have $\{v_1, v_3\}$ a metric basis and hence the metric dimension of G is 2. There is also one more minimum metric generator $\{v_1, v_6\}$.

3 Data Structures for Graph Representation

In this section, we introduce the data structures required to develop the algorithm.

Consider the following graph $G = (V, E)$ as shown in Fig. 1.

3.1 Distinguish Set of a Vertex $\mathbf{v_k}$

Let $v_k \in V$. The distinguish set of a vertex v_k, denoted by $D(v_k)$ is the set of pairs of vertices (v_i, v_j) where v_k distinguishes v_i and v_j.

$D(v_k)$ can be represented by a Boolean $n \times n$ matrix D_k as follows:

$D_k = (x_{ij})$ where

$$x_{ij} = \begin{cases} 1 \text{ if } v_k \text{ distinguishes } v_i \text{ and } v_j \\ 0 \text{ otherwise.} \end{cases}$$

That is,

$$x_{ij} = \begin{cases} 1 \text{ if } d(v_k, v_i) \neq d(v_k, v_j) \\ 0 \text{ otherwise.} \end{cases}$$

For example, the matrix D_3 is shown in Table 4.

With the definition of D_k, we have created a Boolean three-dimensional vector D as follows:

Table 4 Matrix D_3

	v1	v2	v3	v4	v5	v6	v7
v1	0	1	1	1	0	1	0
v1	1	0	1	0	1	1	1
v3	1	1	0	1	1	1	1
v4	1	0	1	0	1	1	1
v5	0	1	1	1	0	1	0
v6	1	1	1	1	1	0	1
v7	0	1	1	1	0	1	0

$$D_{ijk} = \left(x_{ijk}\right) \text{ where } x_{ijk} = \begin{cases} 1 \text{ if } v_k \text{ distinguishes } v_i \text{ and } v_j \\ 0 \text{ otherwise.} \end{cases}$$

3.2 *Notation*

D_{**k} denotes the two-dimensional matrix given below:

$$\begin{pmatrix} x_{11k} & x_{12k} & \dots & x_{1nk} \\ x_{21k} & x_{22k} & \dots & x_{2nk} \\ \dots & \dots & \dots & \dots \\ \dots & \dots & \dots & \dots \\ x_{n1k} & x_{n2k} & \dots & x_{nnk} \end{pmatrix}$$

In other words, D_{**k} is the two-dimensional matrix got by projecting the three-dimensional matrix D in the third dimension with specific value k. For example, for the Graph G shown in Fig. 1, D_{**3} is shown in the matrix in Table 4. In a similar manner, we can define D_{i**} and D_{*j*} also.

4 Breadth-First Search

Consider the *BFS* search starting from v and visiting all the vertices of G. Let *BFS*(G, v) be the spanning tree formed by the breadth-first search starting from vertex v. This partitions V into $V_0, V_1, V_2, V_3, \ldots, V_k$ where V_i is the set of vertices at level i in the spanning tree. Note that the root is assumed to be at level 0. For example, consider the graph in Fig. 2. The *BFS* spanning tree starting from v_1 is given in Fig. 3.

We have the following observations directly derived from the basic definition of BFS search.

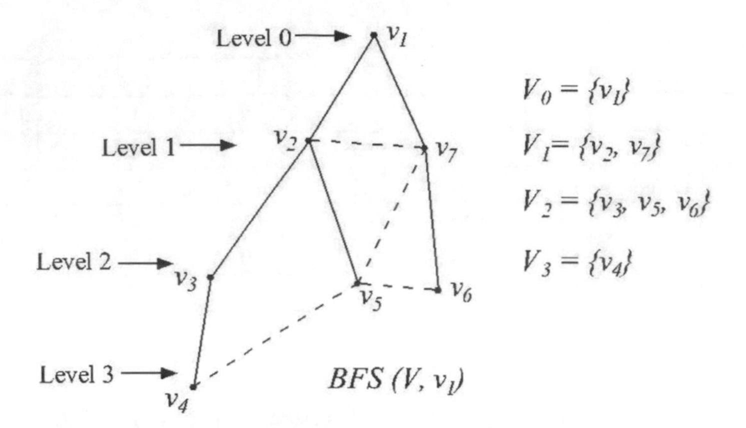

Fig. 3 BFS tree starting from v_1

1. If $u \in V_i$, the shortest distance from the root v to u is i.
2. Obviously, every tree edge joins two vertices at adjacent levels.
3. Every non-tree edge of the graph always joins vertices at the same level or at adjacent levels.
4. Let $v_i \in V_i$ and $v_j \in V_j$, if $|i - j| > 1$ the $(v_i, v_j) \notin E$.

The following is a detailed algorithm that produces the BFS spanning tree with level numbers for each vertex.

4.1 Algorithm 3: Find the BFS Tree with Partition of Vertices

Input

Graph $G = (V, E)$ with n vertices and m edges. Let v_1 be the vertex from which we start BFS.

Data Structures

QUEUE is an ordered list of vertices that works as a queue in the BFS process.

VISITED() is a Boolean array indexed by vertices. *VISITED(i)* is *true* indicating that the vertex is already visited in the BFS process and *false* otherwise.

LEVEL() is also an array of integers indexed by the vertices. *LEVEL(i)* represents the level of vertex v_i in the BFS spanning tree.

Algorithm

1. *QUEUE* is initially empty; *VISITED(i)* is false for all $i = 1$ *to* n;
2. *Insert* v_1 *into the QUEUE; VISITED(1)* = *true; LEVEL(1)* = *0.*
   This means that $V_0 = \{v_1\}$
3. *While QUEUE is non-empty continue the following and exit step 3 when empty:*

3.1 *Delete one entry from the QUEUE and call it u.*
3.2 *For every vertex x adjacent to u such that VISITED(x) is false do the following:*
  3.2.1 *Insert x into the QUEUE*
  3.2.2 *VISITED(x) = true*
  3.2.3 *LEVEL(x) = LEVEL(u) + 1*

Output

LEVEL(i) for i = 1 to n. Note that V_i *is the set of vertices x with LEVEL(x) = i*

Theorem 1 *In the above algorithm, after the end of execution, LEVEL(x) =* $d(v_1, x)$ *for all the vertices x, where* v_1 *is the root node of the tree.*

Proof We prove this result by the method of induction. For the root vertex v_1, the result is obviously true, as $d(v_1, v_1) = LEVEL(v_1) = 0$. Also, for every vertex x adjacent to v_1, $LEVEL(x) = 1 = d(v_1, x)$ as per the algorithm. This means that the result holds for the root and the vertices at level 1. Assume that the result is true for all the vertices up to level $k - 1$. Let x be a vertex with. $LEVEL(x) = k$. We have to show that $d(v_1, x) = k$. As per the algorithm, there is a vertex u such that $LEVEL(u) = k - 1$ and x is adjacent to u. As per induction hypothesis, $d(v_1, u) = k - 1$. As per the algorithm, there is no vertex before u was pulled out from the QUEUE adjacent to x. So, $d(x, v_1) \nless k$. So, the shortest path from v_1 to x is the shortest path from v_1 to u and then edge (u, x). This proves that $d(v_1, x) = k$. This completes the proof.

Let us illustrate the flow of the algorithm in the graph as shown in Fig. 2.

Initial Iteration—Step 1 of Algorithm

QUEUE							
Index	1	2	3	4	5	6	7
Vertex	v_1	v_2	v_3	v_4	v_5	v_6	v_7
VISITED	0	0	0	0	0	0	0
LEVEL	–	–	–	–	–	–	–

Initial Iteration—Step 2 of Algorithm

QUEUE	v_1						
Index	1	2	3	4	5	6	7
Vertex	v_1	v_2	v_3	v_4	v_5	v_6	v_7
VISITED	1	0	0	0	0	0	0
LEVEL	0	–	–	–	–	–	–

Iteration 1—Step 3.1 of Algorithm

QUEUE							
Index	1	2	3	4	5	6	7
Vertex	v_1	v_2	v_3	v_4	v_5	v_6	v_7
VISITED	1	0	0	0	0	0	0
LEVEL	0	–	–	–	–	–	–

Iteration 1—Step 3.2 of Algorithm

QUEUE	v_7, v_2						
Index	1	2	3	4	5	6	7
Vertex	v_1	v_2	v_3	v_4	v_5	v_6	v_7
VISITED	1	1	0	0	0	0	1
LEVEL	0	1	–	–	–	–	1

Iteration 2—Step 3.1 and 3.2 of Algorithm

QUEUE	v_5, v_3, v_7						
Index	1	2	3	4	5	6	7
Vertex	v_1	v_2	v_3	v_4	v_5	v_6	v_7
VISITED	1	1	1	0	1	0	1
LEVEL	0	1	2	–	2	–	1

Iteration 3—Steps 3.1 and 3.2 of Algorithm

QUEUE	v_6, v_5, v_3						
Index	1	2	3	4	5	6	7
Vertex	v_1	v_2	v_3	v_4	v_5	v_6	v_7
VISITED	1	1	1	0	1	1	1
LEVEL	0	1	2	–	2	2	1

Iteration 4—Steps 3.1 and 3.2 of Algorithm

QUEUE	v_4, v_6, v_5						
Index	1	2	3	4	5	6	7
Vertex	v_1	v_2	v_3	v_4	v_5	v_6	v_7
VISITED	1	1	1	1	1	1	1
LEVEL	0	1	2	3	2	2	1

Final Iteration after Iterations 5, 6, 7—Steps 3.1 and 3.2 of Algorithm

QUEUE							
Index	1	2	3	4	5	6	7
Vertex	v_1	v_2	v_3	v_4	v_5	v_6	v_7
VISITED	1	1	1	1	1	1	1
LEVEL	0	1	2	3	2	2	1

Theorem 2 *Let $G = (V, E)$ be a simple connected graph and $v \in V$. BFS (V, v) is the spanning tree got from BFS starting with v. Let $V_0, V_1, V_2, \ldots, V_k$ be the partition of the vertex set such that V_i is the set of vertices in Level i in the BFS Spanning tree. Vertex v distinguishes a pair of vertices u_1, u_2 if and only if $u_1 \in V_i$ and $u_2 \in V_j$ for $i \neq j$.*

Proof Let $u_1 \in V_i$ and $u_2 \in V_j$ where $i \neq j$. By Theorem 1, $d(v, u_1) = i$ and $d(u, u_2) = j$. So, u distinguishes u_1 and u_2.

Conversely, suppose v distinguishes u_1 and u_2. Then $d(v, u_1) \neq d(u, u_2)$ So, by Theorem 1, u_1 and u_2 are at different levels in $BFS(V, v)$. Assume u_1 is at level i and u_2 at level j where $i \neq j$. This shows that, $u_1 \in V_i$ and $u_2 \in V_j (i \neq j)$. This proves the result.

4.2 Algorithm 4: To Find the Dimension and Metric Basis of a Graph

Input

A simple connected Graph $G = (V, E)$ with n vertices and m edges.

Preprocessing

Let the vertices be ordered from as a set $\{v_1, v_2, \ldots, v_n\}$.

Find the distance matrix $d = (d_{ij})$ where d_{ij} = distance from vertex v_i to vertex v_j.

Algorithm

1. For every vertex $v_k \in V$, $k = 1$ to n, do the following:
   1.1. Determine the Breadth-First Search Tree *BFS* (V, v_k) where v_k is the root of the tree.
   1.2. In *BFS* (V, v_k) the set V_t has the vertices at level t where $t = 0, 1, 2, \ldots, h$ where h is the height of the tree.
   1.3. Initialize

$$D_{**k} = \left(b_{ijk}\right) \text{ where } 1, j, k = 1 \text{ to } n \text{ defined by}$$

$$b_{ijk} = \begin{cases} 0 \text{ if both } v_i \text{ and } v_j \in V_t \text{ for some } t = 0 \text{ to } n \\ 1 \text{ otherwise} \end{cases}$$

2. For every (i, j) where $i = 1$ to n and $j = 1$ to n.

$$\text{Calculate } de_{ij=} \sum_{k=1}^{n} b_{ijk}$$

3. Consider the matrix $\left(dc_{ij}\right)_{\substack{i=1 \text{ to } n \\ j=1 \text{ to } n}}$
   Choose the minimum entry in the matrix.
   $k = \min\{dc_{st} \mid s, t = 1 \text{ to } n\}$.
   Assume that $k = dc_{ij}$ for a specific value i and j.

Output

1. The metric dimension of G is k.
2. Consider $M = \{v_t \text{ such that } b_{ijt} = 1\}$. M is the metric basis of G.

References

1. Brigham RC, Chartrand G, Dutton RD, Zhang P (2003) Resolving domination in graphs. Math Bohem 128(1):25–36
2. Cáceres J, Hernando C, Mora M, Pelayo IM, Puertas ML, Seara C, Wood DR (2007) On the metric dimension of cartesian product of graphs. SIAM J Discrct Math 21(2):423–441
3. Estrada-Moreno A, Rodríguez-Velázquez JA, Yero IG (2013) *k*-metric dimension of graphs. arXiv:1312.6840 [math.CO]
4. Chartrand G, Eroh L, Johnson MA, Oellermann OR (2000) Resolvability in graphs and the metric dimension of a graph. Discrete Appl Math 105(1–3):99–113
5. Chartrand G, Poisson C, Zhang P (2000) Resolvability and the upper dimension of graphs. Comput Math Appl 39(12):19–28
6. Chartrand G, Salehi E, Zhang P (2000) The partition dimension of a graph. Aequ Math 59(1–2):45–54

7. Fehr M, Gosselin S, Oellermann OR (2006) The partition dimension of cayley digraphs. Aequ Math 71(1–2):1–18
8. Harary F, Melter RA (1976) On the metric dimension of a graph. Ars Comb 2:191–195
9. Haynes TW, Henning MA, Howard J (2006) Locating and total dominating sets in trees. Discrete Appl Math 154(8):1293–1300
10. Johnson M (1993) Structure-activity maps for visualizing the graph variables arising in drug design. J Biopharm Stat 3(2):203–236, pMID: 8220404
11. Johnson MA (1998) Browsable structure-activity datasets. In: Carbó-Dorca R, Mezey P (eds) Advances in molecular similarity. JAI Press Inc, Stamford, Connecticut, pp 153–170
12. Khuller S, Raghavachari B, Rosenfeld A (1996) Landmarks in graphs. Discrete Appl Math 70:217–229
13. Kuziak D, Yero IG, Rodríguez-Velázquez JA (2013) On the strong metric dimension of corona product graphs and join graphs. Discrete Appl Math 161(7–8):1022–1027
14. Oellermann OR, Peters-Fransen J (2007) The strong metric dimension of graphs and digraphs. Discrete Appl Math 155(3):356–364
15. Okamoto F, Phinezy B, Zhang P (2010) The local metric dimension of a graph. Math Bohem 135(3):239–255
16. Saenpholphat V, Zhang P (2004) Conditional resolvability in graphs: a survey. Int J Math Math Sci 2004(38):1997–2017
17. Sebő A, Tannier E (2004) On metric generators of graphs. Math Oper Res 29(2):383–393
18. Slater PJ (1975) Leaves of trees. Congr Numer 14:549–559
19. Slater PJ (1988) Dominating and reference sets in a graph. J Math Phys Sci 22(4):445–455
20. Tomescu I (2008) Discrepancies between metric dimension and partition dimension of a connected graph. Discrete Appl Math 308(22):5026–5031
21. Yero IG, Kuziak D, Rodríguez-Velázquez JA (2011) On the metric dimension of corona product graphs. Comput Math Appl 61(9):2793–2798
22. Yero IG, Rodríguez-Velázquez JA (2010) A note on the partition dimension of cartesian product graphs. Appl Math Comput 217(7):3571–3574

A Deep Learning Approach in Early Prediction of Lungs Cancer from the 2D Image Scan with Gini Index

Denny Dominic and Krishnan Balachandran

Abstract Digital Imaging and Communication in Medicine (DiCoM) is one of the key protocols for medical imaging and related data. It is implemented in various healthcare facilities. Lung cancer is one of the leading causes of death because of air pollution. Early detection of lung cancer can save many lives. In the last 5 years, the overall survival rate of lung cancer patients has increased, due to early detection. In this paper, we have proposed Zero-phase Component Analysis (ZCA) whitening and Local Binary Pattern (LBP) to enhance the quality of lung images which will be easy to detect cancer cells. Local Energy based Shape Histogram (LESH) technique is used to detect lung cancer. LESH feature extracts a suitable diagnosis of cancer from the CT scans. The Gini coefficient is used for characterizing lung nodules which will be helpful in Computed Tomography (CT) scan. We propose a Convolutional Neural Network (CNN) algorithm to integrate multilayer perceptron for image segmentation. In this process, we combined both traditional feature extraction and high-level feature extraction to classify lung images. The convolutional neural network for feature extraction will identify lung cancer cells with traditional feature extraction and high-level feature extraction to classify lung images. The experiment showed a final accuracy of about 93.27%.

Keywords DiCoM images · ZCA whitening. LBP · Image enhancement · Deep learning

1 Introduction

Lung cancer disease causes over a million deaths consistently throughout the world. It represents all new cancer growth of 13% cases and 19% of disease-related deaths around the world. There were two million new lung disease cases assessed to happen in 2018. In India, lung malignancy establishes 6.9% among all new diseases and 9.3%

D. Dominic (✉) · K. Balachandran
Department of Computer Science and Engineering, Christ (Deemed to be University), Bengaluru, India
e-mail: denny.cmi@gmail.com

D. S. Jat et al. (eds.), *Data Science and Security*, Lecture Notes in Networks and Systems 132, https://doi.org/10.1007/978-981-15-5309-7_11

among all cancer growths related in both genders, and reason for the disease-related mortality rate is more in men [1].

The proper treatment can increase the survival rate [1]. Lung cancer growth varies according to geographic area and ethnicity. A general 5-year endurance pace of lung cancer growth is around 15% in developed nations and 5% in developing nations. Screening by low portion computer tomography (CT) in high hazard populace showed a relative hazard decrease of 20% in lung cancer growth mortality, however, with a bogus positive pace of 96% [1].

Surface grouping as a significant research point in picture handling and example acknowledgment has been broadly examined during the most recent, quite a few years. The early agent surface grouping technique is the co-event framework strategy and separating-based methodologies. These strategies are for the most part delicate to direction and brightening changes [2]. CT scans helped in finding, which is an exploration feeling to modify the clinical symptomatic procedure [3]. The ongoing appearance of profound learning methods has highlighted the probability of consequently revealing highlights from the preparation pictures even among highlights inside the profound structure of a neural network [3].

Deep learning methods are extensively used to identify cancer growth tissues. Deep learning is a method of execution of an artificial neural system with different layers for the imitation of the nodules of the lung. Layers of profound neural network remove different highlights and henceforth give numerous degrees of reflection [4].

Convolutional neural networks (CNNs) can provide better results than Deep Belief Networks (DBNs). Dataset provided to this CNN can achieve a good benchmark for computer vision. Even CNN can be attracted to the best interest in machine learning because it has a very strong representational ability in learning.

2 Literature Survey

The noises in the images are always distracting facts in the image processing [5]. Ilya Levner [6] introduced object marker and topographical function with watershed segmentation. Ginneken [7] has divided the classification either pixel classification-based category or rule-based. The lung images extract into rule-based which is an acceptable approach for the researchers [8, 9]; the rule-based approach is preferred for classification. Clarifications have become a major concern and speculation [10–12]. The LBP distribution of various regions in a similar surface image may be extraordinary. The regular LBP techniques just processed the LBP histogram of the entire image and ignored the LBP measurement highlight in neighborhood locales. So as to utilize maximum capacity of neighborhood paired examples, separate the revolution-invariant LBP histograms in little locales to become familiar with the LBP texton for surface depiction. All the surface pictures were spoken to with the event of textons, not the occurrence of local binary patterns [2].

The LESH feature extraction procedure depends on the histogram of the energy which stores significant data about difference in intensity. It is fit for checking critical example varieties in DiCoM images. The higher degree LESH coefficients relate to the most critical arrangement of highlights when chosen prompts the same order while decreasing the curse of dimensionality [13].

The medical image classification algorithm joins elevated-level component extraction from a coding system with customary image highlights, known as CNMP. Supposedly, these examinations are the first occasion when a deep model had been legitimately used by considering customary image highlights for a group of medicinal image. The trial outcomes demonstrate that our technique can accomplish an exactness of which is the best SVM coding system and include a combination by significant edges. Also, the impact of image augmentation is on the algorithm's accuracy and running time [14].

The central feature extraction is used for benevolent nature of the lung without performing the morphology and surface highlights. For the best information, this is the main research to apply deep learning strategies to the issue of pneumonic nodule characterization. This investigation may fill in as a premise to address the canny nodule identification issue. He accepted that the perfect start for deep learning with application to errands [15].

The proposed strategy has diminished false positives essentially in competitors by utilizing the most discriminative surface highlights. The observational outcomes give the proof that the proposed strategy can productively characterize nodule. Later on, they are intending to utilize developmental calculations so as to look for ideal features [16]. The utilization of wavelet highlights to portray receiver operating characteristic (ROI) [17].

The cutting edge, for example, ResNet, Google Net, or VGG, is utilized for restorative picture investigation [18]. Likewise, new models were structured that performed better for the grouping of cancer growth. 3D profound systems were presented for 3D volumetric restorative picture investigation, recognizing the most separating highlights that characterize the quantity of ailments. The group of CNN models is additionally used to improve the prescient intensity of the model to better forecast of disease.

3 Methodology

Input Image Dataset: This experiment was conducted on lung's digital image dataset. This dataset has a total of 221 images in which the pixel value of one image is 128*128 pixels.

Image Preprocessing: Lungs' images are having low contrast which creates a hindrance to detect nodules in images. That is why we applied Contrast Limited Adaptive Histogram Equalization (CLAHE) for the enhancement of images. In this method, it first divides images into block sand and then calculates the values of histogram for each block using a specific number of gray levels. After that, histogram value will be clipped at a specific threshold. At last, histogram results will map in each region of the images. It interpolates gray-level mapping which helps in reconstruction of the final CLAHE Image.

This work has used Lung Image Database Consortium (LIDC) having 221 lung CT scans of cancer patients. This CT scan provides diagnostic information about suspected patients. With the help of this information, it will be easy to recognize the patient's condition. Using the LIDC dataset, the coordinates of the lung's nodule outline are obtained and a binary image for accessing the segmentation of lung nodules is achieved.

Image Segmentation: The contrast level between background area and lung nodule is very strong. Due to this reason, it is very tough to utilize threshold segmentation to obtain receiver operating characteristic (ROI) of lung nodules. To remove the interference, CNN is used.

For image segmentation, Convolution Neural Network (CNN) algorithm is used. A CNN is having multiple perceptron layers, namely, convolutional layer, Re–Lu layer. Dropout layer, convolutional layer, and pooling layer are normalized layers.

In CNN architecture, generally pooling layer and convolutional layer are used for combination of images. Pooling layers perform two operations such as mean-pooling and max-pooling. The average of mean-pooling calculates neighboring pixels, while maximum of feature points is by max-pooling. Mean-pooling reduces size errors, and max-pooling reduces estimated errors. CNN algorithms are used for image segmentation.

In this way, it completes every CNN. The counting procedures are implemented here, which gives the current output agreed by two CNNs.

LESH Base Feature Extraction: Based on the calculation of the histogram of the local energy pattern for the image of interest, local energies are calculated along with different orientations.

Before calculating phase congruency (PC), the image is first convolved with a bank of 2D log-Gabor filter with different orientations o and scales s. Let $\mathrm{G}_{\mathrm{so}}^{\mathrm{even}}$ and $\mathrm{G}_{\mathrm{so}}^{\mathrm{odd}}$ be the odd-symmetric and even-symmetric filters at scale s and orientation o. Then image result of convolution with the response vector is given as

$$[e_{so}(z),\ o_{so}(z)], \tag{1}$$

where $z = (x, y)$ represents a location. Hence, at a given scale, orientation can be computed by the amplitude of the response. And sensitive phase deviation measure is given as

$$\Phi_m(z) = \cos\bigl(\Phi_m(z) - {}^{-}\Phi_m(z)\bigr) - |\sin\bigl(\Phi_m(z) - {}^{-}\Phi_m(z)\bigr)| \tag{2}$$

Now, local energy is calculated which is $E(z)$. At last, the sum of Fourier amplitude components 2D phase congruency for the image is computed as local energy normalized, where T-noise cancellation factor and $W(z)$- weighting of the frequency spread. A_m and ϕ_m represent the phase angle and amplitude, respectively. Further, the resultant of LESH feature vector is calculated as follows:

$$H_{r,b} = \sum W_r x\ PC(z) \times \delta_{r-b} \tag{3}$$

$$W_r = \frac{1}{\sqrt{2\pi}\sigma} \theta^{[(x-rx0)^2+(y-ry0)^2]/\sigma^0 2} \tag{4}$$

where W_r is the Gaussian weighting function of the region r in the image, PC represents the local energy computed by the equation, and δ_{r-b} represents Kronecker's delta of the orientation.

LBP Feature (Local Binary Pattern): The LBP administrator names the image pixel a nearby example, which is figured out by contrasting its dim worth and those of its neighbors. The rotation-invariant of LBP is used to build LBP image. Every pixel of the image is evaluated. This method is used to represent texture of the image.

Gini Index: It is a correlation-based criterion to estimate the feature's ability. It follows splitting rules in decision tree which examines the decrease of impurity when it is selected for the feature. Impurity relates to the ability of feature to distinguish between the possible images. Gini index can be calculated by

$$\text{Gini}(x) = \sum_i P(X_j) \sum_c P(Y_c|X_j)^2 - \sum_c P(Y_c)^2 \tag{5}$$

where $P(X_j)$—prior probability of feature X having value X_j. $P(Y_c|X_j)$—probability that a random sample of the dataset, whose feature X has value X_j, belongs to class Yc. $P(Y_c)$—prior probability that a random sample belongs to class Y_c and see the result in Fig. 1.

ZCA Whitening Transformation: Whitening transformation is a preprocessing step to remove redundancy. This processing is required to compute zero-mean in covariance matrix.

$$\text{Covariance matrix is calculated as} \sum = E\left(xx^T\right) = \frac{1}{m} \sum_{i=1}^{m} (x_i)(x_i) \tag{6}$$

where m—no. of image patches sampled in the dataset. Now, compute eigenvectors $v_{1,}$ v_2, ... $v_{n,}$ and eigenvalues λ_1, λ_2, ... λ_n of covariance matrix. ZCA whitening follows definition as $x_{ZCAwhite} = W^x_{ZCAwhite}$ as given in Fig. 2.

4 Results

This research work has segmented lungs with and without nodules for classification. The results are evaluated by using classification and accuracy. This experiment was conducted using MATLAB. The X-Validation and Y-Validation are performed after classification. Finally, results are tested and analyzed by performance measures. This work had provided a series of lung images from the dataset to obtain a confusion

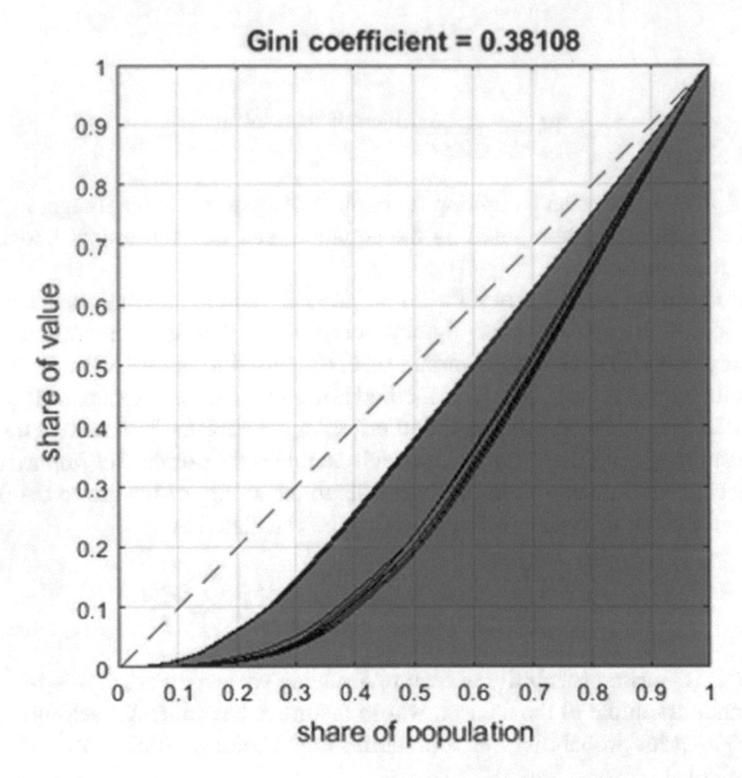

Fig. 1 Gini coefficient

Fig. 2 ZCA whitening of lung image

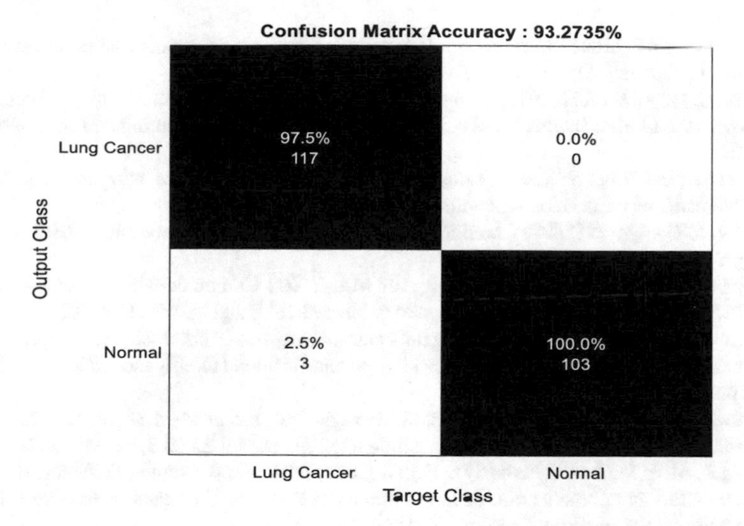

Fig. 3 Confusion matrix

matrix as given in Fig. 3. With the help of that confusion matrix, we calculated the accuracy of 93.2735%. For better performance, receiver operating characteristic (ROC) curve and confusion matrix are employed to analyze a model. Confusion matrix describes the target class and output class. The ROC curve is obtained to compute false positive rate and true positive rate.

5 Conclusion

This research work proposes a novel image classification algorithm for lung cancer from the 2D image scan. We have used convolutional neural network for feature extraction so that it can easily identify lung cancer cells. In this process, we combined both traditional feature extraction and high-level feature extraction to classify lung images. The experiment showed an accuracy of about 93.27%. We have discussed the training progress, and the validation RMSE is 9.9164. Gini index is a correlation criterion used to estimate feature ability.

References

1. Cruzroa A, Caicedo JC, Gonzalez FA (2011) Visualpattern mining in histology image collections using a bag of features. Artifi Intell Med 52(2):91–106
2. He Y, Sang N, Huang R (2011) Local binary pattern histogram based texton learning for texture classification. In 18th IEEE international conference on image processing, 2011

3. Wen-Huang LK (2015) Computer-aided classification of lung nodules on computed tomography images via Deep Learning Technique. OncoTarg Ther
4. Rahul C, Ghansala KK (2018) Convolutional neural network (CNN) for image detection and recognition. In First International conference on secure cyber computing and communication, IEEE
5. Chaudhary A, Singh SS (2012) Lung cancer detectionon CT images using image processing. In International transaction on computing sciences, vol 4
6. Levner I, Zhangm H (2007) Classification driven watershed segmentation. In IEEE transction on image processing, vol 16, no 5
7. Ginneken BV, Romenyand BM, Viergever MA (2001) Computer- aided diagnosis in chest radiography: a survey. In IEEE, transactions on medical imaging, vol 20, no 12
8. Sharma D, Jindal G (2011) Identifying lung cancer using image processing techniques. In International conference on computational techniques and artificial intelligence (ICCTAI'2011), vol 17, pp 872–880
9. Nguyen HT et al. (2003) Water snakes: Energy-driven watershed segmentation. In IEEE transactions on pattern analysis and machine intelligence, vol 25 no 3, pp 330–342
10. Farag A, Ali A, Graham J, Elshazly S, Falk R (2011) Evaluation of geometric feature descriptors for detection and classification of lung nodules in low dose CT scans of the chest. In IEEE International Symposium, Chicago, IL, USA
11. Lin P-L, Huang P-W, Lee C-H, Wu M-T (2013) Automatic classification for solitary pulmonary nodule in CT image by fractal analysis based on fractional Brownian motion model. Pattern Recogn 46(12):3279–3287
12. Farag A, Elhabian S, Graham J, Farag A, Falk R (2010) Toward precise pulmonary nodule descriptors for nodule type classification. Medical Image Computing and Computer-Assisted Intervention. MICCAI 2010. Springer, Berlin, Germany
13. Wajid K, Hussain A (2016) Lung cancer detection using local energy-based shape histogram (lesh) feature extraction and cognitive machine learning techniques. In IEEE 15th international conference on cognitive computing, 2016
14. ZhiFei L, HuiFang D (2018) Medical image classification based on deep features extracted by deep model and statistic feature fusion with multilayer perceptron. 2061516, Hindawi
15. Wen-Huang KL (2015) Computer-aided classification of lung nodules on computed tomography images via Deep Learning Technique. OncoTarg Ther
16. Khan SA, Shunkun Y (2015) Effective and reliable framework for lung nodules detection from CT scan images. OncoTarg Ther
17. Madero H, Guadalupe V (2015) Automated system for lung nodules classification based on wavelet feature descriptor and support vector machine. In Madero Orozco et al. (eds) Biomedical Engineering Online, vol 14, p 9
18. Usma N, Sambyal AS (2018) Advances in Deep Learning Techniques for Medical Image analysis. In 5th IEEE international conference on parallel, distributed and grid computing, 2018

A Survey on Guiding Customer Purchasing Patterns in Modern Consumerism Scenario in India

Sonia Maria D'Souza, K. Satyanarayan Reddy, and P. Nanda

Abstract This proposed work is used to simplify and ease the shopping experience for a customer through the user-friendly android application. Generally, in the shopping malls user does happy shopping but feels tired and sick for standing in the long queues for the billing. Each product can be billed by reading the barcode on each product; thus, a single person scanning thousands of products of hundreds and more customers is a tedious task. Thus, we have come up with the mobile application which helps the customer do easy shopping without long queues for billing.

Keywords Mobile communication · QR code scanner · Wi-Fi · Semantics

1 Introduction

As the technology grows, there is lot of improvement in living standards of the society. Shopping has become the hobby of the people. When it comes to the name of shopping, most of them prefer online shopping; there are so many e-commerce shopping sites, e.g., Flipkart, Amazon, e-bay, Limeroad, Snapdeal, and many more, which use the virtual shopping cart [1]. An Internet shopping basket comprises three sections.

The original version of this chapter was revised: The current affiliation (East Point College of Engineering, Bangalore, 49, Karnataka, India) of the author K. Satyanarayan Reddy is replaced with a revised affiliation "Cambridge Institute of Technology, Bangalore 36, Karnataka, India". The correction to this chapter is available at https://doi.org/10.1007/978-981-15-5309-7_35

S. M. D'Souza (✉) · K. S. Reddy · P. Nanda
Cambridge Institute of Technology, Bangalore 36, Karnataka, India
e-mail: soniapradeepkumar@gmail.com

K. S. Reddy
e-mail: satyanarayan.reddy@gmail.com

P. Nanda
East Point College of Engineering, Bangalore 49, Karnataka, India
e-mail: nandaashwin7@gmail.com

D. S. Jat et al. (eds.), *Data Science and Security*, Lecture Notes in Networks and Systems 132, https://doi.org/10.1007/978-981-15-5309-7_12

1. Inventory,
2. Wish list, and
3. Checkout system.

The upshot list is being composed of all the statistics anticipated to manifest any upshots to the client and to lacquer a trading interchange on the web. Statistics are to be remembered for the upshot directory. A wish list authorizes end users to spoor the upshots they want to foothold. A go-cart also shows the cost of each product they have been selected. For the payment, user can select the online payment or through credit or debit card, and many go with cash on delivery. With this, there is a secure server, as servers are the backbone of the Internet as well as these shopping websites [2]. With the above information, we know how the online shopping is done but everyone does not prefer online shopping; people do shopping in the malls, but they have to wait in the long queues for billing, where only few billing counters and one person at each counter billing thousands of products. Thus with the same barcode scanning and billing, we have come up with the new mobile application which scans the products, does the billing, and at the end gives the QR code. The user can just scan that code, pay the bill, and walk out.

2 Related Work

2.1 Online Shopping Carts/E-Commerce

The online shopping has features like store design, go-cart, inventory, transform options, product forwarding, and decree processing.

Smart shopping using RFID
The "Quick-witted go-Cart" is come into being to abet a people in regular shopping as far as discounted time spent while acquiring an item at the best value accessible.

Major components are

(a) Server Communication component (SCC),
(b) User Interface and Display Component (UIDC), and
(c) Automatic Billing and Inventory Management Component (ABIMC).

The usage of RFID in applied science is used for the ascetic upshot identification inside the go-cart and thus abolishesg end user involvement in the proccss of product reading for remittance. Each product must have RFID tags.

The reader must be able to associate upshots throughout the mezzanine, as trolleys where you can space items from 20 cm above the ground and to about 1.5 m higher [2]. RFID tags are costlier, and it also requires the RFID reader with antenna cart. Which makes the system with the more cost? In this system, when item is blow in or do away from the cart, RFID tag associates upshot and refurbish draft law (Table 1).

Figure 1 gives an overview of the frequency and till what range the tags can be

Table 1 Frequency and range of RFID tags

Frequency	Mode	Range	Transfer rate
125–135 kHz	Passive	Short range (up to 0.5 m)	Low
13.56 MHz	Passive	Medium range (up to 1.5 m)	Moderate
860–930 MHz	Passive	Medium range (up to 5 m)	Moderate to high
433 MHz	Active	Ultra long (up to 100 m)	High
2.45 GHz	Active	Long range (up to 10 m)	Very high

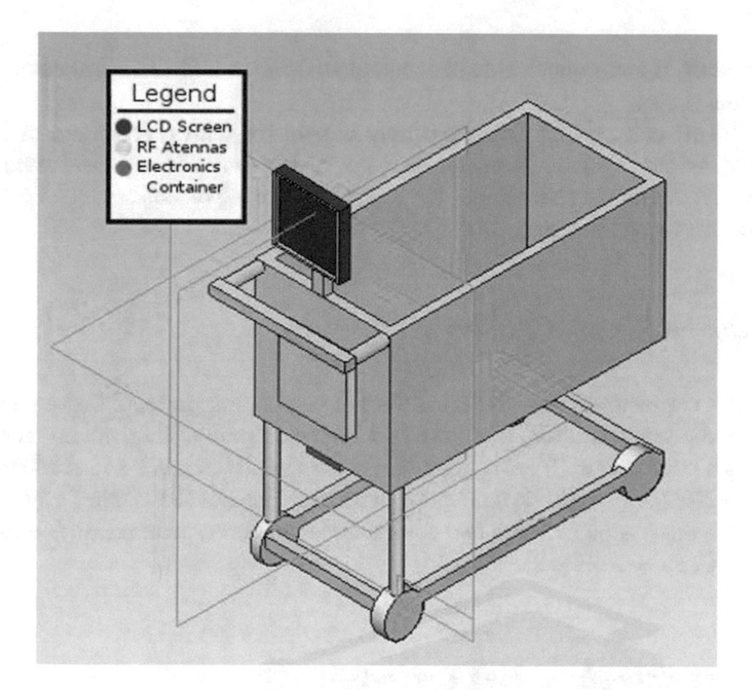

Fig. 1 Implementation of smart shopping cart using RFID tags

read; generally, when the frequency and range increase, the cost also increases, which is a burden to the seller or the owner of the store, while implementing the shopping with the RFID each product, which contains the RFID tags and after billing the tags have to be removed which is a tedious task again. By this practice, it is easy for customer who is shopping but more work for the seller or works in the shop.

The problem or concerns with RFID are as follows:

- Since the proprietor of a thing will not really know about the nearness of an RFID tag and the tag can be perused a way off without the information on the

Fig. 2 NFC tags

individual, it gets conceivable to accumulate delicate information about a person without assent.

- On the off chance that a labeled thing is paid for with MasterCard or related to utilization of an unwaveringness card, at that point it is conceivable to by implication find the character of the buyer by perusing the all-around exceptional ID of that thing (contained in the RFID tag).

2.2 *Digital Shopping Using NFC*

Near-field communication (**NFC**) is a lot of correlation protocol that emancipate two cathodic gadgets. NFC is a lot of short-extend remote innovations, regularly requiring a partition of 10 cm or less. NFC consistently includes an initiator and an objective. NFC is just a single innovation, with Bluetooth and RFID similarly as ready to initiate a discussion between two contraptions; however, there are differentiations inside NFC, as well (Fig. 2).

2.3 *Easy Shopping Using Barcodes*

Shopping is in the blood of all the people; everyone walks into the big shopping malls for shopping and passing time. Most huge shopping centers experienced reproduction and scale extension. Presently, shopping centers are greater with progressively bottomless merchandise and more assortment of products. Individuals are in the quest for top-notch buyer merchandise. To understand the trouble in client shopping, enormous stores have propelled a shopping center format map, Touch Mall shopping framework, and other shopping guides and so forth.

Barcode is a discernible, fathomable, statistics for the most part depicts progressively toward item that grants the standardized tag. Initially, standardized identifications methodically reprimand to facts by shifting the compass and dispersing of equal

Fig. 3 Barcode

Fig. 4 QR code

lines, and might be allocated to as direct or one-dimensional. Later, two-dimensional codes were created, utilizing square shapes and etoiles (Fig. 3).

2.4 QR Code (Quick Response Code)

It is the trademark for a sort of grid scanner platitude. A QR code makes use of four commend conceal vogues to productively stockpile statistics. It accounts for gloom squares organized in a square crisscross on a white basis, which can be perused by an imaging earmark, to give an instance a camera (Fig. 4).

3 Existing System

Predicament: Any handpicked upshot is released into the go-cart; RFID reader reads the tag inside the product and the data of the item is extricated and showed on the LCD screen. Simultaneously, charging data is additionally refreshed.

(1) At the point of customers end with the "start button" which helps to make the framework and afterward the segments to display.
(2) Each item belongs to an RFID label mainly focused on one of the kind, assimilation. These assimilations are sustained in the directory relegated relating upshot [3].
(3) At the point the customer drops any item on to the truck, then the RFID per user peruses the tag. The data item extricated finally showed.

(4) These means are rehashed until the finish of shopping button is squeezed. Once the "End Shopping" button is squeezed, the all-out bill is sent to ace pc through Wi-Fi (ZigBee).
(5) Toward the finish of shopping, the client can straight away take care of the tab and leave.

It is the framework with RFID labels yet which is not actualized in all the shops or shopping centers.

It has a few disadvantages as it is costly; all the products cannot be equipped with the RFID and it requires the hardware items

like the ZigBee RFID antenna and RFID reader [4].

4 Proposed System

It is the mobile application which helps you scan the products; when the customer purchasing, he or she can scan the product by themselves before putting it into the cart. It is like the self-scan of the products before going to the billing counter. Rather than going to the billing counter and waiting for one person to scan all your purchased products, it is better to self-scan the products and just go scan the QR code of the total amount of bill, pay the money, and checkout [5] (Figs. 5 and 6).

5 Design and Implementation

It is the android application where the user has to sign up with the application given the user id and the password. The application opens the screen in which user has to enter the location in which he can search for the nearby malls in a particular location; he gets the list of the malls. The user can now select the particular mall which he wants to visit. If the user does not know the route to the mall, the app itself shows the navigation route to mall. Once the user reaches the mall, he can find which are the different shops present in the mall. After entering the particular shop, he can search for the particular product or the item through the application itself; if the item is there it shows where the item is located, else it will not. In general, everyone does the budget planning; money is very much important to all. While shopping also the laymen keep the budget, and thus we have a budget planner included in the application. Users can set the budget and then start shopping; if the user crosses the particular set budget, the application gives the sound alert saying you have crossed the budget, showing the option RESET the budget or checkout. Now the user RESET the budget and start shopping again or walk to the billing counter to checkout (Figs. 7, 8, 9, 10, 11, 12, 13, and 14).

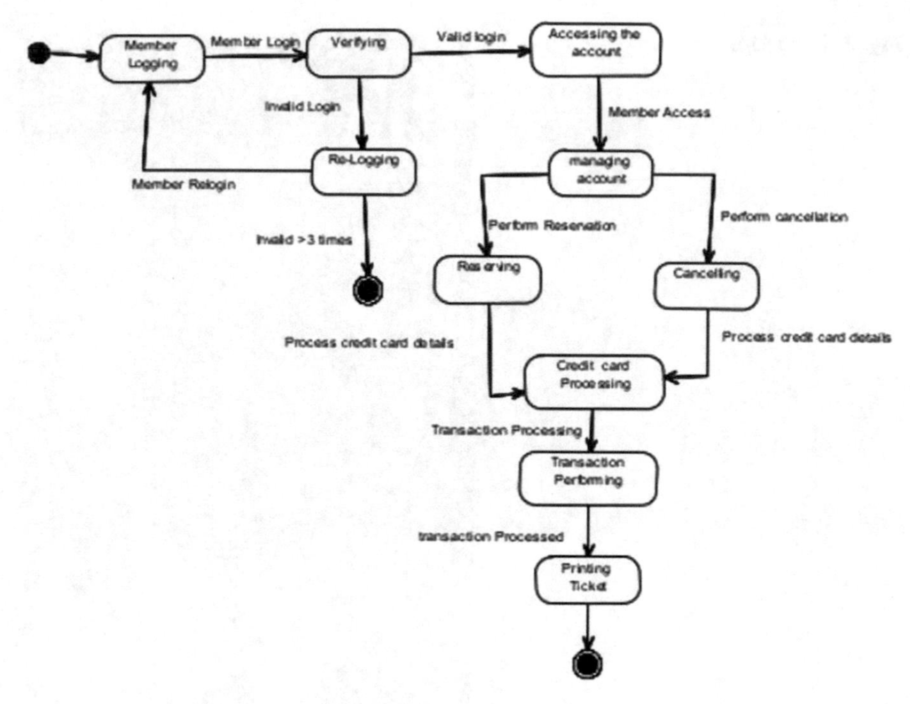

Fig. 5 The proposed system

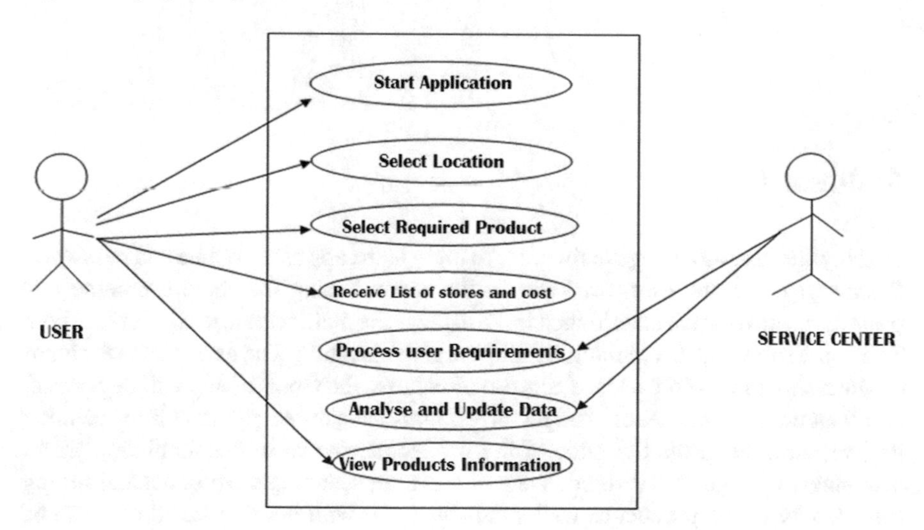

Fig. 6 User interactions

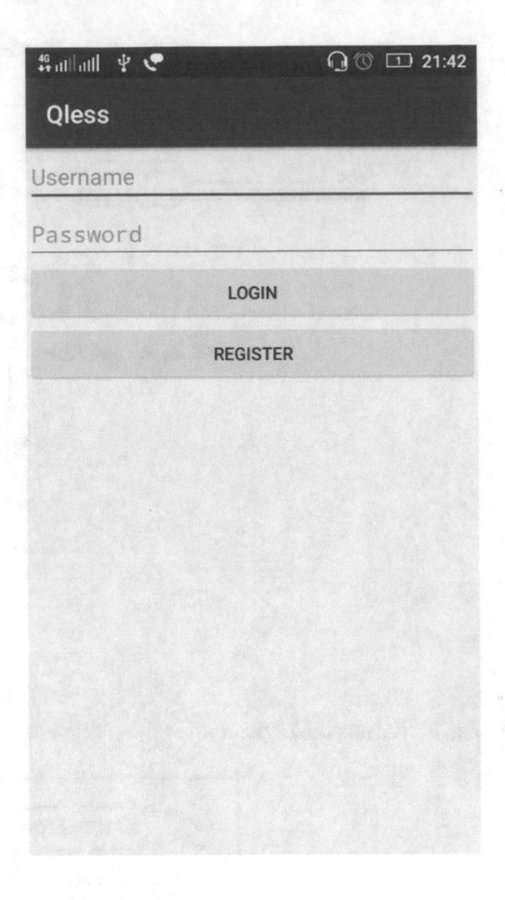

Fig. 7 Login page

6 Results

Initially, the user has to register in order to login in the app as username and password. Once registered, the user can login in the app by using the same username and password. Once the user has logged in, it displays the main screen; it asks for location. The app displays the list of malls near to the given location. The user selects for maps or starts shopping. If the user selects the directions, the Google Map will be opened, and then the user can set the budget. Now, user can start shopping before scanning the cart view. Scan the QR code of the product as shown in the simulation. After scanning, the product is verified. View of cart after scanning the product and setting the quantity of the product as well as updated cart with items scanned can also be displayed in the app. If budget exceeds, alert message with option change budget or checkout for billing is displayed. In this app, budget reset is also possible. Finally, the QR code for the total shopping amount is generated.

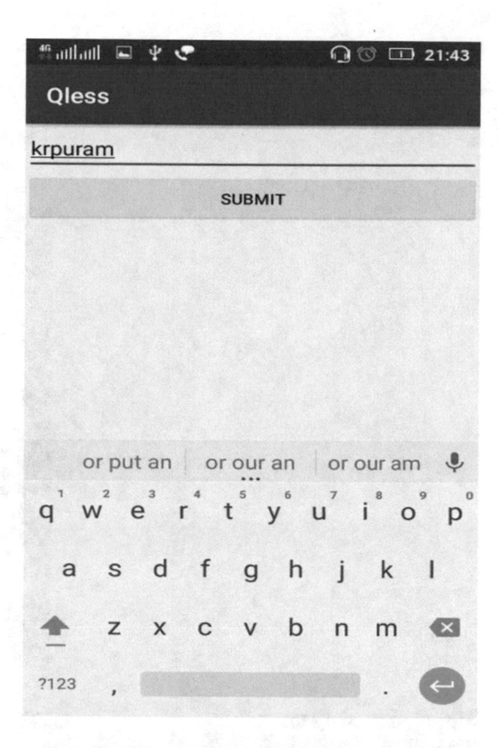

Fig. 8 Entered location name

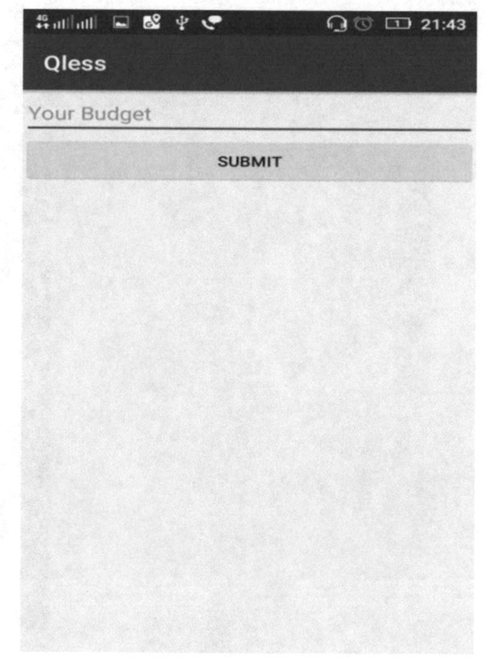

Fig. 9 Budget planner

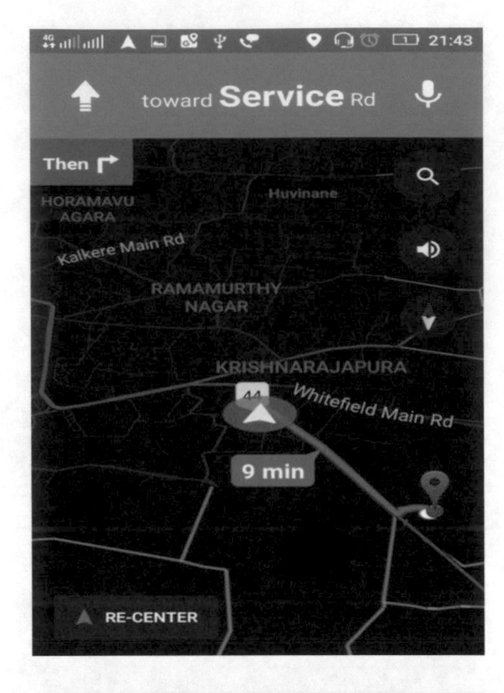

Fig. 10 Navigation routes

Fig. 11 Scanning the product

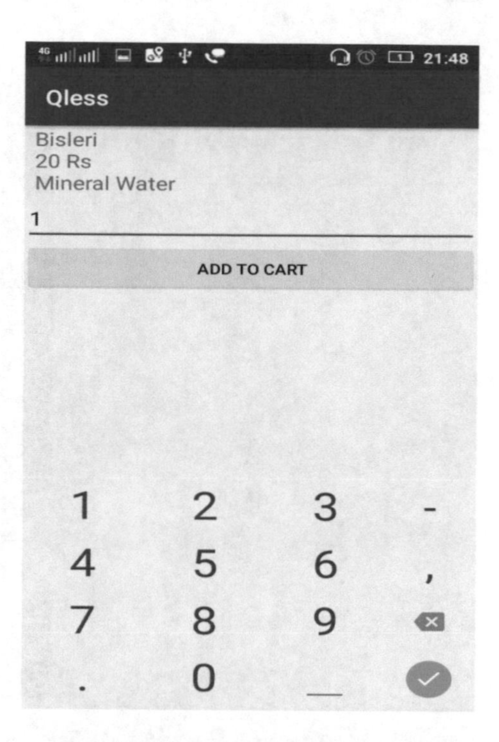

Fig. 12 Quantity of the product

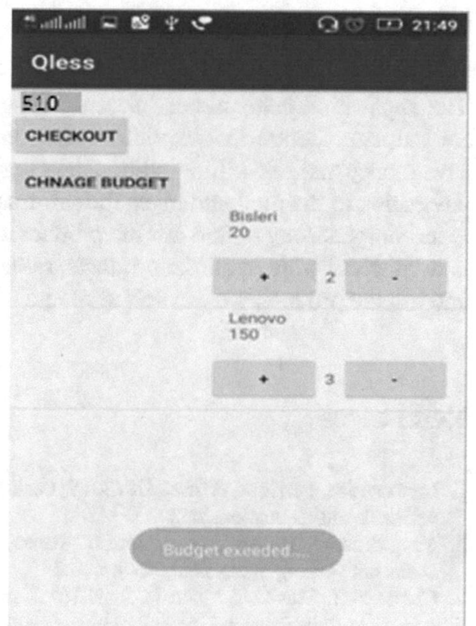

Fig. 13 Budget exceeded alert

Fig. 14 Amount paid through QR code

7 Conclusion and Future Work

The application helps the end user to self-glance the upshot for purchasing, check for the manufacture and expiration date, and do the budget planning which helps save money and self-billing without any long queues, and the application shows the navigation to the mall and gives the information about all the products. As barcode is the simplest way to find out the product information which is less cost and which is used world wide in all the products, rather than waiting for someone else to scan and bill the product, we can self-scan and bill the items.

References

1. McCune JM, Perrig A, Reiter MK (2015) Seeing-is-believing using camera phones for human-verifiable authentication. 3(2)
2. Yang Z, Zhang B, Dai J, Champion AC, Xuan D, Li D (2018) A distributed mobile system for social networking in physical proximity. 2
3. Chang Y-H, Chu C-H, Chen M-S (2016) A general scheme for extracting QR code from a non-uniform background in camera phones and applications. 4 (Canada)
4. American Time Use Survey. http://www.bls.gov/tus/charts
5. Chandra Babu DVS (2017) Wireless intelligent billing trolley for supermarket. Int J Adv Res Technol 3(1)

Diabetic Retinopathy Detection Using Convolutional Neural Network—A Study

Farha Fatina Wahid and G. Raju

Abstract Detection and classification of Diabetic Retinopathy (DR) is a challenging task. Automation of the detection is an active research area in image processing and machine learning. Conventional preprocessing and feature extraction methods followed by classification of a suitable classifier algorithm are the common approaches followed by DR detection. With the advancement in deep learning and the evolution of Convolutional Neural Network (CNN), conventional preprocessing and feature extraction steps are rapidly being replaced by CNN. This paper reviews some of the recent contributions in diabetic retinopathy detection using deep architectures. Further, two architectures are implemented with minor modifications. Experiments are carried out with different sample sizes, and the detection accuracies of the two architectures are compared.

Keywords Diabetic retinopathy · Deep learning · Convolutional neural network

1 Introduction

Diabetic retinopathy (DR) is the most common ophthalmic disease that affects an eye fundus. DR is generally caused due to damage of tiny blood vessels in the retina and may lead to leakage of blood vessels and swelling of eye, and it is one of the main reasons for vision loss in people [1, 2].

"Haemorrhages", "microaneurysms" and "exudates" are the significant symptoms of DR [3]. The deformations in the walls of blood vessels, like a balloon, are the "microaneurysms". It is one of the factors used to specify the stages of DR. "Haemorrhages", on the other hand, are caused when capillaries get damaged and

---

F. F. Wahid
Department of Informtion Technology, Kannur University, Kerala, India
e-mail: farhawahid@gmail.com

G. Raju (✉)
Department of Computer Science and Engineering, Christ (Deemed to be University), Bengaluru, India
e-mail: kurupgraju@gmail.com

D. S. Jat et al. (eds.), *Data Science and Security*, Lecture Notes in Networks and Systems 132, https://doi.org/10.1007/978-981-15-5309-7_13

blood leaks from it. Therefore, the categories of "haemorrhage" are "dot, flame and blot". Dark red spots in the eye indicate dots and are difficult to distinguish from "microaneurysms", but are comparatively larger in size. Flame "haemorrhages" are shaped like a flame and are due to leakage of blood aligned with nerve fibres. Blots are irregular in shape and largest in size [3]. Fluids that leak out of blood vessels to adjacent tissues are referred to as "exudates". Exudates are either hard or soft [4].

The two categories of DR are proliferative (PDR) and non-proliferative DR (NPDR). Proliferative DR occurs when new blood vessels grow, i.e. proliferates, and affect the normal flow of vessels. The absence of proliferation is identified as non-proliferative DR. NPDR indicates the beginning of DR, whereas PDR indicates advanced stages [3]. NPDR can be further categorized as mild, moderate and severe. Mild NPDR indicates the presence of "microaneurysms" in less numbers. The presence of less than 20 "haemorrhages" along with "microaneurysms" indicates moderate NPDR. If the count of "haemorrhage" is 20 or greater, it indicates severe NPDR. Proliferative DR leads to the formation of abnormal vessels and is characterized by the presence of "neovascularization" [3].

Different algorithms and approaches are designed for the early and effective detection of DR from fundus images. In recent years, Convolutional Neural Network (CNN)-based models for DR detection have become popular. In this chapter, a brief survey of CNN models for DR detection is carried out. Based on that, two CNN architectures are chosen, modelled and tested with fundus images from MESSIDIOR database.

2 Related Works

From a large collection of CNN-based DR detection papers, a few are chosen and are summarized below:

In 2018, Wan et al. used CNN for DR detection by image classification [5]. The authors experimented with CNN architectures "AlexNet", "GoogleNet", "VggNet" and "ResNet". "Hyper-parameter tuning" and "transfer learning" were the key focus. Data augmentation and data normalization were carried out on the input images. CNN models were then trained for DR detection by adopting transfer learning. The authors reported the highest accuracy of 95.68% for VggNet model, with hyper-parameter tuning.

A CNN model for DR detection was proposed by Chakrabarty (2018) where normalized grayscale images were employed [6]. The input images were resized to 1000×1000. CNN architecture was defined with three sets of "convolution, ReLU and max-pooling" layers. The generated features were then given to an ANN which formed the dense layer of CNN. With HRF image database, a training accuracy of 91.67% was reported. Validation was carried out using six images which resulted in 100% accuracy.

Burewar et al. (2018) presented a model for classifying retinal images based on the severity of DR [7]. They employed "U-Net" segmentation with CNN and region merging. Initially, green channel of image was enhanced using "CLAHE" algorithm. From the enhanced image, vessels were segmented using "U-Net" and the output was further modified by region merging with the input image. Finally, the images were classified using CNN network which consisted of 13 convolutional layers, 5 pooling and 2 fully connected layers. The proposed architecture could classify severe NPR with an accuracy of 93.64 and the other classes with lesser but comparable accuracy.

Hemanth et al. (2019) designed a diabetic retinopathy detection and classification scheme using deep CNN [8]. The schemes incorporate both hybrid deep learning and image processing concepts for DR detection. The colour input images were resized to 150 × 255, and on each channel of RGB image is enhanced with histogram equalization followed by CLAHE. The enhanced image was given as input to CNN. Experiments with images from MESSIDIOR database reported a recognition accuracy of 97%.

A deep CNN-based DR detection model was proposed by Rubini et al. [9]. Each input colour image was resized to 224 × 244 and given as input to a CNN model that consisted of five convolutional layers and pooling layers. This was followed by dropout layer and three FCN. The output of each layer indicates the weights for DR symptoms which were refined through iterations. Based on experiments using MESSIDIOR and ROC dataset, maximum training accuracy of 99.87% was reported.

Zhang et al. (2019) proposed a DeepDR system for DR identification and its grading [10]. The model directly detected the presence and severity of DR using ensemble learning and transfer learning. It worked based on the combination of CNN and standard deep neural network. The identification and grading model gave accuracies of 97.7% and 96.6%, respectively.

Binocular Siamese-like CNN architecture together with "transfer learning" technique was used for DR detection by Zeng et al. (2019) [11]. The model accepted binocular fundus images and learned their correlation to aid prediction. Cross entropy loss and contrastive loss were adopted to guide gradient descent. An AUC value of 0.951 was reported.

Based on the review carried out, two CNN models are chosen for implementation and explained in detail in the next section.

3 CNN Models for Diabetic Retinopathy Detection

CNN is a common deep learning technique which works based on the principle of image convolution. The basic layers in a CNN are "convolutional layer", "pooling layer" and "fully connected layers". Different CNN models are constructed for DR detection by making changes in the CNN architecture such as changing the number of layers and kernels, varying kernel sizes, changes in activation function, optimizer, learning rate, etc. In this work, two models based on CNN architecture for DR detection are selected.

3.1 CNN Model 1—Deep CNN-Based DR Detection

"Deep CNN-based Diabetic Retinopathy Detection" (DCNN-DRD) model was proposed by Rubini et al. [9]. The model was developed to classify retinopathy images into healthy or defective based on the symptoms of diabetic retinopathy. The model consisted of five convolutional layers and pooling layers each with three fully connected layers preceded by a dropout layer. As CNN is capable of automatically generating features from input images, low-level DR symptoms are learned by initial layers which gradually transforms to high-level symptoms for DR detection [9].

The DCNN-DRD architecture accepts inputs as colour retinal images. Images are resized to 224 × 224 and provided the input layer of CNN with no additional preprocessing. The first convolutional layer contains 32 filters each with kernel size of 5 × 5. A stride of size 1 × 1 is used for the convolutional layers. After the convolutional layer, a pooling layer is provided which reduces the image to half the original size. Max-pooling operation with a stride length of 2 is incorporated. The pooling layer is followed by ReLU function in order to introduce non-linearity to the outputs [9]. The output of the first layer is passed to the next convolution layer which consists of 64 filters. As in the first layer, convolutional layer is followed by max-pooling and ReLU function. The filter size and stride length remain the same as the initial layer. The third and fourth convolutional layers contain 128 and 256 filters, respectively, each of size 5 × 5 with stride length of 1 followed by max-pooling layer of size 2 × 2 with stride length 2 and ReLU function. The last convolutional layer contains kernel of size 14 × 14 with 256 filters followed by a max-pooling layer. The final output is then flattened to form a linear vector and is followed by a dropout layer to avoid overfitting. The output from dropout layer is connected to three fully connected layers (FCN) each with size 2048, 256 and 64, respectively. All the FCN are followed by ReLU activation function. The third FCN is followed by an FCN whose size depends on the class of the problem [9]. This is followed by a softmax layer to perform softmax activation function and a classification layer. In this work, we consider only the detection of DR; the number of classes is only 2 and requires only two nodes in the output layer. The layers in DCNN-DRD architecture are depicted in Fig. 1.

3.2 CNN Model 2—A CNN Model for DR Detection

Chakrabarty proposed a simple CNN model for classifying diabetic retinopathy images into normal or abnormal in 2018 [6]. The input images are subjected to a preprocessing step in which colour retinal images are converted to normalized grayscale images. These images are then passed to the input layer of CNN. For the CNN architecture, three convolutional layers, ReLU and max-pooling layers are used in sequence. Further, flatten layer is used to form a single feature vector. Each convolutional layer consisted of 32 filters (feature detectors) of size 3 × 3. Max-pooling

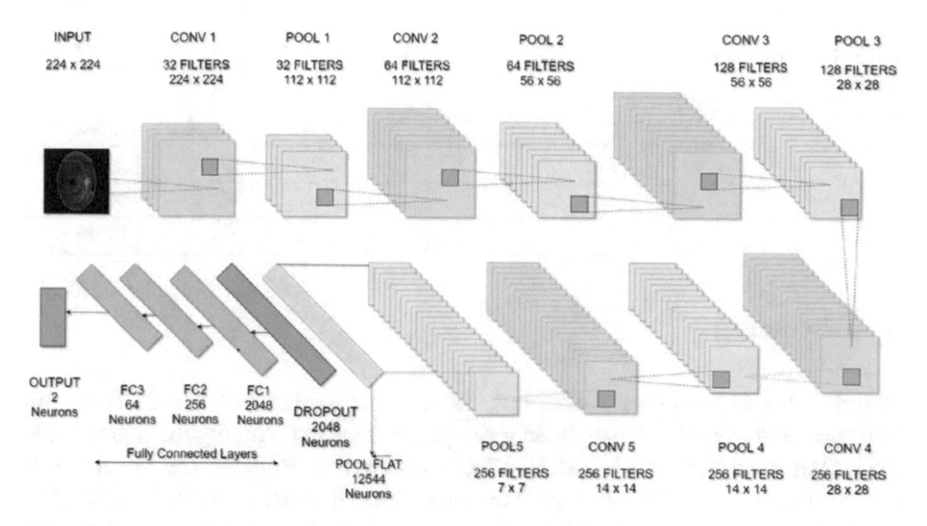

Fig. 1 Architecture of DCNN–DRD model [9]

layer with stride dimension of 2 × 2 is incorporated. One layer of ANN is used with 128 units as the dense layer.

4 Experiments

For the conduct of the study, the models are constructed using deep learning framework available in MATLAB R2019a. The construction of DCNN-DRD model is carried out as explained in Sect. 3.1, whereas for the model discussed in Sect. 3.2, apart from the details mentioned, a softmax layer is introduced following fully connected layers of size 128 and 2, where 2 represents the number of classes in the problem. The flattened layer is avoided for both the models.

The models are evaluated on fundus images from MESSIDIOR database. Three experiments are carried out on both the models. The main difference in the experiments is the number of images included for the training. For experiment 1, a total of 160 images are considered with equal number of normal and retinopathy affected images. For experiments 2 and 3, a total of 320 and 640 images are considered. Data augmentation by replication is carried out to increase the number of input images. In both cases, an equal number of samples from both the classes are included. The learning rate of the network was fixed to 0.0005 for both the experiments. A maximum of 500 epochs is considered for all the experiments. 70% of the entire data is used for training, and testing is carried out using the remaining 30%. Table 1 shows the recognition accuracy obtained for the three experiments using the two models.

Table 1 Recognition accuracy

Experiment no.	No. of samples	Recognition accuracy (%)	
		Model 1	Model 2
1	160	87.50	80.00
2	320	98.95	93.75
3	640	100	97.92

5 Discussion

Based on the experiments carried out, Table 1 reveals that both the CNN models built proved to be effective diabetic retinopathy detection. Among the two models, the performance of model 1—DCNN-DRD is better than model 2. The first model is relatively complex and requires more resources. One of the basic problems with CNN is the requirement of large number of training samples. This fact with respect to the given architectures is highlighted in the results. In the experiments, data augmentation is achieved through replication. But replication is not considered as a good approach and in literature as well as part of tools several data augmentation techniques are available.

6 Conclusion

A study on CNN-based models for diabetic retinopathy detection is carried out. Two existing CNN models are built and experimented with standard fundus images from MESSIDIOR database. The work reported is only a preliminary study with minor modifications in the architectures considered. Further, impact of addition/deletion of layers and fine-tuning of the parameters need to be investigated. A detailed investigation of the impact of different data augmentation techniques is another area to be explored.

References

1. Diabetic Retinopathy: Causes, Symptoms, and Treatments (2019) Retrieved from September 13, 2019, https://www.medicalnewstoday.com/articles/183417.php
2. Diabetic retinopathy—Symptoms and causes—Mayo Clinic (2019) Retrieved from April 16, 2019, https://www.mayoclinic.org/diseases-conditions/diabetic-retinopathy/symptoms-causes/syc-20371611
3. Salamat N, Missen MMS, Rashid A (2019) Diabetic retinopathy techniques in retinal images: A review. Artif Intell Med. https://doi.org/10.1016/j.artmed.2018.10.009
4. Exudate: MedlinePlus medical encyclopedia (2019) Retrieved from September 13, 2019 https://medlineplus.gov/ency/article/002357.htm

5. Wan S, Liang Y, Zhang Y (2018) Deep convolutional neural networks for diabetic retinopathy detection by image classification. Comput Electr Eng 72:274–282. https://doi.org/10.1016/J.COMPELECENG.2018.07.042
6. Chakrabarty N (2018) A deep learning method for the detection of diabetic retinopathy. In 2018 5th IEEE Uttar Pradesh section international conference on electrical, electronics and computer engineering (UPCON). IEEE, pp 1–5. https://doi.org/10.1109/UPCON.2018.8596839
7. Burewar S, Gonde AB, Vipparthi SK (2018) Diabetic retinopathy detection by retinal segmentation with region merging using CNN. In 2018 IEEE 13th international conference on industrial and information systems (ICIIS). IEEE, pp 136–142. https://doi.org/10.1109/ICIINFS.2018.8721315
8. Hemanth DJ, Deperlioglu O, Kose U (2019) An enhanced diabetic retinopathy detection and classification approach using deep convolutional neural network. Neural Comput Appl 1–15
9. Rubini SS, Nithil RS, Kunthavai A, Sharma A (2019). Deep convolutional neural network-based diabetic retinopathy detection in digital fundus images. Springer, Singapore, pp 201–209. https://doi.org/10.1007/978-981-13-3600-3_19
10. Zhang W, Zhong J, Yang S, Gao Z, Hu J, Chen Y, Yi Z (2019) Automated identification and grading system of diabetic retinopathy using deep neural networks. Knowl Based Syst 175:12–25. https://doi.org/10.1016/J.KNOSYS.2019.03.016
11. Zeng X, Chen H, Luo Y, Ye W (2019) Automated diabetic retinopathy detection based on binocular siamese-like convolutional neural network. IEEE Access 7:30744–30753. https://doi.org/10.1109/ACCESS.2019.2903171

ALOSI: Aspect-Level Opinion Spam Identification

Pratik P. Patil, Shraddha Phansalkar, Swati Ahirrao, and Ambika Pawar

Abstract Opinion mining, an upcoming area of computational linguistics, deals with analysis of public sentiment polarity of a product giving rise to a new platform on social media and e-commerce for users to express/explore opinions on products and/or product manufacturers for the betterment of the product development. A tentative product rating can be derived from the product opinions mined. ALOSI takes opinion mining to a deeper level performing opinion mining at aspect level giving well-analyzed results at product component level. But the data received is not always totally relevant leading to false rating of the product or aspect(s). This irrelevant data can be labeled as spam and is important to identify as it may restrict the model. The presented work models a product-based aspect-level sentiment analysis on public reviews with spam identification and filtering, hence, highlighting the difference in the rating of product or aspect(s) with the presence and absence of spam reviews.

Keywords Aspect-level sentiment analysis · Computational linguistics · Machine learning · Opinion spam · Opinion mining · Social media

P. P. Patil · S. Phansalkar (✉) · S. Ahirrao · A. Pawar
Symbiosis Institute of Technology, Department of Computer Science and Engineering, Lavale, Pune 412115, Maharashtra, India
e-mail: shraddhap@sitpune.edu.in

P. P. Patil
e-mail: patil.pratik@sitpune.edu.in

S. Ahirrao
e-mail: swati.ahirrao@sitpune.edu.in

A. Pawar
e-mail: ambika.pawar@sitpune.edu.in

D. S. Jat et al. (eds.), *Data Science and Security*, Lecture Notes in Networks and Systems 132, https://doi.org/10.1007/978-981-15-5309-7_14

1 Introduction

Today, with the advent of digital marketing and Internet industry, there has been a rise in the number and types of platforms for expressing viewers' opinions on a product or services through social media. Nowadays, social media sites like Facebook, Twitter, Pinterest, LinkedIn, etc. and review websites like Amazon, IMDB, Yelp, etc. have democratized public opinions [1]. Analyzing this enormous collection of opinions for sentiment extraction is essential for the manufacturers in realizing the public opinion on their products and for prospective buyers to buy right product.

The user opinions are multimodal but predominantly in text format. The sentiment analysis at aspect-level can further help the manufacturer granulize the public opinion attributed to different parts (components) of the products. Review opinion mining is worsened with spam reviews which might affect the rating of the product with irrelevant, manipulative, fake, and sometimes polarized reviews. In this work, the presented system implements and validates a model which performs aspect-level sentiment analysis and spam detection on a product- or domain-specific dataset. The system is divided into three phases—dataset preparation, aspect-level sentiment analysis, and spam detection. The main goal of this system is to give manufacturer a precise sentimental and spam-free analysis of their product and its aspects. The results will also show the involvement and effect of spam reviews on the ratings of the product.

1.1 Sentiment Analysis

Sentiment analysis (a.k.a. Opinion Mining) is an upcoming field and focusses on study of the opinion, sentiment, or emotion toward an entity [2]. An entity can be anything from an event, individual to any topic. Sentiment analysis is the detection of sentiments from a set of text and is represented in terms of polarities (positive(satisfied), negative (not satisfied), or neutral (nothing to comment)). Sentiment analysis can be seen spread over all domains involving data science, viz., social media, healthcare, management, politics, etc. [3]. Opinion mining has two main methods—sentiment analysis and sentiment classification. Sentiment classification involves using subjective analysis to assign polarity labels to the text documents. Subjective analysis checks if the data involves user's feelings or not. Analyzing these feelings will give the manufacturers a review of their product at much lower level granularity [4].

Sentiment analysis comprises three main classification levels—document-level, sentence-level, and aspect-level [3] (as shown in Fig. 1). This classification is based on the artifacts that are analyzed and their outcomes.

a. **Document-level**: Uses the entire document to find out the overall sentiment. This is applicable when the whole document refers to a single entity, like a mobile phone or a person.

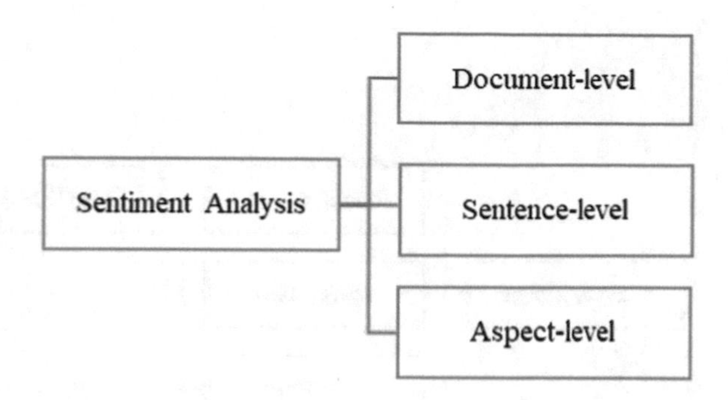

Fig. 1 Levels of sentiment analysis

b. **Sentence-level**: Determines the overall sentiment of a sentence. Every sentence in a document is iterated, and sentiments of multiple entities present can be determined.
c. **Aspect-level**: Here, the sentiment for each topic is evaluated. Aspect-level sentiment analysis is also referred to as entity/feature-level sentiment analysis. Sentiments on different multiple entities can be determined within a single sentence or token.

1.2 Aspect-Level Sentiment Analysis

Aspect-level sentiment analysis is the study of sentiment at the component-level. Aspects can also be called as features or entities. This analysis is carried out in three phases [5] which are identification of sentiment–aspect pairs in textual data, classification of sentiment–aspect pairs with respect to opinion polarity, and lastly aggregation of classified for consolidation overview. Also, it is worthwhile to see that there are some parameters like flexibility, robustness, and speed [6]. Flexibility is when an application tends to work better in certain domains. Robustness deals with the informal writing styles with which people express their sentiments. High-speed performing model is required for better aspect detection and sentiment analysis [6]. Aspect-level sentiment analysis can detect multiple aspects present in the single sentence and find out the respective polarity associated with it.

E.g., *"Phone build quality is great; camera is also good but the battery life is bad."*

From the above example of review, we can manually identify that the aspects present are phone quality, camera, and battery life, and the sentiments are positive (great), positive (good), and negative (bad).

Figure 2 as explained in [6, 7] shows the various approaches in aspect-level sentiment analysis with respect to three methods, viz., aspect detection, sentiment analysis, and hybrid approach. Aspect detection involves the extraction of aspects from textual data. This can be done by checking most frequently occurring words [8], syntax of sentence [9], using various machine learning algorithms like LDA, LSI, naïve Bayes, support vector machine, etc. or a combination of either of these methods. Sentiment

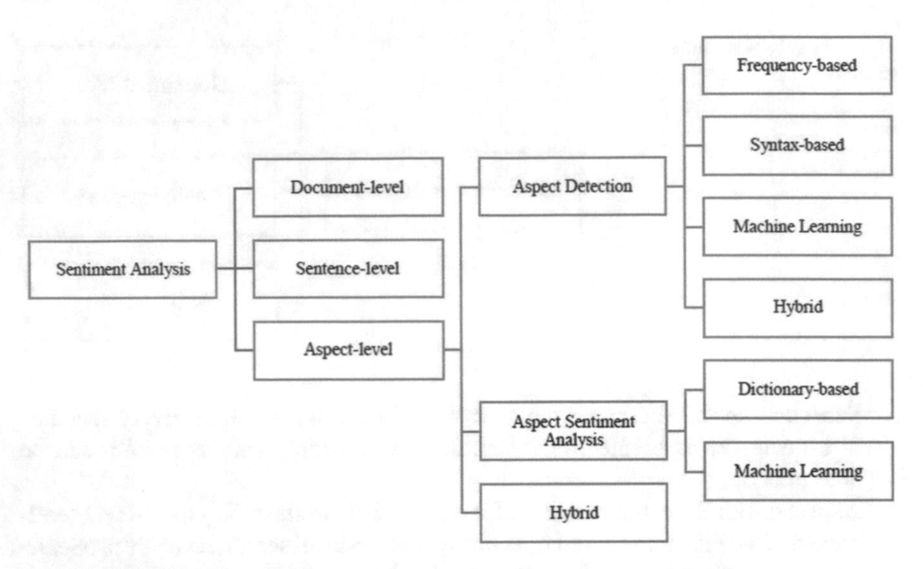

Fig. 2 Classification of aspect-level sentiment analysis approaches

analysis involves extracting sentiment polarity from the aspect text. Aspects can be sentimentally analyzed by maintaining and using a dictionary [10] or machine learning algorithms [11].

1.3 Spam Detection

Spam detection is a significant activity when reviewing a product online. It mostly consists of stuff like fake reviews taking user to other third-party links, irrelevant reviews, highly polarized reviews, etc. Spam reviews can be sourced to an individual or a spammer community. It can be done either to promote or demote a product. Spammers usually post certain polarized reviews and are very manipulative. Our target for this system is to detect spam reviews and to filter them out to present a spam-free authentic product analysis. Not much work has been done on opinion spam detection but a lot of work can be found on email and SMS spam detection using bag-of-words technique to detect spam messages and emails. Spam emails [12] can have features like unsolicited by unknown sender, bulk irrelevant reviews.

Several frameworks have been proposed till date for detecting spam, especially in email and SMS, but not in opinion mining. Hu et al. [13] proposed and explained a framework which does spammer detection on social media platform based on sentiment analysis. On similar note, Shehnepoor et al. [14] proposed a network-based framework called NetSpam for detecting spammers, both individuals and communities. NetSpam uses heterogeneous information networks to map spam detection

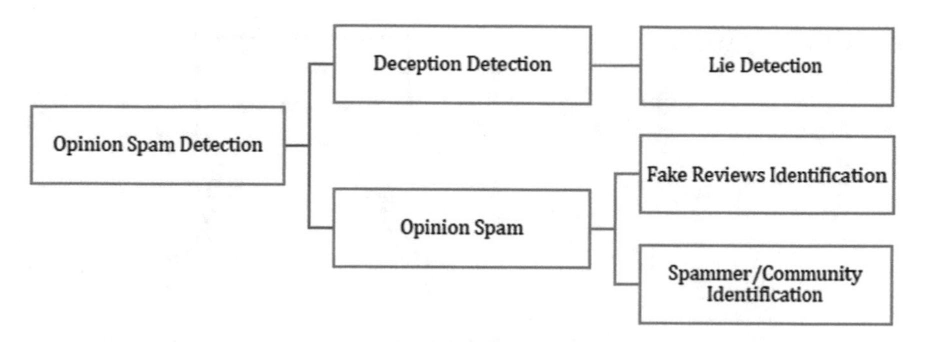

Fig. 3 Taxonomy of spam detection approaches

to classification problem and incorporates metapath connecting the reviews. TopicSpam [15] uses the LDA-based topic modeling approach for detecting truthful and deceptive reviews. Thanigaivelan et al. [16] proposed a distributed anomaly detection system implemented in the world of Internet of Things (IoT) which uses propagation of control messages through a network.

Figure 3 explains the basic taxonomy of opinion spam detection approaches with respect to two main types—deception detection and opinion spam. Deception detection is detection of deceptive opinions or lies expressed in reviews [17, 18]. Opinion spam involves identification of fake reviews, spammer individuals, and spammer communities. Spammers are the people who deliberately post spam/fake reviews of the product. These fake reviews can be used for product promotion or demotion purposes. As discussed in [19], the performance of learning classifier system (LCS), a rule-based machine learning system, is by performing sentiment analysis on twitter dataset and spam detection on SMS spam dataset. Narayan et al. [20] utilized the supervised learning techniques for spam detection in the opinion mining. The presented work uses different sets of features with sentiment score calculated to build models and compute their performance using different classifiers. Section 2 describes the methodology and the result analysis of the Aspect-Level Opinion Spam Identification (ALOSI) model.

2 Methodology and Workflow

Sentiment analysis and spam detection are two very important fields in linguistics. Sentiment analysis helps gain a sentiment out of textual data using polarities, while spam detection is used to remove the anomalies changing the sentiment polarity. In the age of ever-growing social networks and e-commerce, both these fields play a vital role in analyzing textual data and guessing the true personality of review and reviewer. Reviewer expresses sentiments/opinions by making use of certain words or phrases which can be classified as positive, negative, or neutral (Fig. 4).

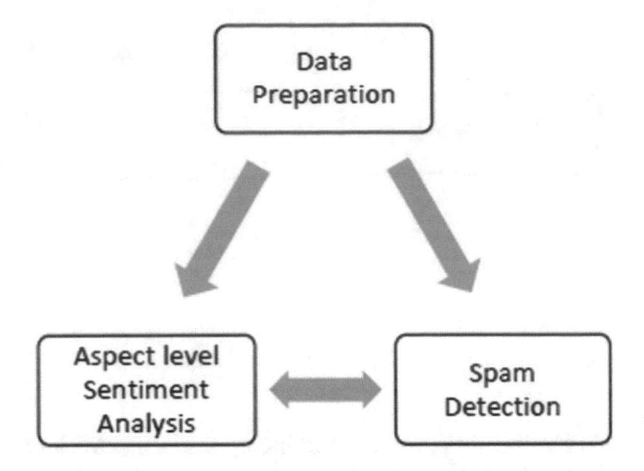

Fig. 4 Phases of ALOSI

I. **Data Preparation**:

(a) **Review Dataset**:

Experimentation is done on product-specific datasets of Nokia 6610 and Canon SD500 [8]. The product dataset has approximately 630 reviews and 300 reviews. The dataset is a labeled dataset and contains reviews labeled either ham or spam. The dataset is in CSV file format with tab as the delimiter. Data preparation involves the steps used to pre-process and clean the dataset making it ready to give as input to the proposed model. Figure 5 shows the workflow of Aspect-Level Opinion Spam Identification (ALOSI).

(b) **Data Pre-processing**:

Data pre-processing involves the cleaning of the dataset. This step involves eliminating all the noisy, redundant, and unwanted data, making data ready to use. Pre-processing also involves tokenization, lemmatization, parts-of-speech tagging, and

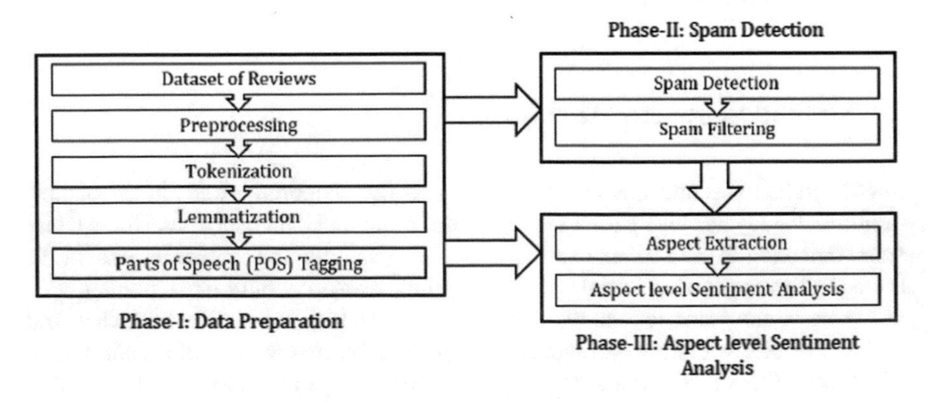

Fig. 5 Workflow of ALOSI

stop words removal, which can be explained further. Furthermore, this data is given as input to the tokenizer.

(c) **Tokenization**:

Tokenization is the process of splitting each sentence into parts delimited by spaces and ignoring special symbols and numbers. The pre-processed data is given as input to the tokenizer which converts the reviews into sentences making it easy for further steps like POS tagging and aspect detection.

(d) **Lemmatization**:

Lemmatization is used to group together the different types of words having a similar meaning. Just like stemming, words are reduced to its parent words (a.k.a. seed words). But unlike stemming the reduced word has a meaning, e.g., "*Data science is being studied by limited people.*" From the example, we can find the difference between stemming and lemmatization. Word "*studied*" is reduced to "*studi*" by stemmer, while lemmatizer reduces it to "*study*" which has more meaning.

(e) **Parts-of-Speech (POS) Tagging**:

Parts-of-Speech tagging takes in the tokenized data as input and performs word tokenization on it independently converting each word into token. Every word tokenized is then tagged for part-of-speech. Here the stop words and punctuations are removed. Stop words removal involve the words like "*a*", "*an*", "*this*", "*I*", etc. This is done to get more refined results for aspect detection and sentiment analysis.

II. **Spam Detection**:

The presented system is domain-independent but requires product-specific reviews. The goal of ALOSI is to present manufacturers with a sentimentally well-analyzed and spam-free result at aspect level. Spam detection involves the removal of any unwanted, irrelevant reviews or reviews contributing no meaning to product reviews. For illustration purpose, the product-specific review data of Nokia 6610 have been considered. The pre-processed data is given as input to the spam detector and eliminator before aspect-level spam detection to get more refined results with respect to the product. Spam detection is done keeping note of few key points. Spam detection model has been developed using multinomial naïve Bayes and Bernoulli naïve Bayes out of which Bernoulli naïve Bayes resulted in better performance. When using ALOSI, spam detection can be considered as an alternative step as it can be bypassed for aspect-level sentiment analysis.

(a) **Spam Key Points**:

The spam detection is done keeping in mind some key points. The definition of spam is relative to domain chosen. Following are the key points:

i. *Hyperlinks redirecting user to a third-party website*: Hyperlinks are often considered to be an indicator of spam. In ALOSI, they are a suspect of spam and may redirect prospective buyers to a third-party webpage. However, some of these webpages can be authentic. Hence, ALOSI extracts all the hyperlinks from dataset and bookmarks the reviews where it appears.
ii. *Reviews talking nothing about aspects of product or product itself*: Done after the aspect extraction, this optional step involves reviews having nothing in common with the product or product aspects itself and is labeled as non-contributive. These aspects are tagged, and the manufacturers can take a call as per which to keep or discard.
iii. *Advertising/Promotional/Bonus fake reviews*: Reviews showcasing a different product or redirect to a different product will be tagged as spam. These reviews usually follow a specific pattern and are in a certain format. These reviews mostly try to demote the product and market their own product. These reviews are usually big (> 50 characters) and involve certain keywords like "buy", "discount", "coupon code", etc.

(b) **Feature Selection for Spam**:

The input data from the tokenizer and lemmatizer is converted to vectors for extracting features so that it is understandable to the machine learning models. This requires essentially three steps, in the bag-of-words model:

i. Calculating term frequency [21] for every word in each message.
ii. Calculating the inverse document frequency [21], so the frequent tokens have lower weights.
iii. Finding the L2 norm [21] which normalizes the vectors to unit length, to get the gist of the original text length.

The data is converted to vectors using count vectorizer, and the bag-of-words and unique words are detected with their frequency for every review message in the dataset. All the unique words are selected and saved in a dictionary. One review message has many dimensions, and processing them individually is difficult. Hence, use of machine learning is done to make this process automated and quick to process. Considering all the vectors, there are in total 1838 unique words in the dictionary. A sparse matrix has been generated for this high-dimensional data. Details of sparse matrix are as follows with dimension in format of number of reviews, unique words.

- *Dimension of sparse matrix: (623, 1838)*
- *Non-zero spaces: 7891*
- *Sparsity: 0.69%.*

(c) **Training model for detecting spam**:

With the review messages converted to vectors, machine learning model can now be developed for detecting ham/spam reviews. Scikit module has been used to train the

Table 1 Classification report for all models

Algorithm(s)	Accuracy	Confusion matrix	Precision	Recall	F1-score
Multinomial naïve Bayes	0.9454253611556982	[[584 0] [26 10]]	0.95	0.95	0.92
Bernoulli naïve Bayes	0.9678972712680578	[[584 0] [14 22]]	0.97	0.97	0.96

model using Multinomial naïve Bayes and Bernoulli naïve Bayes [22]. According to the literature review done on spam detection, most of the models were developed using these classifiers. Hence, these classifiers were selected for model building and performance analysis.

Once the models were trained, a classification report was generated for all the models. The classification report contained the accuracy of the model along with the confusion matrix, precision, recall, and f1-score. The Nokia 6610 dataset has approximately 630 product-specific reviews labeled as either ham or spam. Table 1 shows the classification report for all the models implemented. According to classification report, Bernoulli naïve Bayes gives much realistic results with 96% accuracy, precision, and recall at 97% and f1-score at 96% outperforming multinomial naïve Bayes giving 94% accuracy and 95% precision, recall, and 92% f1-score. Confusion matrix gives a projection of how many reviews were identified as spam or ham using dictionary approach while feature extraction.

(d) **Cross-validation**:

Cross-validation involves reserving specific part of the dataset on which model is tested after the training. Once the models are tested, and accuracy and classification reports are generated, and the models are prepared for real-time spam detection. In order to do so, cross-validation [23] is to be done, and it is to be plotted with a learning curve to gain an understanding of how change in data can change the results. Firstly, the dataset is split into training and testing set with ratio of 80:20. Cross-validate all models and calculate mean and standard deviation. Figure 6 shows the cross-validation results with the mean and standard deviation.

The scores obtained in the cross-validation are quite low as compared to original accuracy result. To infer more information, we represented it with a learning curve. Figure 7 shows the learning curve for multinomial naïve Bayes, and Fig. 8 shows the learning curve for Bernoulli naïve Bayes. We can understand that when the number of spam reviews increases in case of multinomial naïve Bayes, the accuracy tends to fall resulting in a less efficient model. But in case of Bernoulli naïve Bayes, the accuracy tends to be either constant or slightly increases depending on the reviews. However, the cross-validation scores slightly fluctuate around the 94% range for both models.

Once the learning curve is plotted, it is now time to remove all the spam reviews from the dataset. With reference to our keywords, the hyperlinks have been removed from reviews and stored in a separate file from which manufacturer can make a

```
Multinomial Naive Bayes

[0.94       0.94       0.94       0.94       0.94       0.94
 0.94       0.94       0.93877551 0.93877551]

Mean: 0.9397551020408162
Standard Deviation:  0.0004897959183673084


Bernoulli Naive Bayes

[0.94       0.94       0.94       0.94       0.94       0.94
 0.94       0.94       0.93877551 0.93877551]

Mean: 0.9397551020408162
Standard Deviation:  0.0004897959183673084
```

Fig. 6 Cross-validation results with the mean and standard deviation

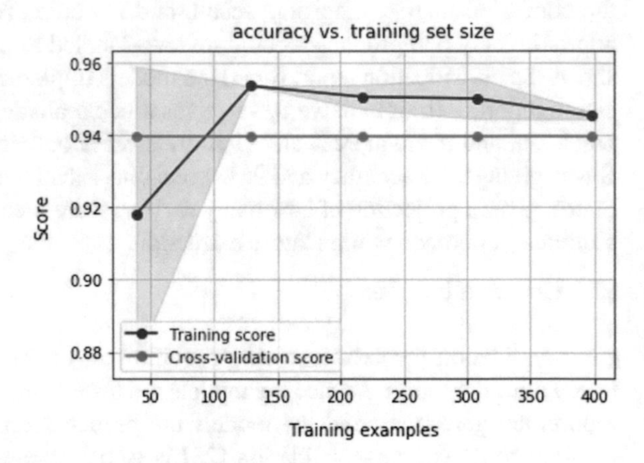

Fig. 7 Learning curve for multinomial naïve Bayes

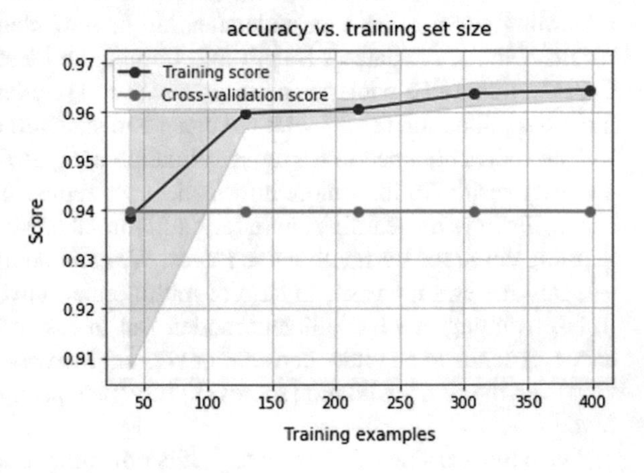

Fig. 8 Learning curve for Bernoulli naïve Bayes

decision which links to save or delete. All reviews in advertising and promoting product will be removed from dataset, thus keeping only relevant reviews. The filtered dataset is then fed to aspect-level sentiment analysis module for aspect detection and sentiment analysis. Aspect detection can then detect aspects from reviews, and those reviews with no relevant aspects shall be removed keeping only reviews with actual aspects which will then be sentimentally analyzed.

III. **Aspect-Level Sentiment Analysis**:
(a) **Aspect Detection**:

Aspect detection for presented model is the process of detection of features of the product. These features are identified using the parts-of-speech tagged words. The words which are *Nouns (NN, NNS)* and *Proper Nouns (NNP, NNPS)* are considered to be aspects. These aspects can be considered as features of product(s), e.g., "*Computers have various features/aspects like screen size, hard-drive capacity, resolution, CPU, GPU, RAM,* etc." Aspect detection is done using parts-of-speech tagging and frequency-based approach. Only aspects having occurrences greater than 1 are considered. This removes any small possibility of false noun word occurrences as aspects giving a more refined result.

(b) **Aspect-Level Sentiment Analysis**:

Aspect-level sentiment analysis deals with finding out the sentiment of each and every aspect in a review detected in aspect detection step. Patil et al. [7] explained a topic modeling approach for aspect-level sentiment analysis. This presented model uses a different approach that of aspect detection. Sentiments are influenced by adjectives occurring near to the aspect word. Hence, in the model, the sentiment is identified using all the adjectives (JJ, JJR, JJS) and adverbs (RB, RBR, RBS) close to the aspect word, e.g., "*The mobile phone has a great battery life but a poor camera.*" In this example, we can manually identify that "*mobile phone*", "*battery life*", and "*camera*" are nouns and hence can be considered as aspects. While words like "*great*" and "*poor*" are adjectives which depict the sentiment involved for the respective aspect. Hence, we get the result as positive for battery life and negative for camera.

3 Conclusion

ALOSI's main goal is to provide the manufacturer with a relevant sentimentally analyzed and spam-free data giving them an idea of the public's opinions on aspects of their product. ALOSI is divided into three phases: the first phase being data preparation, second being spam detection, and finally aspect-level sentiment analysis. The dataset is a product-specific textual review dataset consisting of text reviews without any other attributes. For illustration purposes, Nokia 6610 and Canon SD500 datasets were used.

Spam detection is a very important requirement in computational linguistics for analysis of comments and reviews. Spam reviews have a great impact on the topic being discussed. Using frequency and dictionary approaches, a dictionary of unique words is created. Spam detection model has been developed with multinomial and Bernoulli naïve Bayes, out of which Bernoulli naïve Bayes gave optimum results. This model, placed before aspect-level sentiment analysis, is done so that a more refined result is obtained without going back to it, reducing the time complexity of the model. The spam-free data is then fed to aspect extraction to extract aspects and later perform sentiment analysis.

When working on this project, it was observed that there are far more sarcastic reviews in the dataset. This causes incorrect results when performing sentiment analysis as these reviews are inconclusive for sentiments. Sarcasm detection is a very vast sub-field under sentiment analysis and NLP, and is much required for evaluating sentiments from reviews [24]. In the future, work can be done on including sentiment analysis with trust dynamics. Trust dynamics is a field which determines the impact of trust of a user on the product ratings and how it is influenced by direct experiences [25]. This could be a very interesting and a cross-domain work.

References

1. Tang D, Qin B, Liu T (2015) Deep learning for sentiment analysis: successful approaches and future challenges. Wiley Interdisc Rev Data Min Knowl Disc 5(6):292–303
2. Medhat W, Hassan A, Korashy H (2014) Sentiment analysis algorithms and applications: A survey. Ain Shams Eng J 5(4):1093–1113
3. Shivaprasad TK, Shetty J (2017) Sentiment analysis of product reviews: A review. In 2017 international conference on inventive communication and computational technologies (ICICCT), IEEE, pp 298–301
4. Serrano-Guerrero J, Olivas JA, Romero FP, Herrera-Viedma E (2015) Sentiment analysis: A review and comparative analysis of web services. Inf Sci 311:18–38
5. Tsytsarau M, Palpanas T (2012) Survey on mining subjective data on the web. Data Min Knowl Disc 24(3):478–514
6. Schouten K, Frasincar F (2016) Survey on aspect-level sentiment analysis. IEEE Trans Knowl Data Eng 28(3):813–830
7. Patil PP, Phansalkar S, Kryssanov VV (2019) Topic modelling for aspect-level sentiment analysis. In Proceedings of the 2nd International Conference on Data Engineering and Communication Technology, Springer, Singapore, pp 221–229
8. Hu M, Liu B (2004) Mining opinion features in customer reviews. AAAI 4(4):755–760
9. Guang Qiu, Bing Liu, Jiajun Bu and Chun Chen. Expanding domain sentiment lexicon through double propagation. In *IJCAI* (Vol. 9, pp. 1199–1204), 2009
10. Hu M, Liu B (2004) Mining and summarizing customer reviews. In Proceedings of the tenth ACM SIGKDD international conference on Knowledge discovery and data mining. ACM, pp 168–177
11. Titov I, McDonald R (2008) Modeling online reviews with multi-grain topic models. In Proceedings of the 17th international conference on World Wide Web. ACM, pp 111–120
12. Enrique Puertas Sanz (2008) José María Gómez Hidalgo and José Carlos Cortizo Pérez. Email spam filtering. Adv Comput 74:45–114
13. Hu X, Tang J, Gao H, Liu H (2014) Social spammer detection with sentiment information. In 2014 IEEE International Conference on Data Mining (ICDM), IEEE, pp 180–189

14. Shehnepoor S, Salehi M, Farahbakhsh R, Crespi Noel (2017) NetSpam: A network based spam detection framework for reviews in online social media. IEEE Trans Inf Forensics Secur 12(7):1585–1595
15. Li J, Cardie C, Li S (2013) Topicspam: a topic-model based approach for spam detection. In Proceedings of the 51st annual meeting of the association for computational linguistics (vol 2 Short Papers), vol 2, pp 217–221
16. Thanigaivelan NK, Nigussie E, Kanth RK, Virtanen S, Isoaho J (2016) Distributed internal anomaly detection system for Internet-of-Things. In 2016 13th IEEE annual consumer communications & networking conference (CCNC), IEEE, pp. 319–320
17. Rosso P, Cagnina LC (2017) Deception detection and opinion spam. In A practical guide to sentiment analysis. Springer International Publishing, pp 155–171
18. Rout JK, Singh S, Jena SK, Bakshi S (2017) Deceptive review detection using labeled and unlabeled data. Multimedia Tools Appl 76(3):3187–3211
19. Arif MH, Li J, Iqbal M, Liu K (2017) Sentiment analysis and spam detection in short informal text using learning classifier systems. Soft Comput 1–11
20. Narayan R, Rout JK, Jena SK (2018) Review spam detection using opinion mining. In Progress in intelligent computing techniques: theory, practice, and applications. Springer, Singapore, pp 273–27
21. Jing L-P, Huang H-K, Shi H-B (2002) Improved feature selection approach TFIDF in text mining. In Proceedings International Conference on Machine Learning and Cybernetics, 2002, vol 2, IEEE, pp 944–946
22. McCallum A, Nigam K (1998) A comparison of event models for naive bayes text classification. In AAAI-98 workshop on learning for text categorization, vol 752, no 1, pp 41–48
23. Kohavi R (1995) A study of cross-validation and bootstrap for accuracy estimation and model selection. IJCAI 14(2):1137–1145
24. Maynard DG, Greenwood MA (2014) Who cares about sarcastic tweets? investigating the impact of sarcasm on sentiment analysis. In LREC, pp 4238–4243
25. Falcone R, Castelfranchi C (2004) Trust dynamics: How trust is influenced by direct experiences and by trust itself. In Proceedings of the Third International Joint Conference on Autonomous Agents and Multiagent Systems, 2004 (AAMAS 2004), IEEE, pp 740–747

Adaptive Artificial Bee Colony (AABC)-Based Malignancy Pre-Diagnosis

Sujatha Arun Kokatnoor and Balachandran Krishnan

Abstract Lung cancer is one of the leading causes of death. The survival rate of the patients diagnosed with lung cancer depends on the stage of the detection and the timely prognosis. Hence, early detection of anomalous malignant cells is needed for pre-diagnosis of lung cancer as it plays a major role in the prognosis and treatment. In this work, an innovative pre-diagnosis approach is suggested, wherein the size of the dataset comprising risk factors and symptoms is considerably decreased and optimized by means of an Adaptive Artificial Bee Colony (AABC) algorithm. Subsequently, the optimized dataset is fed to the Feed-Forward Back-Propagation Neural Network (FFBNN) to perform the training task. For the testing, supplementary data is furnished to well-guided FFBNN-AABC to authenticate whether the supplied investigational data is competent to effectively forecast the lung disorder or not. The results obtained show a considerable improvement in the classification performance compared to other approaches like Genetic Algorithm (GA) and Particle Swarm Optimization (PSO).

Keywords Adaptive artificial bee colony · Feed-forward back-propagation neural network · Risk factor and symptoms and modification rate

1 Introduction

Lung cancer is one of the [1, 19] well-known types of cancer disseminating disastrous effects in the modern world. Lung cancer [17] is a deadly disease triggered by the abominable growth of unusual cells metamorphosing into a tumor. The significant features in cancer staging may be broadly grouped into three categories such as primary tumor, regional lymph nodes, and metastasis [5].

Premature recognition is the most significant factor in decreasing the number of casualties on account of lung cancer [2, 7, 20]. In fact, the timely recognition of the

S. A. Kokatnoor (✉) · B. Krishnan
Department of Computer Science and Engineering, School of Engineering and Technology, CHRIST (Deemed to Be University), Bangalore 560074, India
e-mail: sujatha.ak@christuniversity.in

D. S. Jat et al. (eds.), *Data Science and Security*, Lecture Notes in Networks and Systems 132, https://doi.org/10.1007/978-981-15-5309-7_15

lung cancer [8, 14] has been an arduous task [18]. Three approaches to identification of lung cancer such as biochemical diagnosis, clinical diagnostics, and cytological histology diagnosis are currently followed [9]. Erstwhile investigators have unambiguously upheld the unassailable fact that almost all significant categories of lung cancer have been triggered by a singular factor, viz., cigarette smoking [10, 11].

Image processing and data mining methods have appeared on the arena as high-tech quality techniques for augmenting non-automated assessment of the aggressive ailment [13]. In this sense, medical data mining has expressed hopes as a gifted computing tool [15, 16] which is automatically applied to the study of patient data for the purposes of the identification of medically relevant information [12].

In the remainder of the document, Sect. 2 offers a bird's eye view of the relative investigational efforts closely linked to the innovative technique. Section 3 adorns the ins and outs of our eye-catching system launched with much fanfare. While Sect. 4 deftly discharges its duty of well tabulating the cheering test outcomes harvested by the new proposed method, the all-embracing conclusions beautify the contents of the final Sect. 5.

2 Related Work

The paper by Fang et al. is designed in order to recognize and authenticate the factors associated with the genes of the lung cancer to connect all its paths, along with a newly designed, network-based biomarker recognition technique together with an enhancement analysis of genes collection [21]. The result has revealed, using the combined detection method, a collection of fresh and unparalleled gene factors with possible causes of lung cancer by smoking cigarettes.

According to Berni et al., the authors have investigated multiple delays in rapid ambulatory diagnosis (RODP) [22] for presumed patients with lung cancer with those in the literature and instruction manuals, analyzed the effects of the reference route and warning symptoms on interruption, and verified that whether interruption is disrupted.

Nancy et al. [6] researched the network-based method of identifying a six-gene signature linked to smoking that co-expressed as important signaling passages in the non-small cell lung cancer (NSCLC). A smoking-related six signature-based gene that forecasts causes and risks of lung cancer and survival rate was recognized by the pathway-dependent method. This gene signature was identified and predicted for lung cancer in smokers with prospective medical undertones.

Tran [4] investigated the technique of identifying cost-conscious biological markers for the ideal predictive accuracy of NSCLC cancer of the lungs. Only by merging countless genes, the complex nature of cancers has limited forecasts of the status. The authors have found an insignificant group of nine signature-based genes on a sample of 12,600 NSCLC genes, which are identical to the NSCLC lung carcinoma deduction foundation, which are used as genetic markers.

Polat et al., using PCA, preprocessing weighted, and artificial immune recognition system (AIRS) principles [3], identified lung cancer. The technique is completed in three main phases. At first, by the way of theory analysis, lung cancer dataset of 57 factors is reduced to 4 factors. Second, a method constructed on weight-based fuzzy preprocessing system was used before the operation of the key classifier as a pre-dispensing feature. At last, the artificial immune recognition system was used as a classifier.

Bullinaria and AlYahya [2] have proposed a model that explores the efficiency of the Artificial Bee Colony (ABC) algorithm to optimize the link weights of feed-forward neural networks for classification tasks, and authors have provided a more detailed comparison with the conventional Back-Propagation (BP) training algorithm.

In Saadya Fahad Jabbar's paper [1], a new way of classifying cancer based on profiles of gene expression is presented. This procedure combines the mutual information method and the Firefly Algorithm (FA). First, before using the FA, the functional reduction takes place. Ultimately, the vector support machine can be used to separate cancer into its different types.

3 Proposed Methodology

The structural design of our innovative lung cancer pre-diagnosis mechanism is depicted in Fig. 1. The primary objective of this work is to classify lung cancer malignancy based on risk and symptom factors. 52 symptoms, 24 risk factors, and

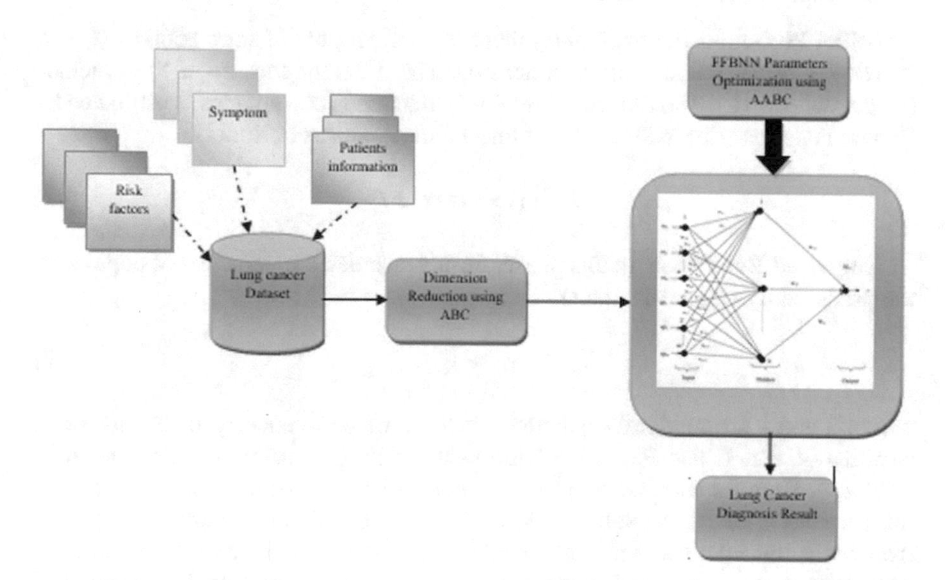

Fig. 1 Proposed architecture for malignancy pre-diagnosis system

44 classified records have been taken for the study. Preprocessing has been carried out by AABC method and fed to the FFBNN model.

3.1 Reducing the Dimension with the Use of AABC

Assuming a particular lung cancer dataset (L) comprises risk aspects and symptoms as the features gathered from the hospital having the dimension of ($I \times J$, where, $I \in 1, 2, \ldots u$, $J \in 1, 2, \ldots v$). Here, I and J represent the number of rows and columns of the dataset correspondingly. The Lung Cancer dataset (L) encompasses data on three different kinds of the patients and their parallel risk aspects and symptoms. They are people suffering from lung cancer, people afflicted with cancer other than lung cancer, and people not affected by cancer.

With the ABC algorithm, the dimensions of the dataset are reduced. ABC algorithm is a meta-heuristic algorithm based upon the swarm, which is inspired by the honey bee's strong fodder behavior. This consists mainly of three modules, viz., employed bees, onlooker bees, and scout bees.

- **Employed bees**: The employed bees are connected with the food sources in the area of the hive, and they transmit information about the nectar value of the food sources exploited by them to the onlooker bees.
- **Onlooker bees**: Onlooker bees watch the dance of the bees in the hive, selecting a single source of food that can be used as per the information given by the bees.
- **Scout bees**: The employed bees whose food sources are deserted change into scout and are on the lookout for fresh food source randomly.

Initial Phase: At the beginning, there is a random set of food sources ($i = 1, 2,..N$). Let N be the population dimension. The different sources of food include the dataset row (I_i). The fitness value of manufactured food sources is estimated by Fitness Function (FF) to evaluate the finest food source as in Eq. (1).

$$FF(j) = \max(Fi) \tag{1}$$

Employed Bee Phase: In this phase, freshly calculated parameters of population are produced by means of Eq. (2):

$$P_{i,j} = x_{i,j} + \gamma_{ij(x_{i,j} - x_{k,j})} \tag{2}$$

where k and j are randomly generated indexes of preference, γ takes any value between –1 and 1, and $P_{i,j}$ is the fresh value of jth parameter is indiscriminately produced. Then for each newly produced population parameter of sources of food, the calculation of fitness value is done. The best population parameter is picked from the estimated fitness value of the entire population. After choosing the efficient population parameter, probability of the picked parameter is predicted and calculated by means of Eq. (3).

$$Pi = \frac{FF_i}{\sum_{i=1}^{d} FF_i} \tag{3}$$

Onlooker Bee Phase: When the estimation of probability of the parameter that was picked is over, then the total number of onlooker bees is found out. Subsequently, fresh solutions ($P_{i,j}$) are engendered for the onlooker bees from the solutions ($x_{i,j}$) according to the probability value (P_j). Thereafter, to obtain a fresh solution, the new fitness function is calculated.

Scout Bee Phase: Now we calculate the parameters which were abandoned during the previous phases for the scout bees. If a parameter which was left out earlier is available, then fresh scouts are replaced with Eq. (3) parameters and determine the fitness value. Thereafter, the best parameters realized till now are committed to memory. Subsequently, the iteration is enhanced, and the process is repeated till the stopping standard is realized. At last, the dwindled dataset (L'= Γx J) is discovered.

3.2 Detection of Anomalous Lung Cells with the Aid of FFBNN-AABC

Number of inputs, hidden devices, and a solitary output unit have been built into neural networks. In this research work, the data is collected from the cancer super-speciality centers. 76 features and 44 patient records are taken for the study. Back-Propagation (BP) is employed as the training algorithm, and 20 numbers of hidden neurons are used.

3.3 FFBNN-AABC ALGORITHM

1. Allot weights arbitrarily for all neurons with the exception of input neurons.
2. The bias function and activation function of the neural network are defined as follows:

$$x(t) = \beta + \sum_{n=1}^{H} (w_m r_{t1} + w_m r_{t1} + \cdots + w_m r_m + w_m s_{t1} + w_m s_{t2} + \cdots + w_m s_m) \tag{4}$$

$$x(a) = 1/\left(1 + e^{-x(t)}\right) \tag{5}$$

   r_1, r_{2}.... are the extracted risk factors in the bias function and s_1, s_2... are the symptoms that are extracted from the reduced dataset. The output layer function for activation is shown in Eq. (5).
3. Learning error is calculated as follows:

$$E = \frac{1}{h}\sum_{n=0}^{h-1} D_n - A_n \tag{6}$$

In Eq. (6), E is the FFBNN network output, D_n and A_n are the expected and the real outputs, whereas the factor h is the total number of neurons in the neural network's hidden layer.

Error Minimization: Arbitrarily selected weights are assigned to covered neurons of the input and output layers. Neurons in the input layer keep their weight constant.

1. Now, the activation function and the bias function are estimated.
2. Thereafter, BP error for each and every node is determined and the weights are revised as shown in Eq. (7):

$$w_{(tn)} = w_{(tn)} + \Delta w_{(tn)} \tag{7}$$

$\Delta w_{(tm)}$ is obtained as

$$\Delta w_{(tn)} = \delta.x(t_n).B \tag{8}$$

where B is the back-propagation error and δ in Eq. (8) corresponds to the learning tempo, which is in the range of 0.2 and 0.5.
3. Thereafter, the steps (2) and (3) are iterated till the back-propagation error gets reduced. The procedure is continued until $B < 0.1$.
4. FFBNN is well guided for executing the testing stage as the error is reduced to a minimum value.

With a view to achieving superior precision and efficient accomplishment in identification of the lung cancer, the FFBNN parameters (w_m, β) are optimized and improved by means of AABC.

3.4 Initial Phase

The initial food sources are produced arbitrarily which are within the range of the boundaries of the parameters by means of the following equation (Eq. (9)):

$$x_{i,j} = x_j^{min} + r(0,1)x\left(x_j^{max} - x_j^{min}\right) \tag{9}$$

x_j^{min} and x_j^{max} are the prearranged boundaries, i= 1,2,.. N, j = 1, 2,.., O where O represents the number of optimization parameters. The fitness function favored in this case is given by Eq. (4). This optimization of FFBNN parameters by means of AABC turns out superior identification outcomes and efficient accomplishment.

Table 1 Performance of our proposed technique with other methods such as FFBNN-ABC, FFBNN-GA, and FFBNN-PSO

Measures	Proposed FFBNN-AABC	FFBNN-ABC	FFBNN-GA	FFBNN-PSO
Accuracy	95	90	60	80
Sensitivity	93	88	77	82
Specificity	100	100	29	67
FPR (False Positive Rate)	0	0.0	71.4	33.3
PPV (Positive Predictive Value)	100	100	67	93
NPV (Negative Predictive Value)	80	60	40	40
FDR (False Discovery Rate)	0	0	33	7
MCC (Matthews' Correlation Coefficient)	86.6	66.4	6.8	39.1

4 Performance Analysis

This lung cancer pre-diagnostic technique algorithm contrasts the output with other technologies like FFBNN- ABC, FFBNN- GA, FFBNN-PSO, and FFBNN (Table 1). The dimension of the specified dataset is 44×76. With the employment of the popular AABC algorithm, the dimension of the specified lung cancer dataset size is brought down, and the dimension of the abridged dataset is 21×76.

Table 1 contains dimensions regarding accuracy, sensitivity, and specificity. From these measures, it is clear from the captioned table that proposed FFBNN-AABC method has been able to achieve a better accuracy of 95%, in relation to those of FFBNN-ABC, FFBNN-GA, and FFBNN-PSO with 90, 60, and 80%, correspondingly. Increased efficiency is due to the adaptive element coupled with the artificial bee colony optimization method incorporated in the preprocessing stage of the data processing.

5 Conclusion

The selection of features is an important preprocessing process for machine learning and data mining tasks that enable the study models to perform better by eliminating redundant and irregular properties. Through this paper, an adaptive artificial bee colony method in the pre-diagnosis stage was explored for lung cancer by means of FFBNN and AABC. This method has shown improved accuracy, sensitivity, and specificity. With an inclusion of biomarkers of lung cancer, this method could further be augmented.

Acknowledgements Authors wishes to acknowledge the technical and infrastructural help rendered by the faculty members of Department of CSE of CHRIST (Deemed to be University), India.

References

1. Jabbar SF (2019) A classification model on tumor cancer disease based mutual information and firefly algorithm. Period Eng Natural Sci 7(3):1152–1162
2. Bullinaria JA, AlYahya K (2014) Artificial bee colony training of neural networks: comparison with back-propagation. Memetic Comput 6(3):171–182
3. Polat K, Gunes S (2008) Principles component analysis, fuzzy weighting pre-processing and artificial immune recognition system based diagnostic system for diagnosis of lung cancer. Expert Syst Appl 34(1):214–221
4. Quoc-Nam Tran (2011) A novel method for finding non-small cell lung cancer diagnosis biomarkers. In: International conference on bioinformatics and computational biology, Las Vegas, NV, USA, pp 18–21
5. Chaudhary A, Singh SS (2012) Lung cancer detection on CT images by using image processing. In: IEEE proceedings of international conference on computing sciences, Phagwara, pp 142–146
6. Guo NL, Wan Y-W (2012) Pathway-based identification of a smoking associated 6-gene signature predictive of lung cancer risk and survival. Artif Intell Med 55(2):97–105
7. Chang W-L, Chang C-C, Chen W-M (2010) Computer-aided diagnosis system for variance estimation of 3D ultrasonography based on gabor filter. In: IEEE proceedings of international conference on biomedical engineering and computer science, Wuhan, pp 1–4
8. Taher F, Werghi N, Al-Ahmad H (2012) A thresholding approach for detection of sputum cell for lung cancer early diagnosis. In: IEEE proceedings of IET conference on image processing, London, pp 1–6
9. Wu Y, Wang N, Zhang H, Qin L, Yan Z, Wu Y (2010) Application of artificial neural networks in the diagnosis of lung cancer by computed tomography. In: IEEE proceedings of the sixth international conference on natural computation, China, pp 147–153
10. Wang Y-B, Cheng Y-M, Zhang S-W, Chen W (2012) A seed-based approach to identify risk disease sub-networks in human lung cancer. In: IEEE proceedings of the 6th international conference on systems biology, Xi'an, pp 135–141
11. Wang C, Li X, Dai L, Zhang G, Gao L, Zhao Q (2011) Analysis of lung cancer gender differences and age structure in the high prevalence areas of Xiamen. In: IEEE proceedings of international conference on spatial data mining and geographical knowledge services, Fuzhou, pp 154– 159
12. Ada, Kaur R (2013) A study of detection of lung cancer using data mining classification techniques. Int J Adv Res Comput Sci Softw Eng 3(3):131–134
13. Ada, Kaur R (2013) Using some data mining techniques to predict the survival year of lung cancer patient. Int J Comput Sci Mobile Comput 2(4):1–6
14. El-Baz A, Gimelfarb G, Falk R, Abo El-Ghar M (2007) A New cad system for early diagnosis of detected lung nodules. In: IEEE proceedings of international conference on image processing, San Antonio, TX, pp II–461–II-464
15. Agrawal A, Misra S, Narayanan R, Polepeddi L, Choudhary A (2011) A lung cancer outcome calculator using ensemble data mining on SEER data. In: Proceedings of the ACM SIGKDD international conference on knowledge discovery and data mining, United States, pp 51–59
16. Zubi ZS, Saad RA (2011) Using some data mining techniques for early diagnosis of lung cancer. In: Proceedings of the 10th WSEAS international conference on artificial intelligence knowledge engineering and data bases, pp 32–37

17. Wan Y-W, Guo NL (2009) Constructing gene-expression based survival prediction model for non-small cell lung cancer (NSCLC) in all stages and early stages. In: IEEE proceedings of international conference on bioinformatics and biomedicine workshop, Washington, pp 338
18. Fatma T, Sammouda R (2011) Lung cancer detection by using artificial neural network and fuzzy clustering methods. In: IEEE proceedings of GCC conference and exhibition, Dubai, pp 295–298
19. Lu L, Wanyu L (2008) A method of pulmonary nodules detection with support vector machines. In: IEEE proceedings of the eighth international conference on intelligent systems design and applications, Kaohsiung, pp 32–35
20. Yu Q, He L, Nakamura T, Suzuki K, Chao Y (2012) A multilayered partitioning image registration method for chest-radiograph temporal subtraction. IEEE proceedings of international conference on computer science and information processing, Xi'an, Shaanxi, pp 181–184
21. Fang X, Netzer M, Baumgartner C, Bai C, Wang X (2013) Genetic network and gene set enrichment analysis to identify biomarkers related to cigarette smoking and lung cancer. Cancer Treat Rev 39(1):77–88
22. Brocken P, Kiers BA, Looijen-Salamon MG, Dekhuijzen PN, Smits-van der Graaf C, Peters-Bax L, de Geus-Oei LF, van der Heijden HF (2012) Timeliness of lung cancer diagnosis and treatment in a rapid outpatient diagnostic program with combined 18FDG-PET and contrast enhanced CT scanning. Lung Cancer 75(3):36–341

A Novel Approach of OTP Generation Using Time-Based OTP and Randomization Techniques

Siddharth Suresh Gosavi and Gopal Krishna Shyam

Abstract There are many ways of generating OTP in and performing user authentication, earlier authentication in a traditional way happened using username and password which became way popular in the digital world. Then, OTP came into the picture, and one-time password is the technique which is two-way authentication. This paper proposes an algorithm that will give an easy way to implement as well as a very secure OTP generation method. This OTP will be a combination of current time value and back-end server-generated value with some random operations on that combined value. This value is 32 bit in size which will give a combination of 8-digit numbers. The proposed OTP algorithm is easier to develop on the other side difficult to reverse engineer and gives efficient performance.

Keywords One-Time Password (OTP) · Authentication · Cloud · Two-factor authentication · Multi-factor authentication · Cloud-based OTP

1 Introduction

In this era of digital world, two-factor verification is very secure and efficient. Though there the level of attacks is increasing day by day, most of the applications are dependent on different verification techniques. In these techniques, one of the widely used techniques is OTP, i.e., One-time password. One-time password is the token sent to authenticating user's mobile number or email address which he enters, and valid entered token reflects that the user is authenticated and genuine. These OTPs are generated using many different techniques such as random OTP, time-based OTP, HOTP, and S/Key OTP generation. This paper shows a study of these OTP generation techniques and their issues, for example, in an organization where developers want a

S. S. Gosavi (✉) · G. K. Shyam
School of C&IT, REVA University, Bengaluru, India
e-mail: siddhugosavi89@gmail.com

G. K. Shyam
e-mail: gopalkrishnashyam@reva.edu.in

D. S. Jat et al. (eds.), *Data Science and Security*, Lecture Notes in Networks and Systems 132, https://doi.org/10.1007/978-981-15-5309-7_16

secure OTP without having difficult implantation along with it secure and irrevocable OTP in order to keep user data safe [1–4].

The study was done to recognize development and user advantages as well as disadvantages of OTP generation algorithms, and it shows that current algorithms are difficult to understand and to implement when developers need an easy to implement an algorithm. Though it is complicated to understand, these algorithms have several issues that are OTP algorithms that can be hacked by an attacker if they have core knowledge about the corresponding algorithm.

The proposed algorithm has several steps that are easy to understand and develop but they are not reversible. The reason the algorithm can't be reversed is that it includes randomization at the array level of the particular digit or number. These algorithms are in different sizes; some are 4 digits, some are 6 or 8 digits, and it is proven that 8 digits (32 bit) is strongest among all because it generates 2^{32} = 4,294,967,296 combinations which are very difficult to predict using any attack when combined with randomization at several levels of the algorithm.

The paper is written in sections where sections start from the introduction of study; the second part is related to the significant work done on core concepts required for development algorithm. The third section shows the proposed algorithm, idea, and needs behind it; the fourth section shows the result of a study done on different algorithms and their issues; final fifth section gives the conclusion of paperwork with acknowledgment and references.

1.1 Our Contributions in This Paper Are as Follows

1. Surveying about one-time password generation algorithms and techniques.
2. Implementation of OTP, easy, secure, and cost-efficient.

1.2 Coverage, Subject

The topics covered in this paper are more related to OTP generation algorithms and random value selection algorithm.

2 Related Works

OTP Technique. OTP is a password system where passwords can only be used once, and the user has to be authenticated with a new password key each time. OTP has much stronger security because the user has to enter a newly created password key even if his key or password is exposed. The OTP is standardized by the IETF and standardized again by verification-related companies [1, 5, 6]. A lot of OTP solutions are secret and/or proprietary; however, some like OATH (usually HMAC-SHA1) are open-source and widely used and supported between OTP providers. OATH uses an event-based OTP algorithm (it can also support time-based; however, this isn't widespread) which usually uses a secret character sequence (a.k.a. seed) only known to the two and the request number and sometimes other data such as customer unique seed, etc. [2].

R (Random Digit). OTP method HOTP is the secure OTP generating used Hash-based MAC. However, HOTP has a fixed digit; it can be vulnerable to dictionary attack and guessing attacks. This method is safe from the attack of another by creating a variable digit of the HOTP. HOTP based on the existing OTP generates the HOTP and the same operation using the SHA-1 algorithm [2, 5].

Fisher–Yates Shuffle. The Fisher–Yates shuffle is an algorithm for generating a random permutation of a finite sequence—in plain terms, the algorithm shuffles the sequence. The algorithm effectively puts all the elements into a hat; it continually determines the next element by randomly drawing an element from the hat until no elements remain. The algorithm produces an unbiased permutation: every permutation is equally likely. The Fisher–Yates shuffle is named after Ronald Fisher and Frank Yates, who first described it, and is also known as the Knuth shuffle after Donald Knuth [7]. This reduces the algorithm's time complexity to O(n), compared to O(n 2) for the naive implementation.

2.1 Issues in Current Algorithms

See Table 1.

3 Proposed Algorithm

3.1 Idea

The idea of proposed algorithm is to generate the time value when exactly the time user tries to initiate login as well as in addition to this the back-end server which will do authentication will pass a random 3-digit value to algorithm and both these values will get combined and randomization, and conversion operation will take place in further steps of algorithm. Finally, the hex or octal converted array of digits will get shuffled and then converted to the final integer of 8-digit desired OTP.

Table 1 Algorithms and their advantages and disadvantages

Name of algorithm	Advantages	Disadvantages
Lamppost's OTP [1] [1981]	• OTP is generated in 60 s • The calculation is performed in 1000 iteration • The user responds to OTP that is easy	• SMS lateness • Cost is high • Incoming messages might be blocked
S/Key OTP [8] [1991]	• It can be easily and quickly added to any kind of operating system • Terminal users are happy to use because of its easiness • Ease of installation	• Time is reduced to some extent
HMAC-based OTP [2] [2005]	• Ability to face the brute force attack • Ensuring frequent use in human time	• Sometimes validation error will come Potentially valid for a long time
Event-based OTP [9] [2005]	• An attacker would require undetected physical access to the device • The OTP value is created at the client's requirements	• An OTP value is valid until another OTP value is used • The client must press a key to create the OTP
Time-based OTP [3] [2011]	• Human-side security can be used in mobile phone applications • OTP is 8 digits and valid for 30 s • OTP is generated randomly	• Sometimes failure in receiving OTP • High cost • Easy to find the secret key

3.2 Algorithm

```
User login attempt time-value in HH:MM: SS format
Backendvalue random generated 3 digit
If time-value and backend-value is present
    Combine time-value and backend-value
          While all the combined values
                Generate random pairs
                If pair digit is more than 127
                      Generate random pair
                Else
                      Convert pair value in hex & octal
          End while
    Select  each  array  of  pairs  and  shuffle  using
    fisher-yates
    Convert List Elements into an int
Else
    Regenerate time-value and backend-value
End if
End
```

Following flowchart diagram shows algorithm working (Fig. 1).

3.3 Discussion

In this section, we describe an algorithm based on TOTP, randomization method, and the proposed OTP algorithm using current time value and back-end server-generated value. Following are the algorithm steps:

Step1—Generation of the Time value

Time value will be in the format of HH:MM: SS, and is the time when the OTP generation needs to happen or we can say it is time when a user is trying to access any service using OTP. For example, the time value is 09:18:59.

Step2—Generation of Random Value from the Back-end server

The server which handles the operation of OTP generation will give its contribution using the random digit method, and it will generate any 3- or 4-digit values. For example, randomly generated value is **654.**

Step3—Combination of Time value + Back-end server-generated value

The combination will be "091859654".

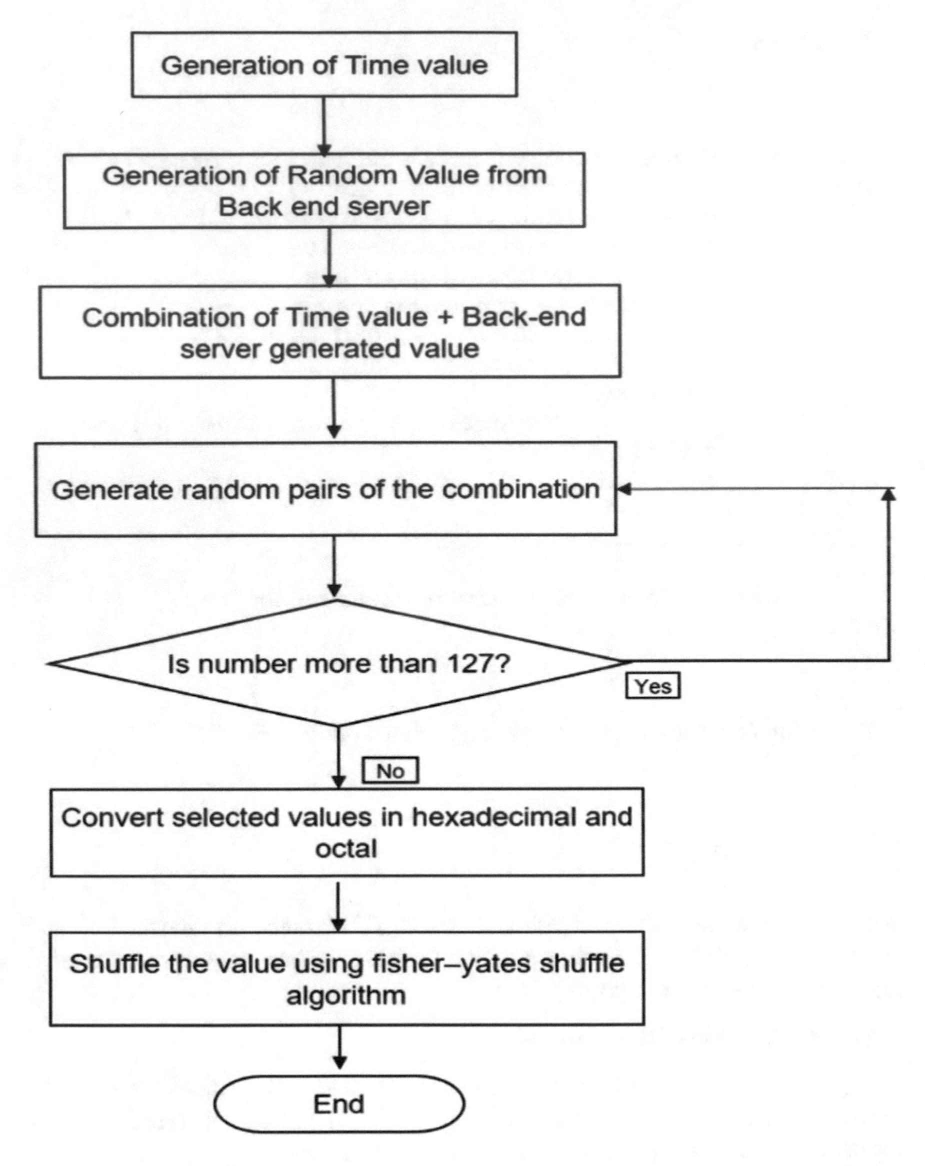

Fig. 1 Algorithm working

Step4—Generate random pairs of the combination

Here, the important process will happen, the algorithm will generate random pairs of 1, 2-, or 3-digit values, and it should follow the following rules:

- 1-digit value ≥ 0 and ≥ 9;
- 2-digit value ≥ 10 and ≥ 99;
- 3-digit value ≥ 100 and ≥ 127.

Table 2 Step4

Random pair	Digit	Value
1	2	09
2	1	1
3	3	859
4	1	6
5	2	36

Table 3 Step4 (b)

Random pair	Digit	Value
1	2	09
2	1	1
3	3	859
3a	2	85
3b	1	9
4	1	6
5	2	36

These rules are based on the ASCII table which will be used in the next step. Till that, let us assume that generated random pairs of the combined value are as follows:

Note: Here, third pair is breaking the rule, and thus algorithm will do step 4 again on these values and we will get new pairs as in Tables 2 and 3.

Step5—Convert selected values in hexadecimal and octal

Here, we will convert this randomly generated paired value into hexadecimal and octal values according to the ASCII table with the following rule:

If the hexadecimal value is alphanumeric, then convert the same selected value into octal. So now we will get a new set of values as now we got a new value as "**91549624**" (Table 4).

Table 4 Step5

Random pair	Digit	Selected value	Hex Value
1	2	09	0 × 9
2	1	1	0 × 1
3	3	859	
3a	2	85	0 × 54
3b	1	9	0 × 9
4	1	6	0 × 6
5	2	36	0 × 24

Table 5 Algorithm comparison table (high = 3, medium = 2, low = 1)

Algorithm	Cost	Development issue	Security issue
Lamppost's OTP	1	1	3
S/Key OTP	3	3	3
HMAC-based OTP	3	3	3
Event-based OTP	2	1	1
Time-based OTP	3	2	3

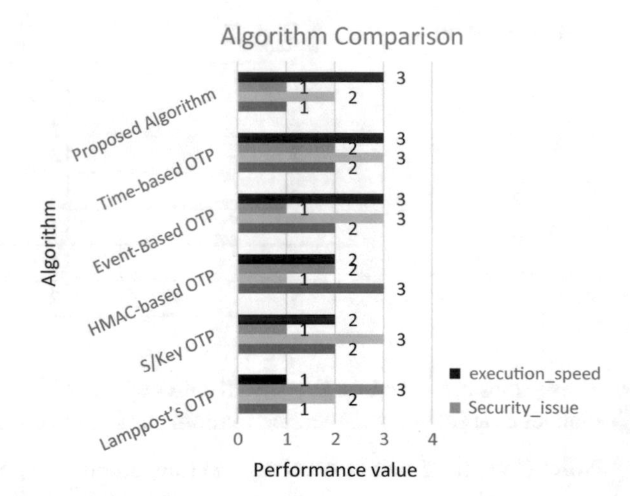

Fig. 2 Algorithm comparisons

Step6—Shuffle the value using the Fisher–Yates shuffle algorithm

Now shuffle will give value as "**14496295**" which will be the generated OTP.

4 Results

The study and analysis of the present algorithms and OTP technique when compared based on cost, development issues for developers and security issues the following result were obtained (Table 5).

Following is the graph which is a direct representation of the table which we created after study on OTP algorithms and techniques (Fig. 2).

5 Conclusion

The proposed algorithm is developed in Python language using Jupyter Notebook and used in the web development application for authentication of a user. Due to log in time value in HH:MM: SS format and back-end random data, the generated algorithm becomes tough to reverse. After several experiments, it shows that the proposed algorithm is less complex to develop but more complex to re-engineer, which is a good advantage for many applications as well as the cost is also efficient in comparison to other algorithms. In future work, the algorithm can be implemented using more complex randomization techniques to make the algorithm more difficult for attackers to recreate.

Acknowledgements This is a matter of pleasure for me to acknowledge my gratitude to the School of Computing and Information Technology, Reva University for giving me an opportunity to explore my abilities via this paperwork. I would like to express my sincere gratitude to our project guide, Dr. Gopal K. Shyam, for his valuable guidance and advice in completing this paperwork. Let me take this opportunity to thank the School Director, Dr. Sunil Kumar S. Manvi for the wholehearted support extended to me throughout the conduct of the study. Last but not the least, I would like to express my sincere thanks to my family members, friends for their immense support and best wishes throughout the academic duration and the preparation of this paper.

References

1. Lamport L (1981) Password authentication with insecure communication. Commun ACM 24(11):770–772
2. Mraihi D, Bellare M, Hoornaert F, Naccache D, Ranen O (2005) HOTP: an HMAC-based one-time password algorithm, document 4226. Internet Engineering Task Force, Fremont, CA, USA. https://www.ietf.org/rfc/rfc4226.txt
3. Mraihi D, Machani S, Pei M, Rydell J (2011) TOTP: time- based OTP algorithm, document 6238. IETF, Fremont, CA, USA. https://www.ietf.org/rfc/rfc6238.txt
4. OTP (2019) http://sites.google.com/site/kalman/OTP. Accessed July 2019
5. Durstenfeld R (1964) Algorithm 235: random permutation. Commun ACM 7(7):420. https://doi.org/10.1145/364520.364540
6. Yassin A, Jin H, Ibrahim A, Qiang W, Zou D (2013) Cloud authentication based on an anonymous OTP. In: Han Y-H, Park D-S, Jia W, Yeo S-S (eds) Ubiquitous information technologies and applications. Springer, Dordrecht, The Netherlands, pp 423–431
7. https://en.wikipedia.org/wiki/Fisher%E2%80%93Yates_shuffle#cite_note-cacm-2. Accessed Aug 2019
8. Haller N (1995) The S/Key one-time password system, document 1760. Internet Engineering Task Force, Fremont, CA, USA. https://www.ietf.org/rfc/rfc1760.txt
9. IETF RFC 2289, A one-time password system, Feb 1998
10. Erdem E, Sandkkaya MT (2019) OTPAAS one time password as a service member. IEEE Trans Inf Forensics Secur 14(3)
11. Eldefrawy MH, Khan MK, Alghathbar K, Kim T-H, Elkamchouchi H (2012) Mobile one-time passwords: two-factor authentication using mobile phones. Secur Commun Netw 5(5):508–516

Classifying Bipolar Personality Disorder (BPD) Using Long Short-Term Memory (LSTM)

Sreyan Ghosh, Kunj Pahuja, Joshua Mammen Jiji, Antony Puthussery, Samiksha Shukla, and Aynur Unal

Abstract With the advancement in technology, we are offered new opportunities for long-term monitoring of health conditions. There are a tremendous amount of opportunities in psychiatry where the diagnosis relies on the historical data of patients as well as the states of mood that increase the complexity of distinguishing between bipolar disorder and borderline disorder during diagnosis. This paper is inspired by prior work where the symptoms were treated as a time series phenomenon to classify disorders. This paper introduces a signature-based machine learning model to extract unique temporal pattern that can be attributed as a specific disorder. This model uses sequential nature of data as one of the key features to identify the disorder. The cases of borderline disorder that are either passed down genetically from parents or stem from exposure to intense stress and fear during childhood are discussed in this study. The model is tested with the synthetic signature dataset provided by the Alan Turing Institute in signature-psychiatry repository. The end result has 0.95 AUC which is an improvement over the last result of 0.90 AUC.

Keywords Borderline personality disorder (BPD) · Long short-term memory (LSTM) · Recurrent neural network (RNN) · Gated recurrent unit (GRU) · AUC

S. Ghosh · K. Pahuja (✉) · J. M. Jiji · A. Puthussery · S. Shukla
CHRIST(Deemed to be University), Bengaluru, India
e-mail: kunjpahuja@gmail.com

S. Ghosh
e-mail: gsreyan@gmail.com

J. M. Jiji
e-mail: jomonjiji@gmail.com

A. Puthussery
e-mail: frantony@gmail.com

S. Shukla
e-mail: samiksha.shukla@christuniversity.in

A. Unal
Digital Monozukuri, Stanford, USA
e-mail: unalayn@gmail.com

D. S. Jat et al. (eds.), *Data Science and Security*, Lecture Notes in Networks and Systems 132, https://doi.org/10.1007/978-981-15-5309-7_17

1 Introduction

In this modern age of fast-paced work–life and intensive workload, our society is slowly crawling toward a mentally ill dystopia. Borderline Personality Disorder (BPD) is one of the most common mental disorders among them and proliferates at a tremendous rate among the youth these days.

Statistics show that about 9.1% of the psychiatric population in the US suffers from personality disorders, out of which 1.4% constitutes of the BPD. The women-to-men ratio [1] suffering from BPD is 3:1. In Bengaluru, India, the technology capital of the country, a statistical report by the National Crime Records Bureau, shows that 35 out of every 100,000 people commit suicide due to mental disorders. Most of the diagnosis [2] done for psychiatric disorders has been hindered based on the retrospective approaches which include studying the abnormal behavior of the patient and the difficulty they experience with their mood swings. The shortcomings of the retrospective diagnostic categories and the severity of these disorders have prompted several pieces of research to enhance current diagnostic methods. In the span of the past few years, the technology has foreseen a tremendous evolution, and with the advancements in mobile phones everyday has opened our world to a whole new level of advancements. It has also spread its roots to the domain of psychopathology and enabled it to take more precise measures for the subjective concepts. The challenging task is analyzing the data produced, since most of the data generated are sequential in nature where most of the information is contained in the order at which different events occur.

This paper discusses Long Short-Term Memory (LSTM) model developed to re-analyze the synthetic data provided in GitHub repository for signature-based machine learning model for bipolar disorder and borderline disorder [3]. This research sought to classify the borderline personality disorder in patient on the basis of their mood patterns over a period of time. This research uses LSTM an artificial Recurrent Neural Network (RNN) architecture, as it provides us with the feedback connections and can process entire sequences of data.

The paper is organized as follows: Sect. 2 addresses the literature survey on Borderline Personality Disorder (BPD), Sect. 3 gives an overview of the data, Sect. 4 gives an overview of the methodology used in the proposed research, Sect. 5 details the discussion and results, and Sect. 6 presents the conclusions and limitations of the research work.

2 Literature Survey

In the paper [4], the author has mentioned bipolar disorder as Major Depressive Disorder (MDD) which affects the quality of life of most of the patients suffering from the borderline disorder. Traditionally, the treatment of MDD depends on the criteria mentioned in the Diagnostic and Statistical Manual of Mental Disorders

(DSM) which offers certain steps and criteria to classify borderline disorders. In this paper, it is shown how these criteria can sometimes be vague and ineffective in treating the borderline disorders faced by a set of patients. So to tackle this problem the researches of this paper tried the concept of neuroimaging which provides various functions of the brain and its structure which can be very useful for getting the state of a patient's brain when they are suffering from disorders and which can further help to detect future symptoms. In this paper, the researcher has made use of machine learning classification algorithms and using the data collected from 63 MRI machines.

In the paper [5], the main problem of borderline disorders being wrongly diagnosed or not diagnosed at all is pointed using the traditional method of consultation. The problem is that most of the time professional consultants are not able to properly diagnose the borderline disorder being faced by a patient, and consultants cannot take many other constraints to detect the early stages of borderline disorders, and hence an artificial intelligence approach is taken to classify unipolar and bipolar depressive disorders. In this paper, the researchers used to methodology, one Particle Swarm Optimization (PSO) algorithm for Feature Selection (FS) process for the elimination of less informative and fewer discriminant features, and Artificial Neural Networks (ANN) for the training process. And the result of this paper shows the vital performance of the ANN–PSO approach expressed that it is conceivable to segregate 31 bipolar and 58 unipolar subjects utilizing chosen highlights from alpha and theta recurrence groups with 89.89% in general order precision.

In paper [6], the author has explained about the bipolar disorder, a brain disorder that affects the brain of the patients which is also well known as manic depression. When a patient is suffering from bipolar disorder, it noted that therein is the symptoms of mood swings and deep state of depression in the patient's behavior. This paper also puts forward how many studies and research have shown that genomics is capable of the detection and treatment of bipolar disorder. Researchers of this study used many deep learning algorithms and came up with an end-to-end deep learning model called DeepBipolar which is used to predict bipolar disorder using genomic data. DeepBipolar embraces the Deep Convolutional Neural Network (DCNN) design that concentrates on highlights from genotype data to anticipate the bipolar phenotype.

In [7], the original work from where our data was taken the authors used random forests to distinguish between healthy, borderline, and bipolar personality disorder, taken in groups of 2. The authors achieved an AUC of 0.919 for the code results of the synthetic data provided in the Alan turing GitHub website and an accuracy of 93% as mentioned in the paper. In [8, 9], MRI data has been used to classify individuals between borderline personality disorder and healthy controls, obtaining an accuracy of 80% [8] and 93.55% [9], although the small sample size should be kept in mind for the latter. In [10], the author has used a non-linear predictive model based on multilayer perceptron to predict BPD.

3 Data

The data was obtained from Alan turing GitHub repository [11] named "A signature-based machine learning model for bipolar disorder and borderline personality disorder". The data used in the proposed work is synthetic data obtained from the work in [9]. By synthetic data we mean data that was generated to exhibit the same statistical properties as the original data, without containing the original entries. The synthetic data in this repository was derived from the signatures of the original mood score data and is therefore in signature form itself. Each dataset contains mood score signatures and their associated diagnostic classification. Data was captured from 130 individuals who took part in AmoSS study. The dataset originally had 48 patients with bipolar disorder, 31 patients with borderline personality disorder, and 51 healthy volunteers. Data was collected over a period of 3 months, although 61 out of the total 130 patients gave data for 12 months. For the proposed work, we were mainly focused on borderline personality disorder and healthy volunteers.

4 Methodology

In this section of the paper, we will be discussing the proposed model, control measures taken to prevent the overfitting of model, Bi-GRU, attention, Bi-LSTM, and cyclic learning rate (Fig. 1).

4.1 The Model

The model proposed to distinguish between borderline personality disorder and a healthy person is a GRU (bidirectional)–LSTM(bidirectional)–attention model trained with cyclic learning rate and Adam optimizer with a maximum learning rate of 0.005 and a minimum learning rate of 0.001. The first layer is 128-node LSTM (Long Short-Term Memory) layer, followed by a 64-node GRU (Gated Recurrent Unit) layer. This GRU layer further goes through attention mechanism which finally gives its output to a 64-node dense layer, followed by a single-node layer with sigmoid activation for classification.

4.2 Control Overfitting

Keras model checkpoint was used to control overfitting. Keras model checkpoint callback method saves the model after every epoch, and with the save_best_only

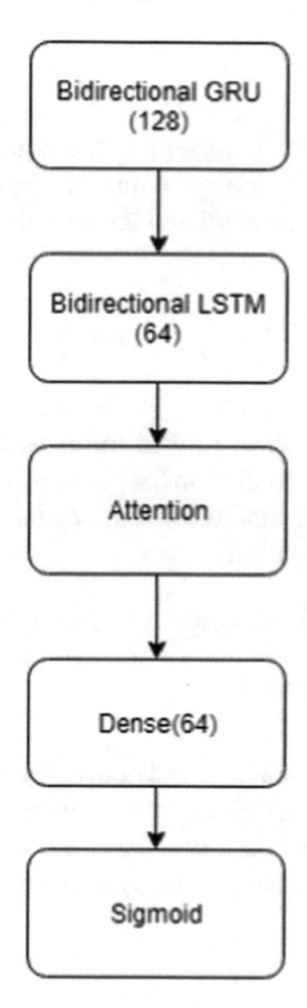

Fig. 1 LSTM–GRU–Attention model (LGA)

hyperparameter the best model is not overwritten. Later for prediction the best model after the training can be called and used for future predictions.

4.3 Bi-LSTM

LSTM, commonly known as long short-term memory networks, learn long-term dependencies in the data. A bidirectional LSTM learns bidirectional long-term dependencies between each time step in time series or sequential data. The learning algorithm is fed once from the beginning to the end of the layer and once from the end to the beginning of the layer. This way, the LSTM learns faster and better.

4.4 Bi-GRU

GRU, commonly known as gated recurrent unit, is a modification over LSTM which has just two gates, namely, the reset gate and the update gate. Bi-GRU similar to Bi-LSTM learns through both forward and backward propagations.

4.5 Attention

The attention mechanism has been proved to be quite useful in domains such as machine translation, sequence predictions, and so on. The attention mechanism basically helps the model to understand the totality of the sentence by disregarding the noise and focus on what's important.

4.6 Cyclic Learning Rate

Cyclic learning rate is used to train a neural network with a learning rate that changes in a cyclic way for each batch of training instead of setting it to a constant value. The learning rate schedule varies between two preset bounds. This helps to come out of a saddle point or local minima faster and leads to convergence at much lesser epochs.

5 Results and Discussions

The model was trained for 50 epochs on an i7 CPU with Keras model checkpoint callback to avoid overfitting. The final AUC attained was 0.95, and accuracy attained was 98%. In the following ROC curve, x-axis denotes false positive rate and y-axis denotes true positive rate. The ROC curve tells us how well the model can distinguish between both the classes. The AUC or area under the curve is 0.95 which states that the model can clearly distinguish between normal and BPD (Fig. 2).

6 Conclusion

The proposed model is trained on synthetic time series data and has verified its performance. The AUC achieved is appreciable over the other existing model. A similar architecture can be extended to other clinical problems with signature-based temporal data. A deeper model with more layers might lead to overfitting. However, perfectly tuned callbacks and appropriate number of training epochs with tuned

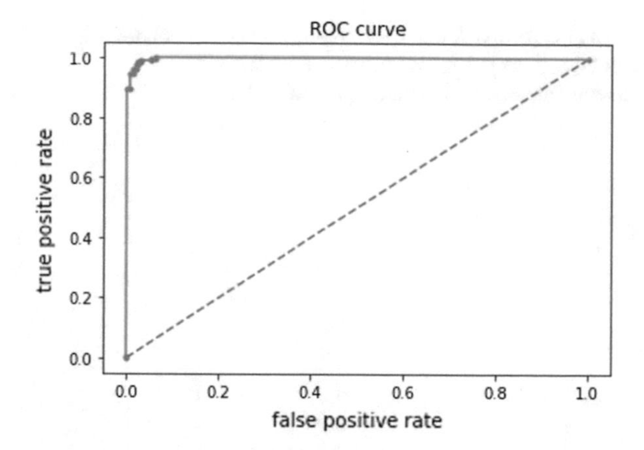

Fig. 2 Performance analysis (AUC-ROC Curve)

drop out after every layer might lead to better results. Modern methods like neural architecture search can also be used to tune several hyperparameters of a deep learning model leading to better results which remains a part of our future work. This paper also shows the ability of deep learning techniques to distinguish between BPD and healthy using signature-based mood data. Extending the proposed work and similar deep learning architectures to distinguish between borderline disorder and BPD and borderline disorder and healthy individual remain a part of the future scope.

References

1. Arribas IP, Saunders K, Goodwin G, Lyons T (2017) A signature based machine learning model for bipolar disorder and borderline personality disorder. arXiv:1707.07124
2. https://www.nimh.nih.gov/health/statistics/personality-disorders.shtml
3. The Economic Times. https://economictimes.indiatimes.com/magazines/panache/world-mental-health-day-nearly-half-of-india-inc-employees-suffer-from-depression/articleshow/66119215.cms
4. Gao S, Calhoun VD, Sui J (2018) Machine learning in major depression: from classification to treatment outcome prediction
5. Erguzel TT, Sayar GH, Tarhan N (2016) Artificial intelligence approach to classify unipolar and bipolar depressive disorders
6. Laksshman S, Bhat RR, Viswanath V, Li X (2017) DeepBipolar: identifying genomic mutations for bipolar disorder via deep learning
7. A signature-based machine learning model for bipolar disorder and borderline personality disorder
8. Sato JR et al (2012) Can neuroimaging be used as a support to diagnosis of borderline personality disorder? An approach based on computational neuroanatomy and machine learning. J Psychiatr Res 46:1126–1132
9. Xu T, Cullen KR, Houri A, Lim KO, Schulz SC, Parhi KK (2014) Classification of borderline personality disorder based on spectral power of resting-state fMRI. In: 2014 36th annual international conference of the IEEE engineering in medicine and biology society, Chicago, IL, pp 5036–5039

10. Maldonato NM et al (2018) A non-linear predictive model of borderline personality disorder based on multilayer perceptron. Front Psychol 9(2018):447
11. https://github.com/alan-turing-institute/signatures-psychiatry

An IoT-Based Fog Computing Approach for Retrieval of Patient Vitals

J. Hilda Janice and Diana Jeba Jingle I

Abstract Internet of Things (IoT) has been an interminable technology for providing real-time services to end users and has also been connected to various other technologies for an efficient use. Cloud computing has been a greater part in Internet of Things, since all the data from the sensors are stored in the cloud for later retrieval or comparison. To retrieve time-sensitive data to end users within a needed time, fog computing plays a vital role. Due to the necessity of fast retrieval of real-time data to end users, fog computing is coming into action. In this paper, a real-time data retrieval process has been done with minimal time delay using fog computing. The performance of data retrieval process using fog computing has been compared with that of cloud computing in terms of retrieval latency using parameters such as temperature, humidity, and heartbeat. With this experiment, it has been proved that fog computing performs better than cloud computing in terms of retrieval latency.

Keywords Internet of Things (IoT) · Adafruit IO · Cloud computing · Fog computing · Python · Temperature · Humidity · Pulse rate · DHT11 · MAX30100 · Pulse oximeter sensor · Linux · VNC viewer

1 Introduction

Internet of Things (IoT) is a technology which involves the connection of sensors and devices, that tend to respond, to the internet and the respective data can be stored and retrieved. IoT is a giant network of connected things and people—all of which collect and share data about the way they are used and about the environment around them. Fog computing extends the services of cloud computing in an efficient way. It is also known as edge computing or fogging, since it connects to the devices nearby and

J. Hilda Janice (✉) · D. J. Jingle I
Department of Computer Science and Engineering, CHRIST (Deemed to be University), Bengaluru, India
e-mail: hilda.janice@mtech.christuniversity.in

D. J. Jingle I
e-mail: diana.jebajingle@christuniversity.in

D. S. Jat et al. (eds.), *Data Science and Security*, Lecture Notes in Networks and Systems 132, https://doi.org/10.1007/978-981-15-5309-7_18

also to the cloud through the edge of the connectivity. The services of fog computing are, it enhances the computation processes, network, and storage services between the devices and the data centers. IoT generates a lot of data of various varieties and by the time the data reaches the cloud, it might lose the moment. The data collected from the devices using IoT show minimized latency when analyzed through fog. It also reduces network traffic and stores only the needed data in the cloud and not all available data. The existing method connects the patient with the doctors in the hospital and the family at home through the cloud. It takes more time to retrieve data and update the concerned people.

2 Related Work

Existing researches have done a lot of work on fast retrieval of time-sensitive data using cloud and fog computing methods. Chen et al. [1], proposed an Adaptive Fog Configuration (AFC) algorithm for the industrial IoT which configures the fog nodes using the power of batteries. Since the capacity of the battery is limited, it creates a confusion in the decision that it is to be made to make it work efficiently. The AFC algorithm minimizes this limitation and the service hosting and certain other important decisions are optimized. It requires the data that are present at the moment and the performance also tends to be more or less precise. Al-Joboury and Hemiary [2], proposed a method enabling doctors and the family of the patient to monitor the vitals of the patient using an application that is connected to the cloud. It explains the devices, middleware, and the other layers used in a simulated method using Cisco Packet Tracer. Mayuresh and Yuan [3] proposed a Secure Mutual Authentication Protocol (SMAP) that avoids session keys to be repeated and also avoids the master secret keys to be stored in order to make it more secure and efficient. It analyzes both the mathematical and experimental ways and compares them both. Zhang et al. [4] proposed an algorithm named as Fair and Energy-Minimized Task Offloading (FEMTO) for fog-enabled IoT networks. It takes the consumption of energy, history of the average energy of the fog nodes, and the priority of each of the nodes, a metric is introduced for scheduling the fairness. Eren and Gitlin [5] proposed a method for optimizing the number of fog nodes for Cloud–Fog–IoT Networks. The necessary number of nodes is determined using wireless channels that work under certain conditions and an accurate number of nodes connected to the network. It is understandable that the selection of the number of fog nodes must be precise and with high capability for computing the path loss in the deployed set of networks.

3 Design and Implementation

This paper proposes a more efficient way to retrieve the vitals of a patient using fog computing [6]. The vitals of the patient that are concerned are temperature, humidity, and the heart rate. The temperature and humidity data are collected through temperature and humidity sensors [7]. The heart rate and pulse oximeter sensor (MAX30100) [8]collects the pulse rate data of the patient. These sensors are connected to the Raspberry Pi [9] which in turn connects to the Adafruit.IO application [10]. This application stores the data and displays the vitals of the patient in two different pages so as to compare the updates of the vitals of the patient both from the cloud as well as fog. The time of updates is noted down and compared to prove their efficiency. The complete architecture of the proposed system is shown in Fig. 1.

The main objectives of the proposed work have been defined based on addressing these problems precisely and clearly as follows: (i) to set up a cloud and fog infrastructure connecting to a Raspberry Pi, (ii) to collect the data from the sensors and process it using the Raspberry Pi, and (iii) to display the data from cloud and fog and compare them with the retrieving time. To implement this proposed method, few applications must be installed and certain guidelines must be followed.

Raspberry Pi: Raspberry Pi 3 Model B has been used for the implementation of this paper. It has a 16-GB hard disk and a 2-GB RAM. Connect the temperature and humidity sensor and the heart rate sensor to the Raspberry Pi with the help of jumper cables. It would be better to glue them to avoid loose connections during implementation. It is more appreciated if the Raspberry Pi is mounted on a flat board, a little bigger than its surface to avoid tampering during work and to give a stable mount [9, 11].

Adafruit.IO: Adafruit.IO tends to be a service for cloud computing. It manages all the data stored in the cloud through it efficiently using the internet. They display

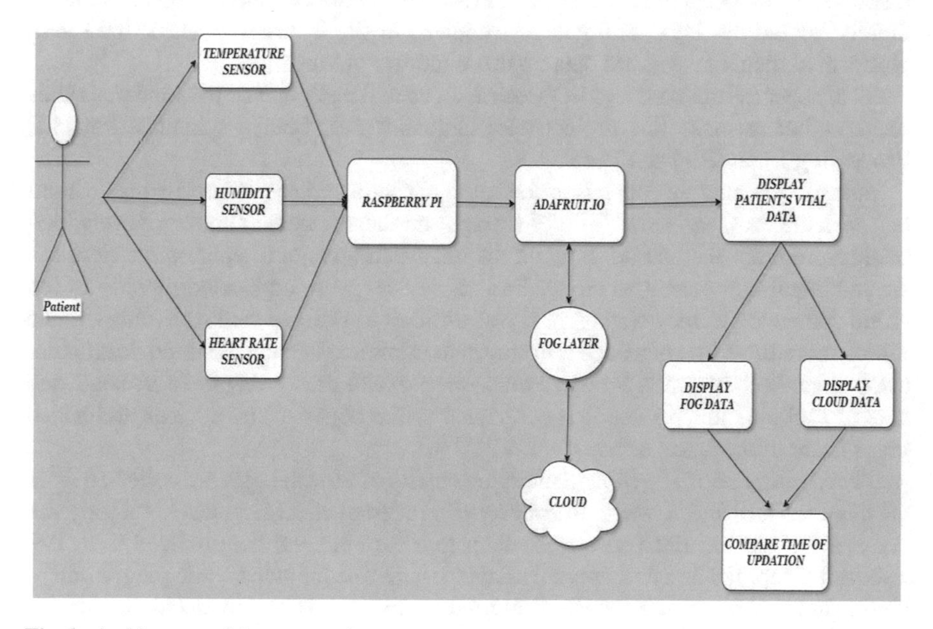

Fig. 1 Architecture of the proposed system

data processed through real time and all other projects that were connected through it. It also tends to give automatic updates in any communication form. It has a feature called Dashboards providing workspace for the client. Dashboards can be accessed as the same as the cloud. There is no location limit or network-related issues in this. Triggers are used to send alerts to the client regarding the devices connected to the dashboard. When the limit set by the client crosses, it sends an alert message and lets the user know the criticality of the issue immediately. All the data stored in Adafruit.IO are kept private and secure. Adafruit.IO supports a certain protocol called MQTT (Message Queue Telemetry Transport) for communicating with the devices that are connected to it. This protocol is used to send and receive data feed from the IoT devices [10].

VNC Viewer: VNC stands for Virtual Network Computing. VNC Viewer is an application which is a graphical system enabling us to share desktop using Remote Frame Buffer Protocol to control another desktop inside ours. It is a platform-independent. A viewer which connects to a port on the server is the usual normal method. But there are ways in which a browser can connect to a server and a server also, in turn, can connect to a viewer through a given port number. An SSH or a VPN connection may also be used to connect to a VNC Viewer since it provides an extra secure connection which has an encryption that is stronger than the normal one. One should also be aware that using VNC Viewer, many others can also access other desktops. So change the privacy to only authorized persons and in the case of a connection from a trustable person, then one may give the credentials for the authorization. This paper is implemented in python for both cloud and fog connecting the raspberry. Now when the code is run, the Adafruit dashboard shows updates immediately since it displays the fog data. Another website is designed inside the Linux platform to display the cloud data on the Windows platform.

This paper is implemented in three modules and each one of the modules relies on the other to make this project work efficiently: (i) Setup—Cloud & Fog, (ii) Processing Data, (iii) Data Display.

Setup: This module consists of the setup of cloud and fog infrastructures. Cloud has been the most extensive form of storage and there is not yet another form which could store more than cloud. Fog, on the other hand, is not a separate structure but an additional component to cloud. Earlier, devices were connected directly to the cloud and hence latency was high. It was difficult to retrieve data from cloud, since it took more time. Fog has been introduced to minimize latency. It would be efficient to combine both together. So when data are retrieved or is going to be updated, fog acts as cache memory since it is connected to the edges of the devices and hence gives faster updates and access.

Processing Data: The second module consists of how the data are collected from all three sensors and how the Raspberry Pi processes it. This involves majorly the code of the project. Each sensor is assigned a variable and is initialized to 0. The Adafruit IO application and its credentials must also be included while programming having in mind that it is connected externally and need to be connected directly. To compare the rate at which cloud and fog give result, it is programmed in such a way that both processes start simultaneously and not one after other—multiprocessing.

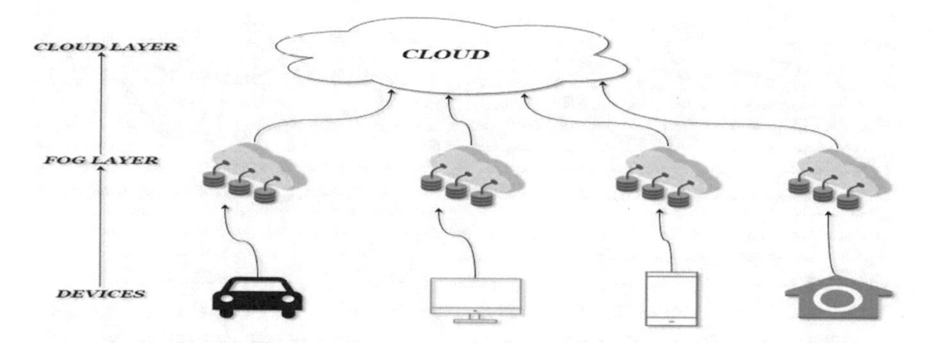

Fig. 2 Setup of cloud and fog infrastructure

Import Client, Feed, and Adafruit DHT. The client is the Adafruit client and the feed is the client's data which is both to and fro. The Adafruit DHT is the module to be imported for the temperature and humidity sensor to work (Fig. 1).

Displaying Data: The data are updated to Adafruit IO, since the VNC viewer is connected with the Raspberry Pi through its AIO key. A fork and join function is used in the main function to start the two processes simultaneously. The pulse has to be continuously monitored and hence it is programmed in such a way that it keeps performing the function and never stops until the power is off. The variables assigned must be the same in Adafruit IO, python code in VNC Viewer and the cloud webpage, to be connected to each other. Only if the variables match, then the Adafruit feeds would be synced. A client object is created to call the functions and a member function is used to send data. The feeds include heartbeat rate, temperature, and humidity of the patient. The AIO key provides security and hence it can be changed anytime. Change the AIO key in the script also to ensure connectivity and security.

4 Results and Discussion

The project was implemented and the variations in updates of fog and cloud were compared by a value "time." The tabulations show the rate at which cloud and fog update the vitals of the patient. As seen in the above graphs, the cloud data updates the values slower than the fog data. Thus, it is seen that fog computing is efficient than cloud computing (Fig. 3) [12].

5 Conclusion and Future Work

This paper was proposed to enhance the already available technology using cloud to be made practical with fog computing. Fog Computing serves to be an add-on and helps collect data faster and efficiently. Like cloud computing, fog computing also

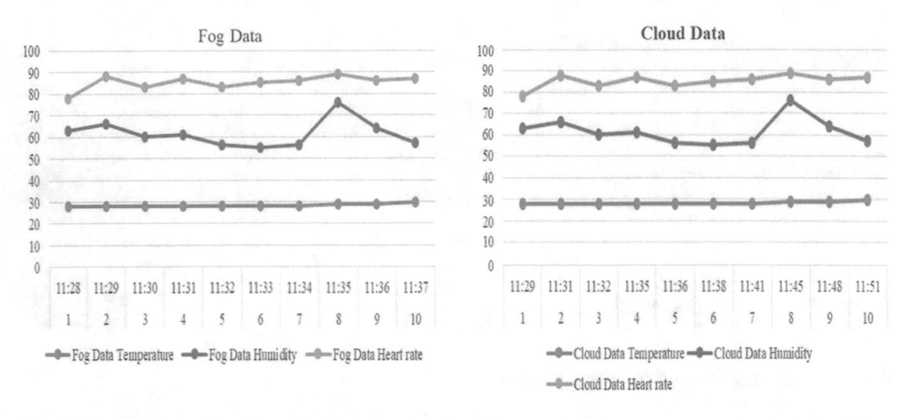

Fig. 3 Comparison of retrieval latency in cloud and fog computing

has its own pros and cons. Fog is more considerable in the case of latency, bandwidth, power efficiency, and user experience but cloud is more considerable in the case of storage, performance, processing, and costs. Fog computing tends to be the only technology which can handle the data of the increasing devices that are connected to the internet. Since it also uses local resources, it makes processing and performance easier and efficient. The latest technologies include IoT as one of the elements in a package rather than a separate technology. In the future, fog computing may also be proved to be the only technology to be able to satisfy the demands during the development of 5G solutions, Artificial Intelligence, and IoT.

References

1. Chen L, Zhou P, Gao L, Jie X (2018) Adaptive fog configuration for the industrial internet of things. IEEE Trans Ind Inf 14(10):4656–4664
2. Al-Joboury IM, Hemiary EH (2018) Internet of things architecture based cloud for healthcare. Iraqi J Inf Commun Technol 1(1):18–26
3. Pardeshi MS, Yuan S-M (2019) "SMAP fog/edge: a secure mutual authentication protocol for fog/edge." IEEE Access 7:101327–101335
4. Zhang G, Shen F, Liu Z, Yang Y, Wang K, Zhou M-T (2018) FEMTO: fair and energy-minimized task offloading for fog-enabled IoT networks. IEEE Internet Things J 6(3):4388–4400
5. Balevi E, Gitlin RD (2018) Optimizing the number of fog nodes for cloud-fog-thing networks. IEEE Access 6:11173–11183
6. https://www.cisco.com/c/dam/en_us/solutions/trends/iot/docs/computing-overview.pdf
7. https://www.mouser.com/datasheet/2/758/DHT11-Technical-Data-Sheet-Translated-Version-1143054.pdf
8. https://datasheets.maximintegrated.com/en/ds/MAX30100.pdf
9. https://static.raspberrypi.org/files/product-briefs/Raspberry-Pi-Model-Bplus-Product-Brief.pdf

10. https://cdn-learn.adafruit.com/downloads/pdf/welcome-to-adafruit-io.pdf
11. https://www.ti.com/lit/ds/symlink/lm2596.pdf
12. https://www.sam-solutions.com/blog/fog-computing-vs-cloud-computing-for-iot-projects/

Recommender System for Analyzing Students' Performance Using Data Mining Technique

Shruti Petwal, K. Sunny John, Gorla Vikas, and Sandeep Singh Rawat

Abstract The increasing amount of data in educational databases demands the use of data mining techniques in order to extract hidden and useful student information that is otherwise ignored. Placement cell in any technical education system always plays an important role in institutions or universities for their recognition. Multinational companies are looking for students who are good at Logical, Quantitative Aptitude, Coding, and Verbal subjects. Therefore, it is highly necessary to analyze their marks in the mock tests and assess their strengths and weaknesses before appearing for the final placement tests. This paper presents the application of performing mining on student marks data collected from Talentio [1] at Anurag Group of Institutions, Hyderabad. A distinctive technique like K-Means Clustering Algorithm was applied to categorize the students into different clusters depending on their performance levels. This method can help students a great deal by letting them as well as their instructors identify areas for improvement, especially in large classes, and take appropriate action in a timely manner.

Keywords Data mining · Clustering · Students' performance · Education

S. Petwal · K. Sunny John · G. Vikas · S. S. Rawat (✉)
Department of CSE, Anurag University (Formerly known as Anurag Group of Institutions), Hyderabad, Telangana, India
e-mail: srawatcse@cvsr.ac.in

S. Petwal
e-mail: shrutipetwal.98@gmail.com

K. Sunny John
e-mail: sunnyjo98k@gmail.com

G. Vikas
e-mail: vikasgorla1111@gmail.com

D. S. Jat et al. (eds.), *Data Science and Security*, Lecture Notes in Networks and Systems 132, https://doi.org/10.1007/978-981-15-5309-7_19

1 Introduction

The highly competitive field of education requires students and graduates to be highly skilled all-rounders. Educational institutions undoubtedly have a major role in making sure that their students are skilled enough to thrive in such a competitive scenario. Getting placed in good organizations is a goal of all graduate students. Besides students and their families, getting maximum students placed in good companies as a part of campus recruitment is a challenge to colleges and universities. Student Classifier aims to help the colleges in boosting their students' performance by analyzing the major aspects of marks scored by them in mock exams conducted to help them in campus placement.

Students seeking jobs or higher education opportunities have to clear some competitive exams. Being students of Computer Science background, we have considered the pattern of exams conducted for Software engineering and related job aspirants. These exams test the candidates' grip over aptitude, reasoning, coding, and communication skills. A mock test was conducted after a series of practice tests by Talentio Solutions for students of our batch to help with our preparation. The marks of students were collected in the form of an excel sheet consisting of the scores of 382 students from different branches of college. Data mining was performed on the same to analyze the performance of students and to divide them into batches based on their performance in various sections of the test. Each group of students comprises of students who are weak at one or more areas in the test. This is an efficient way of making batches of students as each batch of students can be trained in the area that they are lagging at and hence focuses on the overall improvement of the performance of students.

The rest of the paper is organized as follows. Section 2 discusses the related work. Section 3 features the recommended framework used for clustering the students based on their mock tests. Section 4 throws light on the observed experimental results. The paper ends at Sect. 5 which mentions the likely future enhancements of the system.

2 Literature Survey

Amirah Mohamed Shahiri et al. [2] have discussed which method is suitable for mining the students' data and predict their performance considering the most used methods under classification. The algorithms they considered are Decision Tree, Neural Network, Naive Bayes, K-Nearest Neighbor, and Support Vector Machine. The data set they considered has the CGPA and the internal marks of students.

Baradwaj and Pal [3] discussed dividing the students to identify the students who need special attention to reduce fail ration and taking appropriate action for the future examinations using classification. They proposed a solution to improve the division of the students using decision trees under the classification method in data mining. Their data set consists of the previous semester marks, class test grades, seminar, and general proficiency performance, attendance, lab work, and end semester marks of the students in a class.

Saa and Ajman [4] tested multiple decision tree techniques and selected the most suitable one based on efficiency. They analyzed and noticed that some algorithms worked better with the considered data set than the others, in detail, CART had the best accuracy of 40%, CHAID and C4.5 were next with 34.07% and 35.19%, respectively. The least accurate one was ID3 with 33.33%. Their data set contains students' gender, nationality, first language, high school percentage, location, discount, transport, family size, family income, parents' qualifications, parents' occupation, and previous semester's GPA.

El Din Ahmed and Elaraby [5] discussed how to improve the students' performance based on the total marks they acquired and enable them to perform well the next time using ID3 Decision tree algorithm under classification technique. The data set considered consists of students' attendance, mid-term marks, lab marks, and final grades.

Yassein et al. [6] discussed how data mining can be used to improve a student's performance using different techniques and analyzing which technique is suitable. They used feature reduction and classification techniques to reduce the error rate and determined that the factor that most affects the performance of the students is their attendance in class in addition to final and mid-exam grades. They considered clustering, classification, and association. The data set consists of practical, assignment, mid and final exam marks, and success rate.

Srivastava and Srivastava [7] used classification, prediction, clustering, and IBM Intelligent Miner to monitor the students' requirements and their academic progression in order to fill the gap between the rates of enrollments for higher education and the number of higher institutions being established. The data set consisted of students' marks and interests.

Oyelade et al. [8] have discussed analyzing students' results based on cluster analysis and used K-Means to arrange their scores data according to the level of their performance. It can help academic planners of higher institutions in effective decision-making to improve and monitor the students' performance. The data set considered contains students' marks and interests.

Al Hammadi and Aksoy [9] discussed analyzing the importance of data mining at different levels in the field of education like students' level, academic guidance level, and decision-makers' level. The algorithms used for this are Decision trees, K-Means, and Association techniques. The data set used contains the finance, achievements, success, and tailored guidance by academic counselors.

3 Recommender System

The students' performance in each subject is assessed and they are trained to improve only in those subjects in which they are weak. This helps in the overall development of the students and also saves time for them instead of spending time on subjects

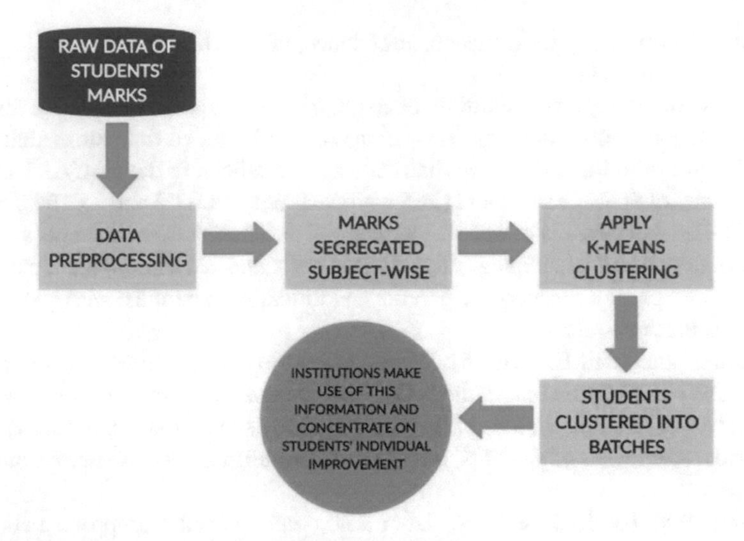

Fig. 1 Recommended architecture of the Student Classifier

in which improvement is not needed. In order to implement this system, K-Means Clustering algorithm has been used to group the students. The recommended system also clusters the students considering the marks acquired by them in a combination of subjects they are weak in. This helps in identifying the students who are weak in more than a single subject. The proposed architecture for the Classifier is shown in Fig. 1.

K-Means Clustering [10] algorithm is one of the most suitable algorithms for Educational Data Mining. It is a type of unsupervised learning which can be easily applied even on large data sets. K-Means partitions a set of data into k number of clusters based on the specified criteria. The working of K-Means begins with choosing k number of clusters and randomly selecting k-centroids C1, C2, …, Ck. At each step, every data point Xi is allocated to the centroid it is closest to, and at the end of each iteration, the new centroid is calculated as the mean of all points assigned in that cluster. The above steps are repeatedly performed until convergence is reached.

4 Experimental Results

The data set used in the Student Classifier is a 27-KB-sized csv file consisting of 382 rows representing marks of 382 students in a mock test conducted by Talentio [1]. It consists of the students' names, roll numbers, marks secured in the four subjects Logical, Quantitative Aptitude, Coding, and Verbal sections. The data set has been collected from the examination branch of Anurag Group of Institutions, Hyderabad. After applying preprocessing on marks, the records are narrowed down to 190.

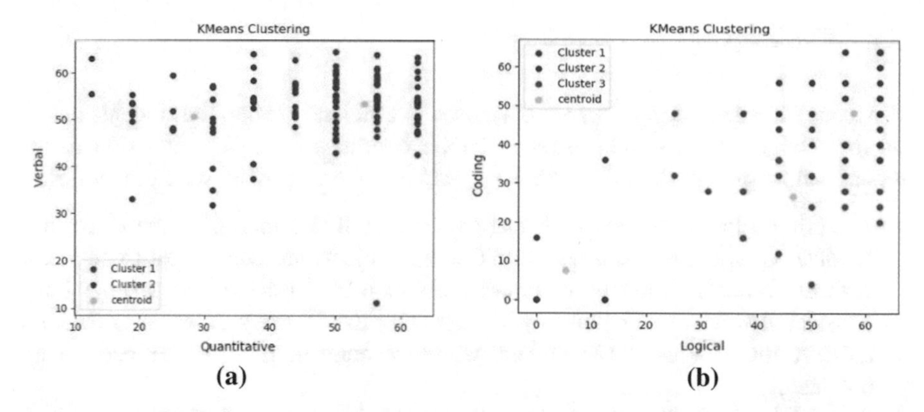

Fig. 2 **a** Cluster visualisation of students with marks <60% in verbal and 65% in the quantitative aptitude. **b** Cluster visualization of students who secured marks <65% in coding and logical subjects

Preprocessing includes eliminating students who have an overall score of less than 5 as most of these students would have just started and ended the test without actually attempting it. Clustering is then performed on the 190 students which results in clusters of students with a varying range of marks. Since the clustering is random, the obtained clusters could sometimes consist of unwanted data points. These unwanted data points or outliers are taken care of by filtering the clusters using required boundary conditions. The results obtained are graphically represented in Fig. 2a, b.

The result of clustering using 65 marks criteria for the Aptitude section and 60-mark criteria for the Verbal section can be viewed graphically in Fig. 2a and the result of clustering using 65 marks criteria for Logical Reasoning section and 60-mark criteria for the Coding section can be viewed graphically in Fig. 2b. The instructor can, therefore, train these students majorly on the subjects which they are weak in for improving their performance.

To verify the accuracy of the proposed method, the manual method of dividing students was also performed by applying formulas on the excel sheet and repeatedly filtering the retrieved rows until the required set of students was obtained. The results obtained using the manual method are the same as those obtained using Student Classifier. The proposed method clearly is better at performing the categorization of students as it is simple and does not require the user to do anything but run the program on the data set. It takes very less time of around 3–4 min to perform the task compared to the manual method which could take hours together for the same.

4.1 Findings

K-Means Clustering algorithm has been used to efficiently group the students based on their subject-wise scores instead of their overall score. We could find many recommendations for the instructor or department. Among them, we can mention:

- The criteria have been set such that students with less than 65% marks for the Logical, Quantitative Aptitude, and Coding subjects are considered to be weak at those subjects. Students with marks less than 60% for the Verbal subject are considered weak at verbal ability. On applying the 65-mark criterion in Quantitative Aptitude, a list of 127 students with low scores in the Aptitude section was obtained.
- On applying both 60-mark condition for the verbal subject and 65-mark condition together, we get a list of 58 students weak at both the sections.
- Similarly, students with low scores in the Logical section are found followed by those weak at coding section using clustering. Then, on applying 65 marks criteria for the Coding and Logical sections together, we get a list of 38 students who are weak at both the sections.

5 Conclusion

This paper discusses the way how data mining has been applied on the data set to extract groups of students having similar marks, which in turn helps the management take appropriate steps that help students in performing better. The result obtained through the Student Classifier is the list of the students after dividing them into batches based on their performance. The college management can refer to this list and allot different venues for different groups of students and help them improve by providing them with special subject-wise training. This helps the students to clear career-based tests with good scores and get jobs in well-reputed companies. It also helps in narrowing down the failure percentage. The proposed solution is simple and helps rule out the drawbacks of the existing method of performance analysis. The fact that this solution doesn't require internet connectivity for its working makes it even more effective. Besides being helpful to students, this solution helps the placement department to be more effective by improving its overall performance.

This solution could be enhanced in the future by designing a simple-to-use front-end interface that can be used by teachers, students, and course administrators. This interface could simply collect marks of students and display the analysis for each area of the test and suggest measures to be taken. This can be done by connecting the front-end to the existing Python code at the back-end. Future works may also include applying this system on regular academic subjects instead of just career-based exams. This will help students improve even their academic percentages. Linking marks of students with their attendance percentages would be another add-on that would help in the analysis.

References

1. https://talentio.online
2. Shahiri AM, Husain W, Rashid NA (2015) A review on predicting student's performance using data mining techniques. In: The third information systems international conference, School Comput Sci, Universiti Sains Malaysia, 11800 USM, Penang, Malaysia
3. Baradwaj BK, Pal S (2011) Mining educational data to analyze students' performance. Int J Adv Comput Sci Appl (IJACSA) 2(6)
4. Saa AA, Ajman (2016) Educational data mining & students' performance prediction. Int J Adv Comput Sci Appl (IJACSA) 7(5)
5. El Din Ahmed AB, Elaraby IS (2014) Data mining: a prediction for student's performance using classification method. World J. Comput. Appl. Technol. 2(2):43–47
6. Yassein NA, Helali RGM, Mohomad SB (2017) Predicting student academic performance in KSA using data mining techniques. J Inform Tech Softw Eng (an open access journal) 7(5)
7. Srivastava J, Srivastava AK (2014) Data mining in education sector: a review. In: Special conference issue: national conference on cloud computing & big data
8. Oyelade OJ, Oladipupo OO, Obagbuwa IC (2010) Application of k-means clustering algorithm for prediction of students' academic performance. Int J Comput Sci Inf Secur (IJCSIS) 7(1)
9. AlHammadi DAA, Aksoy MS (2013) Data mining in higher education. Period Eng Natl Sci 1(2)
10. Qiao J, Zhang Y (2015) Study on K-means method based on data mining. In: Chinese Automation Congress, 2015

Classification of Financial News Articles Using Machine Learning Algorithms

Anshul Saxena, Aishika Banik, Chirag Saswat, M. Joan Jose, and Vandana Bhagat

Abstract The opinion helps in determining the direction of the stock market. Information hidden in news articles is an information treasure which needs to be extracted. The present study is conducted to explore the application of text mining in binning the financial articles according to the opinion expressed inside them. It is discovered that using the tri-n-gram feature extraction process in conjugation with Support Vector machines increases the reliability and precision of the binning process.

Keywords Classifications · n-grams · Stock market · Support vector machine · Text mining

1 Introduction

Nowadays, a plethora of data is generated in a structured and unstructured format. Be it smart devices like mobile phone or tablet, and users are generating different types of data in the form of text and images. In essence, data capturing and storage capacities have given birth to plenty of opportunities to do data analysis. One of the significant data types of unstructured data is text. According to CNN [1], one-third of US citizens prefer to communicate via text rather than speaking overcall. Mining this form of data gives a considerable opportunity to research companies to analyze, summarize, and voice the opinion of customers. The words "Bullish" and "Bearish" tells about the actual state of the market, that it is gaining or losing the value.

This paper presents one such use case where classification labeling of financial articles according to the opinion expressed in them. This categorization of financial articles can form the basis of designing the reference management and document management system for finance professionals. Automation of extraction, classification, and labeling of financial articles helps stock analysts in preparing the reliable buy/sell recommendation reports in a short time.

A. Saxena (✉) · A. Banik · C. Saswat · M. Joan Jose · V. Bhagat
CHRIST (Deemed to be University), Bengaluru, India
e-mail: anshul.saxena@christuniversity.in

D. S. Jat et al. (eds.), *Data Science and Security*, Lecture Notes in Networks and Systems 132, https://doi.org/10.1007/978-981-15-5309-7_20

Section 2 describes the review of the significant prototypes used for text mining of financial data. Section 3 gives an overview of the techniques used as part of the text mining process. Section 4 talks about the experiment design and results obtained from the experiment, followed by a conclusion and future works.

2 Related Works

Schumacher et al. [2] have used Support Vector Machine (SVM) method to statistically predict future stock prices while comparing it with linear regression. They have found that Noun Phrase representation performs better than Bag of Words. In the future, they are planning to use Relevance Vector Regression which gives better accuracy and lesser vectors in classification.

Hierarchical knowledge map (NewsMap) [3] could be generated automatically from the finance and health sections extracted from the online Chinese news. Such a map can be used to find hidden information in news articles about business intelligence and medical knowledge. They tried to create high-quality hierarchical knowledge and to suggest effective map-based visualizations.

The authors have reviewed 9,121 [4] financial news articles and 10,259,042 stock quotes using different textual representations, i.e., Bag of Words, Noun Phrases, and named entities by using the method of machine learning to predict the stock market. This article also shows that Bag of Words achieves the best performance in closeness to actual future stock price.

According to a review paper written by Nassirtoussi et al. [5] which has chronicled articles about market prediction based on online text mining for market prediction. From their literature review, authors have found that there is no proper technical and theoretical framework available for predictability of financial market through the interpretation of sentiment in online data. The article shows a comparative analysis of all available methods which are having a theoretical and technical foundation. The paper will help researchers to identify the significant strategy.

3 Philosophical Underpinnings

3.1 Financial News Article Source

A plethora of articles is available online, which are generated based on the performance figures reported by the company. The primary source of data is the news briefing organized by company management. Also, they report the yearly and quarterly performance of the organization through the balance sheet. According to Indian Accounting Standards, Indian companies need to maintain three types of documents: Balance sheet, statement of cash flows, and statement of profit and loss. Apart from the company website, independent news agencies and analyst reports are an excellent source to comprehend financial news. News articles and analyst reports are used for opinion mining based on the views expressed in the same.

3.2 Textual Representation

Over the period, numerous methods have been discovered to analyze the news article. In the proposed experiment, we have introduced a new system for feature extraction known as Enhanced Feature Selection Framework (EFSF). Under this setup which is implemented using Rapid Miner tool, we are proposing to enhance the feature selection by using n-grams technique which enables the process to generate similar tokens. We have generated a set of three similar types of tokenized words to generate the features which in turn will help the model in achieving higher accuracy by recognizing the sentiment prevailing on Dalal Street.

3.3 Machine Learning Process

The multitude of Machine learning algorithms is available to generate and confirm the list of keywords which can be used to recognize the sentiment carrying by the news article. Two families of machine learning algorithms are deployed for the sentiment classification. Keywords like "Growth," "Up," "down," and "Loss" can lead to predictable outcomes. These outcomes are used to classify the articles as Bullish or Bearish.

Research Question: How can we classify and label the financial articles based on an opinion?

Given that prior research has been done to predict the stock markets using financial text, focus of the current work is to classify the financial articles based on the opinion expressed. It is possible to classify the labeled articles into the Bearish and Bullish segment based on the opinion expressed in news articles. We need to use the abstraction levels to achieve the correct classification scheme for the binning of news articles. We expect to generate the list of keywords which can help in labeling the news articles based on the opinion expressed in it.

4 Experimental Design

The experiment was conducted to analyze news articles by enhancing the feature selection by using n-grams technique. The design of experiment is divided into three partitions, namely data extraction, feature selection, and classification. Data extraction

This phase handles extraction of the data from various data sources. News related to five companies is scrapped from various newspaper websites like Yahoo News, etprime.com. To consider loss and gain of shares, news articles related to companies which have lost and gained the share have analyzed.

Share prices of companies like Adani green and HDFC Asset Management Company limited were the top gainers for the year 2019 according to the ET Now. On the other hand, Yes Bank and IIFL holdings have lost much of the share value for various reasons. Also, to get the input related to the performance of one company, performance of TCS were measured for January Quarter 2019 and January Quarter 2020. In total 120 articles were used as an input to train the system on labeling the articles based on the shareholder sentiment as Bearish or Bullish. The proposed model has automated the entire process using Rapid Miner data mining tool. After extraction, the articles are manually labeled as Bullish or Bearish as per the content of the article.

4.1 Data Preprocessing and Feature Selection

Under the process undertaken, all the 120 articles (Sample size) are tokenized using a tokenize operator. As a part of data preprocessing 80% of the articles are taken as Training dataset and rest of the articles are used as test dataset. All the words extracted are split into one word using the non-letter method, thus forming a word vector. Once the token is generated, words have been transformed from upper case to lower case for better character recognition. Afterward, non-Noun words and verb agreements are removed using the filter stop words. This operator allows the system to filter the keywords which may be useful in recognizing the market sentiment related to the stock. In the end, small length token is removed further using the operator "Filter tokens by length," and all the one word or two n-grams are removed hence paving the way for machine learning algorithms to determine the keywords which can be used for parsing the news articles into "Bull" and "Bear" sentiment the article.

4.2 Classification

In this phase, model trains itself to classify the given article into any of the targeted class as per the features it contains. Amongst various machine learning algorithms,

Proposed model has tested eight machine learning model to check the accuracy of prediction.

Linear models usually deploy regression techniques to identify the keywords. Logistic regression use odds ratio to classify and predict binary variable. Similarly Generalized Linear Model used a log-likelihood method to predict the variable. Naive Bayes method uses the Bayesians classification technique to recognize the pattern. Similarly, Support vector machine uses a non-probabilistic binary linear classifier method to categorize the sentiments. Fast large margin algorithm builds on the linear support vector learning scheme to classify the opinion expressed in a news article. On the other hand, Random forest uses decision tree techniques for opinion mining. Traditionally decision tree uses entropy principle and recursive classification techniques to produce results. Modern enhancements like Random Forest and Gradient boost methods use some other set of features to generate the results. Random forests randomize the split of nodes and generate test cases looking for the best possible results. Gradient boost is a forward selection method which uses classification methods to gradually improve upon estimation methods.

4.3 System Design

Steps of Algorithm as per the phase:

The proposed model is implemented in two phases, namely Learning phase (Training) and Operational (Test) phase as depicted in Fig. 1.

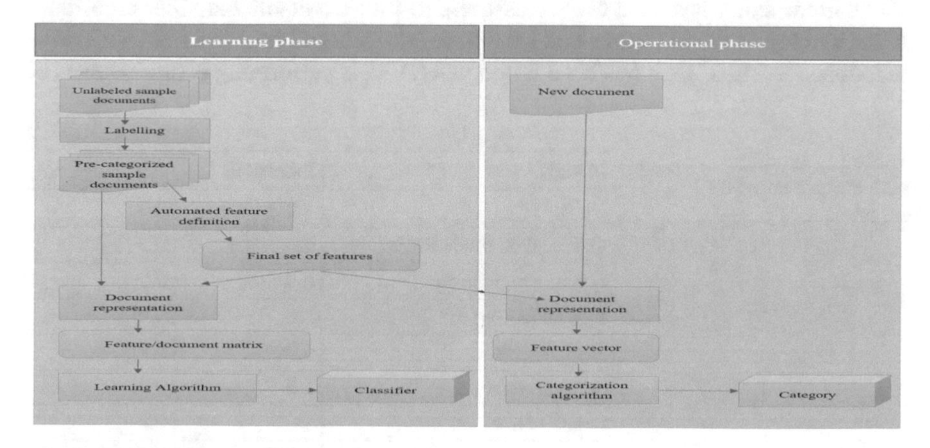

Fig. 1 Text mining systems for market response to news: a survey [4, 5]

Learning Phase

Extraction: Financial articles will be extracted from relevant source.
Laling: Assigns the articles as Bullish or Bearish and generates Pre-categorized sample document (Labelled Article). Labeled data will help the training algorithm to identify the prevailing sentiment.
Feature Definition: Dictionary of tri-n-grams under each category will be created which will generate the list of keywords. These words are deemed as classifiers after performing feature extraction process in the initial stage and hence have been confirmed afterward through the classification algorithm. (Bullish sentiment keywords—"growth," "rise," "increase," Bearish sentiment keywords—"decrease," "fall").
Model Building: Various algorithms will be applied and the model will train itself as per the features.
Feature matrix: Matrix of keywords gets generated based on the features (mentioned in Feature Definition phase) from target document.
Classification: As per the matrix, the training data is classified into categories.

Operational Phase

Test Data: Unlabeled new articles are passed in the operational phase.
Machine Learning: The unlabeled articles will be exposed to the pre-defined features and an appropriate machine learning algorithm is applied to generate Feature vector.
Classification: As per the result, the test data is categorized under the appropriate categories.

5 Experimental Findings and Discussion

The current study has used eight classifiers to find a suitable classifier algorithm. After looking at the performance of the linear models and decision tree models (as mentioned in Table 1), it has been found that SVM is performing better in terms of

Table 1 Algorithm performance based on experiment

Model	Precision	Accuracy
Naive Bayes	1.0	0.8
Generalized linear model	1.0	0.9
Logistic regression	0.9	0.9
Fast large margin	1.0	0.9
Decision tree	0.9	0.8
Random forest	1.0	0.9
Gradient boosted trees	1.0	0.8
Support vector machine	1.0	0.9

accuracy and precision in comparison to other algorithms. Schumacher et al. confirm similar results. Hence, it can be concluded that SVM can be used as a classifier algorithm to label the documents. It has been possible due to the ability of SVM to build a decision boundary in hyperplane using a limited amount of data.

6 Discussion and Conclusion

This article has three contributions: First, to identify and analyze the keywords that can help in automatic labeling of financial news articles based on share market sentiment. This classification can help the financial analyst to look for relevant information in a short time. Second, a novel feature engineering method termed as "Enhanced Feature Selection Framework" (EFSF) is discovered and developed, which in the future can help to perform text analytics on a variety of documents. A combination of three n-grams has resulted in better feature selection. Third, to get the best result from the classifier, eight algorithms are evaluated. SVM, Random Forest, and GLM has given better results. These algorithms can be used further (1) to process high dimensional data (2) is less biased (3) it gives good robustness and stable results to the model.

The study has been limited to the financial news articles related to Indian stocks in the English language. It can be expanded further to other geographies and languages. Further application of the process can be found in the domain of healthcare and other knowledge-oriented processes where the classification of documents is one of the primary concerns.

References

1. Gahran A (2011) One-third of Americans prefer texts to voice calls, Updated 2247 GMT (0647 HKT)
2. Schumaker RP, Zhang Y, Huang CN, Chen H (2012) Evaluating sentiment in financial news articles. Decis Support Syst 53(3):458–464
3. Ong TH, Chen H, Sung WK, Zhu B (2005) Newsmap: a knowledge map for online news. Decis Support Syst 39(4):583–597
4. Khadjeh Nassirtoussi A, Aghabozorgi S, Ying Wah T, Ngo DCL (2015) Text mining of news-headlines for FOREX market prediction: a multi-layer dimension reduction algorithm with semantics and sentiment. Expert Syst Appl 42(1):306–324
5. Mittermayer M, Knolmayer GF (2006) Text mining systems for market response to news: a survey. Inf Syst J 41(184):571–579

OCR System Framework for MODI Scripts using Data Augmentation and Convolutional Neural Network

Solley Joseph, Ayan Datta, Oshin Anto, Shynu Philip, and Jossy George

Abstract Character recognition is one of the most active research areas in the field of pattern recognition and machine intelligence. It is a technique of recognizing either printed or handwritten text from document images and converting it to a machine-readable form. Even though there is much advancement in the field of character recognition using machine learning techniques, recognition of handwritten MODI script, which is an ancient Indian script, is still in its infancy. It is due to the complex nature of the script that includes similar shapes of character and the absence of demarcation between words. MODI was an official language used to write Marathi. Deep learning-based models are very efficient in character recognition tasks and in this work an ACNN model is proposed using the on-the-fly data augmentation method and convolution neural network. The augmentation of the data will add variability and generalization to the data set. CNN has special convolution and pooling layers which have helped in better feature extraction of the characters. The performance of the proposed method is compared with the most accurate MODI character recognition method reported so far and it is found that the proposed method outperforms the other method.

Keywords Convolutional neural network · Data augmentation · MODI script · Handwritten character recognition

1 Introduction

Character recognition is one of the most active topics in the field of pattern recognition and yet challenging because of the presence of several handwritten scripts with each of them having different writing styles. The focus of this study is character recognition

S. Joseph (✉) · A. Datta · O. Anto · S. Philip · J. George
CHRIST (Deemed to Be University), Bengaluru, India
e-mail: solley.joseph@res.christuniversity.in

S. Joseph
Carmel College of Arts, Science and Commerce for Women, Nuvem, Goa, India

D. S. Jat et al. (eds.), *Data Science and Security*, Lecture Notes in Networks and Systems 132, https://doi.org/10.1007/978-981-15-5309-7_21

of MODI Script, which was used to write Marathi until the twentieth century. The reason behind working with the script is that research and development are necessary to extract the information from MODI manuscripts which are stored in various parts of the country and abroad. As such, very less work is done toward MODI script OCR. MODI script is one such script in which character recognition is still in its infancy. MODI script was developed in Devagiri in the twelfth century. Hemadri Pandit also known as Hemad Pant, a famous leader of Yadav Dynasty, introduced MODI script. It is observed that various libraries and temples in India still have a large collection of MODI script [1]. In this study, an implementation of Convolutional Neural Network (CNN) along with the Data Augmentation Method, for the character recognition of MODI script, is reported.

The MODI script has 46 distinctive letters, of which 36 are consonants and 10 are vowels. MODI was easy to write and was commonly used as an official script for writing Marathi until 1950. Because of the difficulty in printing MODI script, its usage was stopped and Devanagiri is now used for writing Marathi. A large number of documents and correspondence during Chatrapati Shivaji's time was written in MODI scripts [2].

MODI character recognition is a very complex task due to its variations in the writing style of individuals, shape similarity of characters, and the absence of word stopping symbol in documents.

The advances in the field of Artificial Intelligence (AI) and Machine Learning (ML) have greatly contributed to the success of various character recognition processes. One of the deep learning approaches (CNN) along with data augmentation is implemented in this study, to classify and recognize the characters of ancient MODI Script.

The rest of the paper is organized as follows; the related work is narrated in Sect. 2. The theoretical background is explained in Sect. 3. In Sect. 4, the methodology is explained. Experimental study and the results are given in Sect. 5 followed by the conclusion in Sect. 6.

2 Literature Review

Compared to other Indian languages very little research has been done toward MODI character recognition [1]. A review of the literature indicates that only 13 published works are available on the character recognition of MODI script. Sadanand et al. have implemented Zernike and Zernike's complex moments in combination with the Euclidian distance classifier for MODI script recognition, and an accuracy of 94.78% was achieved [3]. Otsu's Binarization method and Kohonen neural network classifier were used for MODI character recognition and an overall recognition rate of 72.6% was reported [4]. A two-layer feed-forward neural network and SVM were used for MODI character recognition and a recognition rate of 73.5% was reported [5].

Deep learning-based algorithms are used in various pattern recognition tasks including character recognition. Convolutional Neural Network (CNN) is effectively implemented for character recognition and is one of the best performing deep learning models. Najadat et al. performed the recognition of Arabic characters using CNN and reported an accuracy of 94.9% [6]. CNN architecture based on feature fusion and spatial pyramid pooling was experimented by Keserwani et al. for Bangla character recognition, and an average accuracy of 99.76% was achieved [7]. An experiment on deformed Kannada script by Shobha et al. using CNN-based method has resulted in an accuracy of 92% [8]. Kumar et al. have implemented CNN on Hindi digit recognition and an accuracy of 98.85% has been obtained [9]. CNN based method is implemented by Hossain & Ali, for handwritten digits recognition and an accuracy of 99.15% was achieved [10]. CNN has also been used by Maitra et al. for Oriya numeral recognition and Telugu numeral recognition and the accuracy achieved was 97.2% and 96.55, respectively [11]. An SVM (Support Vector Machine)-based method is used in combination with RBM (Restricted Boltzmann Machine) for the recognition of handwritten multilingual numeral data set [12]. RBM is used for dimensionality reduction and it was observed that complexity was reduced using this method.

It is observed that there is no standard MODI character data set available and the data set is specially generated by individual researchers at the time of each experiment. But the old MODI manuscripts consist of ornamental characters and the recognition of those scripts is a complex task, therefore implementations of more efficient methods are needed to unveil the historic information written in them.

3 An Overview of CNN Architecture

CNN, which is a special kind of deep learning technique, consists of input and output layers with many hidden layers between them. It uses the convolution concept of linear operation from mathematics. The hidden layers include the convolution layer, the pooling layer, and the fully connected layer [13].

The convolutional layer is the layer that does the feature extraction of the sharp features of input images such as corners, edges, and endpoints. It will help in reducing the size of the images. A set of convolution kernels constitute the convolution layer [3]. The feature mapping function is given as follows:

$$F_{i+1} = \alpha(Fi * K + b) \quad (1)$$

where α is the activation function, b the bias, * is the convolutional operation.

The neural network uses various activation functions to determine its output. The widely used activation functions are sigmoid, Tanh, Rectified Linear Unit (ReLU), Softmax, etc. ReLU is generally used in order to add non-linearity to the convolution layers. It converts all the negative values to zero. The function is given by

$$\mathrm{ReLU(Fi)} = \max(\mathrm{Fi}, 0) \quad (2)$$

The output image is then passed on to the pooling layer, which is used to reduce the spatial volume of the input images. This layer comes between the convolutional layers. Average and max-pooling are the most commonly used pooling layers.

The fully connected layer connects neurons in one layer to neurons in another layer. It is used for image classification by training the images. The visual features after the convolution and pooling layer are being transferred to fully connected layers and the function is given by,

$$F_{i+1} = \alpha(Fi.W + b), \tag{3}$$

where '.' represents matrix multiplication, α is the activation function, b the bias, and W the weight matrix.

The last fully connected layer consists of *m* neurons, where *m* is the number of predicted classes [7].

4 Methodology

In this work, a method called ACNN (augmented CNN) is proposed for the recognition of handwritten MODI script, which uses both data augmentation and CNN. The augmentation of the data will add variability and generalization to the data set, which will result in better accuracy while training the network. The steps of the proposed method are shown in Table 1.

The original image (60*60 pixels) is given as the input and data augmentation method will be applied to it. After the data augmentation, the data are passed through the CNN for the recognition of characters, as shown in Fig. 1.

Table 1 The steps of the proposed ACNN method

Step 1. Input the original image
Step 2. Apply image augmentation
Step 3. Generate randomly transformed batch of images
Step 4. Train the network (i) Apply convolutional filter on the input image with ReLU activation function (ii) Apply max pooling for down-sampling (iii) Repeat steps (i) & (ii) with varying number of filters counts and kernel sizes (iv) Apply fully connected layer (two hidden layers) on the extracted features (v) The output layer is constructed with 46 nodes to get the results for 46 characters

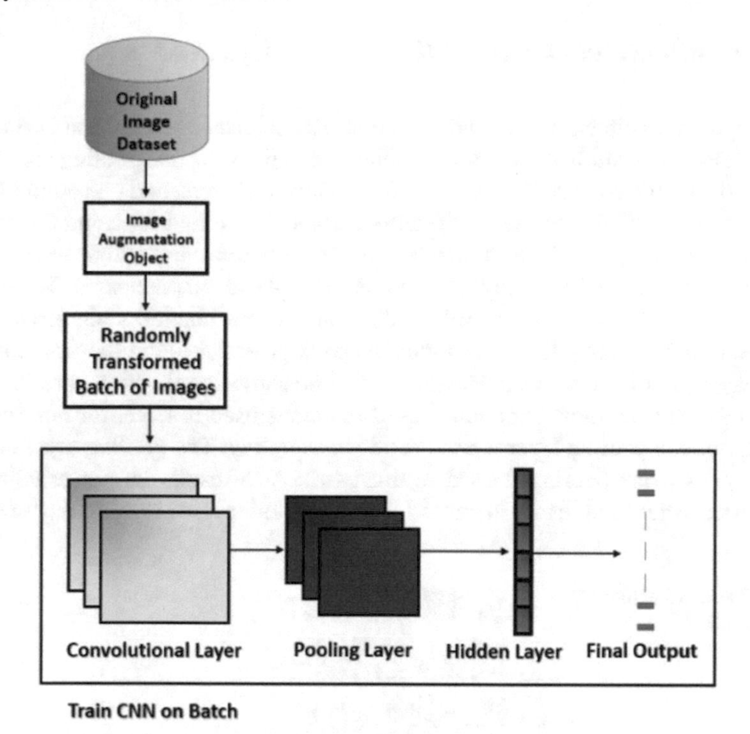

Fig. 1 The proposed model

4.1 Data Augmentation

Data augmentation is a way to artificially create new training data from an existing set of data. Augmentation is essential in deep learning because a huge amount of data is required and it is difficult to get that much data for processing. Several augmentation techniques include rotation, flipping, scaling, adding salt and pepper noise to the data set and many more.

Data augmentation can be applied in different ways. There are two types of data augmentation methods that are commonly used while applying deep learning. In the first type which is called a data set expansion method, the data are augmented by transforming existing images to the new set of images for training and testing. The problem with this type is that the model will not be able to generalize to the unseen data while encountering a small set of data for training. The second type is the in-place/on-the-fly data augmentation method. Instead of generating new data, the augmentation process will transform the existing data and the new set of data will be generated. Thus, the network sees new variations of the data at each and every epoch during the training of the data. On-the-fly method for data augmentation is used in this experiment (for flipping, horizontal, and vertical rotation of 45°).

4.2 Architectural Details of the Proposed Method

The proposed architecture for handwritten MODI character recognition is shown in Fig. 2. The classification model has ten convolutional layers, five pooling layers (2*2 sub-matrices), three dense layers (two fully connected layers, and one output layer). The convolutional layer is used for feature extraction and by increasing the number of convolutional layers, the accuracy of feature extraction can be improved.

This architecture takes a grayscale image as input (60*60 pixel size). The on-the-fly data augmentation is performed on the image using ImageDataGenerator (flip, horizontal and vertical, 45°). The input image is passed through the first convolutional layer (32 filters of window size 2×2). The activation function used is ReLU, which is one of the most commonly used functions used in CNN for nonlinearity. Subsequently, a pooling layer is used for downsampling. The Pooling layer has 2*2 sub matrices (Max pooling is used in this case). As a result, the image volume is reduced and filter samples are increased. The convolution and Max pooling operation

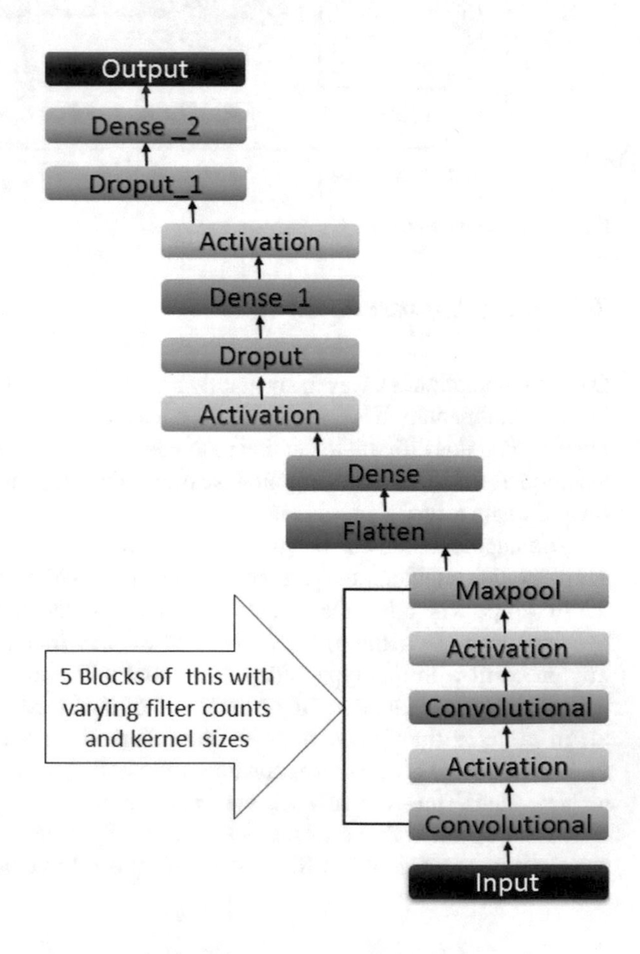

Fig. 2 The architecture of proposed CNN

is repeated serially with the varying numbers of filter counts (32, 64,128, 256,512). A flatten layer is added to convert the pooled feature map to a single column that is passed to the fully connected layers. The features are then passed through two dense layers of 256 nodes each. The ReLU activation function is used for the dense layers. Two dropout functions are added between the two hidden layers. Dropout functions are used for preventing overfitting. The output layer is constructed with 46 neurons for the classification of 46 classes in the MODI character set. The sigmoid activation function is used in the output layer.

5 Experimental Study and Results

The experiment is conducted on the Intel i7 processor with 16-GB RAM, using Python programming.

The MODI script has 46 characters, of which 36 are consonants and 10 vowels. It is a Brahmi-based script used mainly for writing Marathi. The basic character set is represented in the following Fig. 3. As the initial data set, 4600 MODI characters are used in the experiment. 3220 are used as train samples and 1380 as test samples (70:30 ratios). Handwritten characters written by different persons were used for the study.

Data augmentation is performed using ImageDataGenerator function (of Keras), which performs on-the-fly data augmentation. ImageDataGenerator does real-time data augmentation. All the original images are just transformed at every epoch (more data will not be added to the data set) and then used for training, and this way the learned model tends to become more robust and accurate as it is trained on different variations of the same image. The randomly generated data are then subjected to training using CNN. The proposed model has 10 convolutional layers so that more features can be extracted, which leads to better accuracy. The accuracy is predicted with the help of the recognition rate and error rate. Training accuracy and training loss is depicted in Fig. 4.

Fig. 3 Basic Character Set of MODI Script

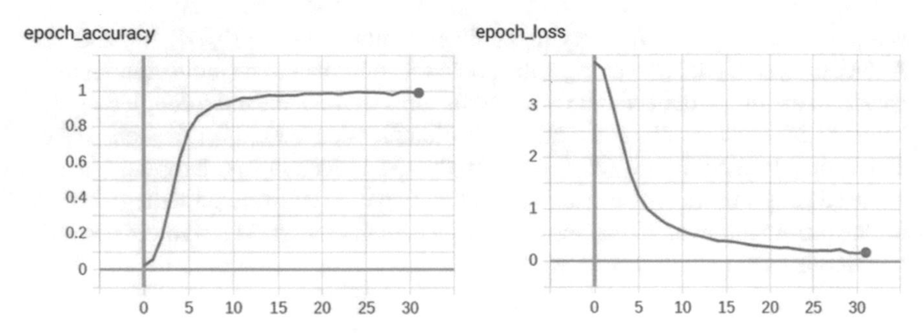

Fig. 4 Training Accuracy and Training loss of the ACNN Network

Table 2 MODI script recognition accuracy using the proposed method and a comparison with the existing method

Sr. no	Author/method	Feature extraction & classification	Data set	Accuracy (%)
1	ACNN method	CNN	MODI character set of 46 characters	99.78
2	Sadanand et al. [3]	Zernike moments & euclidean distance	MODI character set of 46 characters	94.92

The proposed ACNN model fetched a remarkable accuracy of 99.78% on the test data set. The performance of the proposed method is compared with another method (which is the reported highest accuracy method so far), as shown in Table 2. The accuracy obtained using the ACNN method is better compared to any other MODI character recognition method reported so far. It is observed that a similar-looking character gave false predictions, as the system was not able to distinguish some images.

As the number of convolutional layers increases the more features can be extracted and that leads to better accuracy of the recognition system. But at the same time, it is to be noted that an increase in the number of layers leads to an increase in the usage of memory and CPU; hence, from the feasibility point of view, it is necessary to strike a balance to keep the model reliable. The performance of the model can be further increased by making use of the parallel processing power of GPU machines.

6 Conclusion

Handwritten character recognition for MODI script is a promising research area as very less work is done in this field. Deep learning methods are extensively used in character recognition as it gives a better performance. The character recognition of handwritten MODI script using a CNN-based method is performed in this study.

Based on the literature survey and as per the best of our knowledge, a CNN-based study is not conducted on MODI script recognition so far. The proposed method called ACNN is implemented with on-the-fly data augmentation and CNN architecture. The data augmentation is performed using the horizontal and vertical flip of 45°. The on-the-fly data augmentation method equips the network to see a new set of data each time and thus the efficiency of the system is increased. The CNN architecture of the proposed method is designed with appropriate parameters for good accuracy of the MODI script character recognition. Experimentation with various parameters such as the number of convolution layers, filter size, type of activation function, pooling layer size, number of fully connected layers, etc., helped to decide which parameters perform best for MODI script and an accuracy of 99.78% was achieved.

References

1. Joseph S, George J (2019) Feature extraction and classification techniques of modi script character recognition. Pertanika J Sci Technol 27(4):1649–1669
2. Joseph S, George JP, Gaikwad S (2020) Character recognition of MODI script using distance classifier algorithms. Fong S, Dey N, Joshi A. ICT Anal Appl Lect Notes Networks Syst 93, Springer, Singapore
3. Sadanand K, Borde PL, Ramesh M, Pravin Y (2015) Offline MODI character recognition using complex moments. Proce Comput Sci 58:516–523
4. Anam S (2015) An approach for recognizing Modi Lipi using Otsu's Binarization algorithm and kohenen neural network. Int J Comput Appl 111(2):28–34
5. Besekar DN (2012) Special approach for recognition of handwritten MODI script' s vowels. Int J Comput Appl, 48–52
6. Najadat HM, Alshboul AA, Alabed AF (2019) Arabic handwritten characters recognition using convolutional neural network. In: 2019 10th Int. Conf. Inf. Commun. Syst. ICICS 2019, no. January, pp 147–151
7. Keserwani P, Ali T, Roy PP (2019) Handwritten Bangla character and numeral recognition using convolutional neural network for low-memory GPU. Int J Mach Learn Cybern 10(12):3485–3497
8. Shobha Rani N, Chandan N, Sajan Jain A, Kiran HR (2018) Deformed character recognition using convolutional neural networks. Int J Eng Technol 7(3):1599–1604
9. Kumar Reddy RV, Srinivasa Rao B, Raju KP (2019) Handwritten Hindi digits recognition using convolutional neural network with RMSprop optimization. In: Proc 2nd Int Conf Intell Comput Control Syst ICICCS 2018, no. Iciccs, pp 45–51
10. Hossain MA, Ali MM (2019) Recognition of handwritten digit using convolutional neural network (CNN). Glob J Comput Sci Technol 19(2):27–33
11. Sen Maitra D, Bhattacharya U, Parui SK (2015) CNN based common approach to handwritten character recognition of multiple scripts. In: Proc Int Conf Doc Anal Recognition, ICDAR, vol. 2015-November
12. Solley T (2018) A study of representation learning for handwritten numeral recognition of multilingual data set. Springer Lect Notes Networks Syst 10:475–482
13. Jana R, Bhattacharya S (2019) Character recognition from handwritten image using convolutional Neural networks, March. Springer, Singapore

Robot Path Planning–Prediction: A Multidisciplinary Platform: A Survey

D. D. Diana Steffi, Shilpa Mehta, K. A. Venkatesh, and Sunil Kumar Dasari

Abstract In recent times, there is an impressive progress in the field of automation and robotics. Google driverless car or intelligent Unmanned Air Vehicle (UAV) is the latest in the research works aiming for a high degree of autonomy. In this research field of automation and robotics, there is a mandatory requirement of continuously improving path planning algorithms. Path planners aim in finding an optimal and collision-free path in the work environment. The purpose of path planning is to find a kinematical optimal path with the least time complexity as well as model the environment completely. In this paper, we discuss the most successful robot path planning algorithms which have been developed in recent years from different field of science, making it a multidisciplinary approach and concentrate on universally applicable algorithms which can be implemented in aerial robots, ground robots and underwater robots. The algorithms are analysed from an optimality and completeness area perspective. In our study, we have included the Dynamic Programming Planning approach also which none of the review papers have covered.

Keywords Path planning · Optimal path · Source node · Destination node

D. D. Diana Steffi (✉) · S. Mehta · S. K. Dasari
Department of Electronics and Communication Engineering, Presidency University, Bangalore, India
e-mail: steffiseelan@gmail.com

S. Mehta
e-mail: shilpamehta@presidencyuniversity.in

S. K. Dasari
e-mail: mr.sunilkumardasari@gmail.com

K. A. Venkatesh
Department of Mathematics and Computer Science, Myanmar Institute of Information Technology, Mandalay, Myanmar
e-mail: ka_venkatesh@miit.edu.mm

D. S. Jat et al. (eds.), *Data Science and Security*, Lecture Notes in Networks and Systems 132, https://doi.org/10.1007/978-981-15-5309-7_22

1 Introduction

Path planning is the task of determining a path that passes over all points in the area of interest while avoiding obstacles. This task is very important in many robot applications, such as vacuum cleaning robots, painter robots, autonomous underwater vehicles, demining robots, lawn movers, automated harvesters, window cleaners, Google driverless cars, inspection of complex underwater structures and many more.

Path planning targets for moving robots from their initial locations to the destination location by their own actuators and strategies, and the robots must always be able to avoid obstacles to maintain in the process of reaching the destination. Different strategies of path planning have already been tested in robots such as underwater robots, wall-climbing robots and micro air vehicles. These methods can be read from [1–3]. However, they are analysed in a general view, without any specific perspective. Thus, we need to provide a comprehensive analysis on path planning covering latest works. The path planning algorithm is assessed based on completeness, optimality and complexity [4].

The paper is organized as follows: Section II gives a detailed insight into the problem statement, basic definitions and terms. Section III focuses on solutions adapted with real time environment for a high level of autonomy in each of the classified path planning methods as in Fig. 2.

2 Problem Statement

The primary problem in all autonomous robot systems is navigation. Navigation includes the robot locating its own position and the position of the goal and being able to reach the destination without colliding with any obstacle. To put it in a nutshell, the robot must answer these three questions,

1., Where am I?
2. Where is my goal?
3. How can I reach to my goal?

Positioning, mapping and path planning give answers to all the above questions. Positioning: It helps the robot to determine its location in the environment. A number of positioning methods are available [4–7].

Mapping: The robot needs a map to know about the environment; in some cases, the maps are fed in the robot's memory and few cases the map is built as and when the robot discovers the new environment. Map guides the robot to know directions and positions in the environment.

Path Planning: In order to reach the goal, the optimal path must be found. Each robot requires an appropriate addressing scheme to know the goal position. The addressing scheme indicates where it will go from its starting position. There are many issues involved in path planning as shown in Fig. 1. Many researchers have

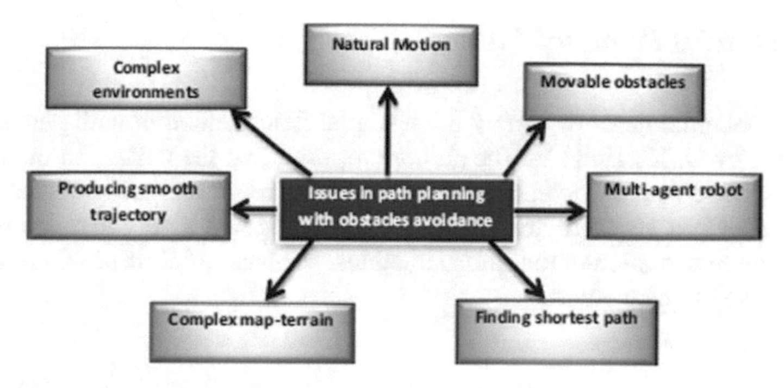

Fig. 1 Issues in path planning

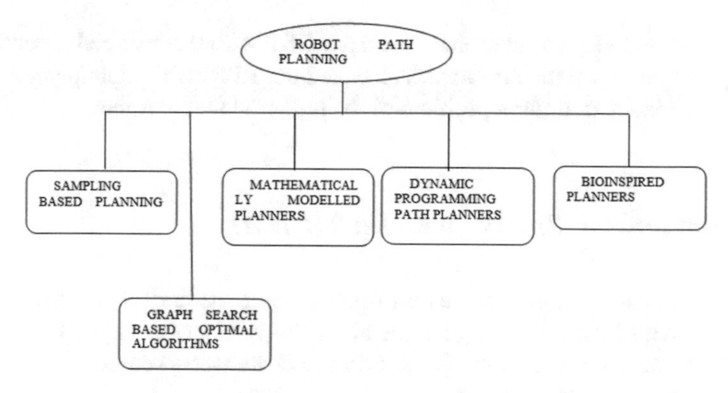

Fig. 2 Classification of path planning in robotics

experimented in finding the shortest optimal path and they have moved in to compute the shortest path in less computational time and also to improve smooth flight [8]. Complex environment [9] and multi-agent [10] issues are also looked into in recent times. There may be many issues in path planning, but the three important metrics which cannot be ignored are completeness, optimality and complexity as mentioned before.

3 Sampling-Based Planning

Sampling-Based Planning (SBP) is a type of path planning in which the configuration space(C-space) is sampled randomly. SBP gives the connectivity available in the c-space by sampling it.

3.1 Artificial Potential Fields

Artificial potential fields or simply the potential field method of path planning was introduced by O. Khatib [11]. The author proposed that the target and the obstacle in the C-space create the potential field in which the robot is considered as a moving particle. This is a very simple path planning method and a very fast one which is easy for implementation in real-time situations, but local minima problem is a huge disadvantage in this method.

3.2 Randomized Potential Planner (RPP)

Carpin and Pillonetto [15] combined the basic SBP with the artificial potential field planning and proposed the new approach called the RPP. RPP implemented random walks to escape local minima problem of the potential field planner.

3.3 Probabilistic Roadmap Method (PRM)

PRM algorithm and its variants are multi-query-based methods that construct a road map that provides with the complete detail of the workspace and its collision-free space. Then the algorithm finds the shortest path between source and destination through the constructed roadmaps.

PRM* Frazzoli and Karaman [12] in their work introduced PRM* which is an incremental PRM algorithm that guarantees asymptotical optimality, as basic PRM algorithm just finds a path from the start to goal by establishing road maps.

PRM-RL Aleksandra et al. proposed a new approach for robot navigation over a long distance with a dynamic unstructured environment [13]. They combined PRM and reinforcement learning to produce PRM-RL which was proved to be very successful in a complex unstable dynamic large environment

3.4 Rrt Series

Rapidly Exploring Random Trees(RRT) LaValle et al. [14] propose a method of path planning with holonomic, nonholonomic and kino dynamic constraints. RRT finds a path between the initial node and the destination node through rapid search in the configuration space. A new node is sampled in every step when the extension from the sampled to the nearest

	node succeeds. RRT finds a path from the initial node to the destination node using Monte Carlo random sampling, which biases the already explored region without considering obstacle region thus, leading to inappropriate sampling which in turn increases the time. So, the main disadvantage is that RRT will consume more time in a cluttered environment and the path is not optimal. RRT is faster than PRM as there is no need to sample the C-space and construct road maps.
RRT*	It is the tree version of RRG, but it holds the asymptotically optimal property and rapid exploration of RRG and RRT. RRT* algorithm rewires the tree as better paths are discovered. This algorithm removes all unwanted connections between the nodes thus optimizes the solution to be less expensive [12].

4 Graph Search-Based Optimal Algorithms

These algorithms plan with nodes and arcs, sometimes called as grid. These planners explore through the nodes and expand them in the process of finding a path between the start and the goal, which is optimal. A network or a graph is generated and search occurs.

4.1 Dijkstra Algorithm

The alma mater in the graph traversal planning is this algorithm proposed by E.W. Dijikstra in 1959 [16]. The algorithm finds the shortest path in the graph where the weights of arcs are pre-known. It can also be said as a special kind of dynamic programming and it is a BFS method. The time complexity is more which leads to huge variants of Dijkstra's algorithm.

4.2 A and Its Variants*

In 1968, Pet et al. [17] extended the Dijkstra's algorithm, in which the number of nodes expanded is reduced with the use the heuristic. Thus, A* planner will be much faster and, in worst case, degenerates. This planner gives an optimal path when applied to the environment which is visibility graph-type modelled and an acceptable solution when applied to the environment which is grid-type modelled.

5 Mathematically Modelled Planners

Linear algorithms and optimal control algorithm come under the mathematically modelled planning category. These planners model the environment for kinematic constraints and the system for dynamic constraints and bound the cost function with kinematics and dynamic constraints which are inequalities or equations to get the optimal solution. Some of the successful mathematically modelled planners are as follows

Chamseddine et al. [18] proposed the flatness-based method that uses differential flatness along the reference path to ensure control flatness. This planner is simple as it linearizes the nonlinear kinematics and dynamic constraints.

Linear Programming — Georgios and Jeff in their paper [26] used linear programming for an optimal path for a group of autonomous vehicles in an adversarial environment.

MILP — Mixed-Integer Linear Programming is very strong in modelling and explains almost all information in a 3D environment [19]. In this planner, the obstacle avoidance issue is mainly looked into and the results are achieved by simplifying and decoupling the robot dynamic model by restricting admissible control. MILP provides an optimal solution for complex environment considerably quick

6 Dynamic Programming Path Planners

Dynamic Programming is an optimization approach that solves a complex problem by breaking them into sequence of simpler problems. This can be said as multistage approach for optimization. Dynamic programming provides a general framework for analysing many types of problems. The category of robotic path planning is not looked in with interest by many researchers. The remaining part of this section clearly highlights the part of dynamic programming in the field of robotic path planning.

- In 2008, Alton et al. presented an efficient dynamic programming algorithm that solves the problem of optimality in multi-location multiple robot rendezvous. In the author's work, a tree structure is exploited with dynamic programming to get an optimal solution [22].
- Hierarchical Dynamic Programming for robot path planning was presented by Bakker et al. in their work [23]. The authors combined Markov Decision Process (MDP) with dynamic programming approach and proposed a path planner for the stochastic task. The advantage of this algorithm is reduced memory complexity.
- In 2006, Willms et al. proposed an efficient dynamic programming planner for real-time applications, where the targets and the obstacles move in the environment. This algorithm provides a collision-free optimal path [24].

- A modified dynamic programming algorithm with an accelerating node was proposed, which provides an optimal path, even when there is a sudden blockage in the path [25]. In the previous work of Williams, the robot continue to move in the same path even when there is a blockage. This presented planner quickly re-plans a new path when there is a blockage.

7 Bio-inspired Planners

In the case of bio-inspired algorithms, the planners should think and behave as a human being in solving the NP-hard problem in path planning and provide an optimal path. There are two main categories in bio-inspired planners, they are evolutionary and neural network algorithms [20, 21].

8 Conclusion

This paper presented an exhaustive overview of the available algorithms in robot path planning. We have included the dynamic programming approach algorithms also, which is widely neglected by many researchers. From the study of the available algorithms, we can say that robot path planning is a multidisciplinary platform, in which all disciplines of science have contributed. The path planners were categorized into five groups: 1. Sampling-based planners, 2. Graph search-based planners, 3. Mathematically modelled planners, 4. Dynamic programming planners, 5. Bio-inspired planners. The following was observed in each category of planners.

- Sampling-Based planners have good time efficiency in a dynamic environment.
- Graph search-based planners provide an optimal solution with little higher computational time.
- Mathematically modelled planners provide optimal solution with lesser computational time when compared to graph search-based planners.
- Dynamic planners have computational complexity but provide a collision-free path.
- Bio-inspired planners are most widely used in recent times. They provide optimal, time-efficient solutions.

Though enough work has been done in the robotic path planning, there is a continuous need for improvement, paving path for many researchers to continuously present new better work.

References

1. Choset HM (2005) Principles of robot motion: theory, algorithms, and implementations, MIT Press
2. LaValle SM (2006) Planning algorithms, Cambridge University Press
3. Sebbane YB (2012) Lighter than air robots: guidance and control of autonomous airships, vol 58. Springer
4. Russell S, Norvig P (2003) Artificial intelligence, a modern approach. Pearson Education, London, UK
5. Visvanathan R, Mamduh SM, Kamarudin K, Yeon ASA, Zakaria A, Shakaff AYM, Kamarudin LM, Saad FSA (2015) Mobile robot localization system using multiple ceiling mounted cameras, IEEE Sensors, 1–4
6. Saito T, Kuroda Y (2013) Mobile robot localization by gps and sequential appearance-based place recognition. In: IEEE/SICE international symposium on system integration (SII), 25–30
7. Forouher D, Große Besselmann M, Maehle E (2016) Sensor fusion of depth camera and ultrasound data for obstacle detection and robot navigation. In: 14th international conference on control, automation, robotics and vision (ICARCV), pp 1–6
8. Baglivo L, Bellomo N, Miori G, Marcuzzi E, Pertile M, De Cecco M (2008) An object localization and reaching method for wheeled mobile robots using laser range finder. In: 4th International IEEE conference intelligent systems, IS'08, vol 1, pp 5–6
9. Nayl T, Mohammed MQ, Muhamed SQ (2017) Obstacles avoidance for an articulated robot using modified smooth path planning. In: International conference on computer and applications (ICCA), pp 185–189
10. Uriol R, Moran A (2017) Mobile robot path planning in complex environments using ant colony optimization algorithm. In: 3rd international conference on control, automation and robotics (ICCAR), pp 15–21
11. Dewangan RK, Shukla A, Wilfred Godfrey W (2017) Survey on prioritized multi robot path planning. In: IEEE international conference on smart technologies and management for computing (ICSTM), pp 423–428
12. Khatib O (1986) Real-time obstacle avoidance for manipulators and mobilerobots. Int J Rob Res 5:90–98
13. Karaman S, Frazzoli E (2010) Optimal kino dynamic motion planning using incremental sampling-based methods. In: Proceedings of the 49th IEEE conference on decision and control (CDC'10), pp 7681–7687
14. Aleksandra F, Kenneth O, Oscar R, Anthony F, Lydia T, Marek F, James D (2018) PRM-RL: long-range robotic navigation tasks by combining reinforcement learning and sampling-based planning, pp 5113–5120
15. LaValle SM, Rapidly-exploring random trees a new tool for path planning. Tech Rep, 98–11
16. Carpin S, Pillonetto G (2005) Merging the adaptive random walks planner with the randomized potential field planner. In: Proceedings of the fifth international workshop on robot motion and control, pp 151–156
17. Dijkstra EW (1959) A note on two problems in connexion with graphs. Num Math 1:269–271
18. Hart P, Nilsson N, Raphael B (1968) A formal basis for the heuristic determination of minimum cost paths. IEEE Trans Sys Sci Cyber 4:100–107
19. Chamseddine A, Zhang Y, Rabbath CA, Join C, Theilliol D (2012) Flatness-based trajectory planning/replanning fora quadrotor unmanned aerial vehicle. IEEE Trans Aeros Elec Sys 48:2832–2848
20. Yue R, Xiao J, Joseph SL, Wang S (2009) Modeling and path planning of the city-climber robot part II: 3D path planning using mixed integer linear programming. In: Proceedings of the IEEE international conference on robotics and biomimetics, pp 2391–2396
21. Aghababa MP (2012) 3D path planning for underwater vehicles using five evolutionary optimization algorithms avoiding static and energetic obstacles. Appl Ocean Res 38:48–62

22. Yang SX, Luo C (2004) A neural network approach to complete coverage path planning. In: IEEE transactions on systems, man, and cybernetics, part b: cybernetics, vol 34, no 1, pp 718–725
23. Ken A, Ian MM (2009) Efficient dynamic programming for optimal multi-location robot rendezvous. In: Proceedings of the IEEE conference on decision and control, pp 2794–2799
24. Bakker B, Zoran Z, Krose B (2005) Hierarchical dynamic programming for robot path planning. In: IEEE transactions on pattern analysis and machine intelligence—PAMI, pp 2756–2761
25. Willms AR, Yang SX (2006) An efficient dynamic system for real-time robot-path planning. In: IEEE Trans Sys Man Cyber—Part B Cyber 36(4):755–766
26. Kala R, Tiwar R (2012) Robot path planning using dynamic programming with accelerating nodes, Paladyn. J Behav Robot 3(1):23–34
27. Georgios C, Jeff S (2005) Linear-programming-based multi-vehicle path planning with adversaries. In: Proceedings of the American control conference, pp 1072–1077

Genome Analysis for Precision Agriculture Using Artificial Intelligence: A Survey

Alwin Joseph, J. Chandra, and S. Siddharthan

Abstract Precision agriculture is a farm management technique which uses the help with the help of information technology to ensure that the crops and soil receive exactly what is required for optimum health and productivity. Genome analysis in plants helps to identify the plant structure and physiological traits. The identification of the right plant genome and the resulting traits help to optimize the cultivation of the plant for better productivity and adaptability. Genome analysis helps the biologist edit the plant genetic makeup structure to make the plant to adapt to the current conditions and thereby reducing the use of fertilizers. For precision agriculture, artificial intelligence techniques help to understand the relationships between plant genome and soil nutrient conditions that help in precision farming effectively reducing the usage of fertilizers by modifying the plants to adapt with the current soil characteristics.

Keywords Precision agriculture (PA) · Genome analysis · Artificial intelligence (AI) · K-Nearest neighbor · Random forest · Next-generation sequencing (NGS) · Sequence analysis · DNA and RNA sequencing · Transcriptomics

1 Introduction

Precision Agriculture (PA) is a technique of farm management, which makes use of the advancements in the Information Technology (IT) to make sure the plants receive the exact amount of nutrition for better health and productivity. The main objective of PA is to ensure profitability, sustainability, and protection of the environment. PA is obtained with the help of specialized equipment's, software's, and IT services [1].

A. Joseph (✉) · J. Chandra · S. Siddharthan
CHRIST (Deemed to Be University), Bangalore, India
e-mail: alwin.joseph@mca.christuniversity.in

J. Chandra
e-mail: chandra.j@christuniversity.in

S. Siddharthan
e-mail: siddharthan.s@christuniversity.in

D. S. Jat et al. (eds.), *Data Science and Security*, Lecture Notes in Networks and Systems 132, https://doi.org/10.1007/978-981-15-5309-7_23

The PA approach is based on processing the real-time data of crops growth, soil and air conditions, along with other relevant information like weather predictions, labor costs, and equipment availability to make smart decisions during farming. Agricultural research centers, like ICAR in India, play their role in integrating different environmental sensor information, along with other relevant information and the Artificial Intelligence (AI), to help the farmers to identify the sections of their fields that require special treatment and to determine the proper amount of water, fertilizers, and pesticides. PA prevents the wastage of resources, by monitoring the soil and plants to have the proper additives for their optimum growth, by minimizing the environmental pollution [2].

Genome analysis helps in identification, measurement, and comparison of genomic features. The features like the DNA sequence, structural variation, gene expression, regulatory elements, and functional elements of a particular organism are compared and studied under genome analysis [3]. Genomic analysis identifies the DNA sequence, from a representative genome of the given species, the species can be plant or animal. The sequence of an organism opens the door to numerous possibilities. Analysis of genomes from the same species can be used to create a statistical picture of the genetic variation, within the populations of that particular species [4].

Artificial Intelligence (AI) is the process of simulating the human intelligence to the computer systems. The processes include learning, reasoning, and self-correction of the decisions they obtain during the processing [5]. There are different methods that help in the analysis of data and to obtain meaningful conclusions from the data. AI technologies can perform genomics data processing faster and effectively, which can give quick and effective insights that help to make better decisions on treatment, spot future possible mutations as well as predict possible future vulnerabilities from any organism's genome data [6]. Effective implementations of the AI techniques in the identification of traits and learning from this can predict the changes effectively.

2 Literature Survey

Edo D'Agaro [7] has surveyed recent research, which proves that machine learning is used efficiently in the analysis of genomic data for the identification of novel gene functions and regulatory regions. The Convolutional Neural Networks (CNN) and Recurrent Neural Networks (RNN) are the best tools for solving tasks in genomics. The paper suggested that the CNN and RNN, used in deep learning, can be applied to determine the protein coding regions, protein–DNA interactions, regulatory regions, and functional RNA gene applications.

Zou et al. [8] have identified the use of deep learning techniques to determine patterns in large datasets of genomic data. The techniques are used in the field of regulatory genomics, variant calling, and pathogenicity scores. The analysis of genomic data uses the help of large training datasets. The main architectures include feed-forward, convolutional, and recurrent networks that are corresponding to different assumptions about data. Deep learning techniques will give high accuracy, but the interpretation of results is challenging.

Pudumalar et al. [9] have surveyed various data mining techniques that can be used with PA to analyze the various biotic and abiotic factors with respect to a particular field. PA uses the information about the soil characteristics, yield data and suggests the farmers the right crop based on their field-specific parameters; this approach reduces the wrong choices about the plant and increases their productivity. A recommendation system with the help of random forest, CHAID, KNN, and Naive Bayes, to recommend a crop for the particular field with respect to the field-specific parameters with high accuracy and efficiency was suggested.

Schuster et al. [10] surveyed the advances in sensing technologies and global positioning system that improve agricultural productivity with the use of data. They identified that there is a significant variation in yield and concentration of pests and plant diseases with reference to geographical location. PA uses the help of various types of sensing information to formulate the specific treatments for the field-specific management of crops. The use of various IT in PA needs effective modeling approaches.

Zhang et al. [11] concluded that genome editing tools helps targeted modification in an organism's genome and to characterize gene functions. The genome editing tools help to improve agricultural traits to improve the crop. The crop improvement helps the plant for better adaptation with the soil. Plants face huge amount of variability due to spontaneous mutations, and the changes due to the chemical mutagens and physical irradiations. Genome editing, is an advanced biological technique that can produce precise and targeted modifications in genome.

Andolfo et al. [12] surveyed and identified the greatest challenges for agriculture is to improve the yield and plant stability without harming the environment. Genome editing technologies are made effective in breeding plants resistant to diseases. Targeted genome engineering provides precise modifications to the plant genome to breed new varieties. Genome editing, modify the crucial players of the plant immunity system. Genome editing strategies help to improve resistance by rewriting the genome sequence of the target genes involved in the plant resistance. Genome editing technologies helps synthetic biology to obtain a reinforced plant defense system.

Bagchi [13] identified that the use of latest technological solutions to make farming more efficient, still remains as a huge challenge. AI has a huge application on the PA sector, it brings a paradigm shift in the farming today. AI-based solutions help farmers to do more with less, with improved quality and faster go-to-market crops. AI changes the agriculture landscape completely. The application of image processing, Genome analysis and genome editing techniques helps PA to improve the outcomes. The future of farming depends on adoption of cognitive solutions in the farms.

The analysis of plant genomic structure helps to identify the genetic structure and traits of a particular plant. The use of these data in the agriculture helps to improve the cultivation of the plants that adapt the existing soil conditions. Precision agriculture that incorporates artificial intelligence techniques can use the process of genome analysis and editing to produce effective crop that suits the soil. This will help reduce the use of fertilizers in farming and to make the cultivation eco-friendly by reducing the impact of chemicals in soil.

3 Methodology

Precision agriculture focuses on the improvement of crop productivity. The information about the agricultural land is to be obtained from the various sources that help in identifying the field characteristics. The identification of the field characteristics helps in determining the appropriate modifications to the plant. The modifications in the plant genome structure helps in the better adaptability of the plant to the soil. The selection of appropriate plant in the soil will help to maximize productivity and also helps to reduce the use of fertilizers and pesticides.

The following flow diagram represents the overall workflow that can be used to choose the appropriate plant in the field to maximize the productivity in the farm.

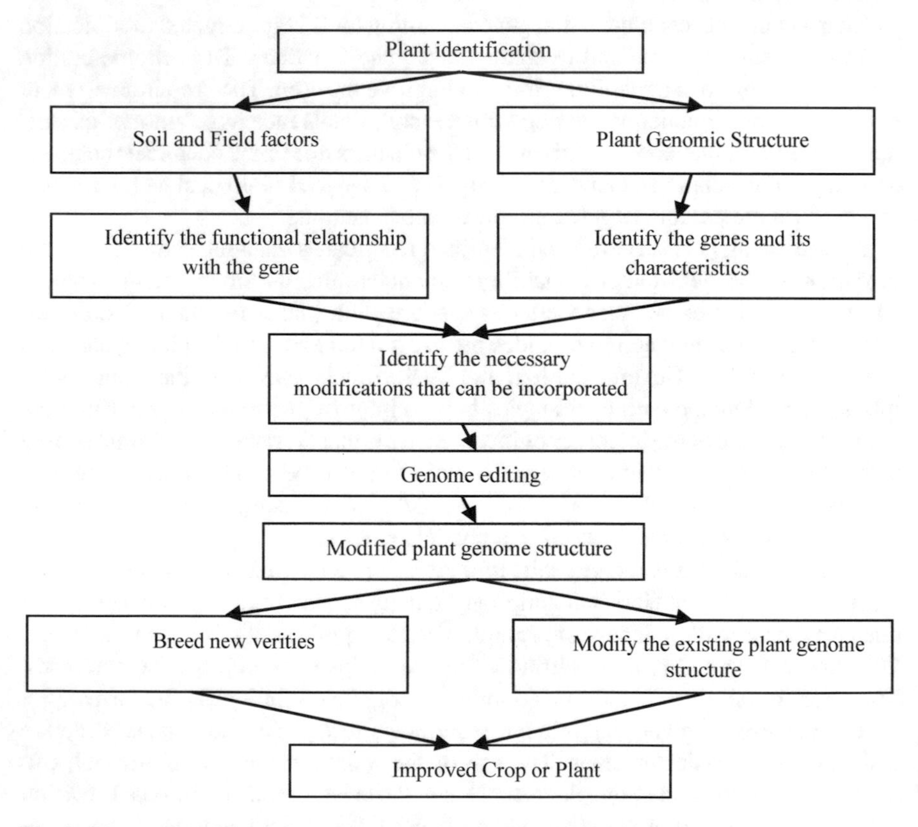

Fig. 1 Proposed workflow of precision agriculture using genome analysis

Figure 1 depicts the proposed workflow of precision agriculture along with the genome analysis. The process starts with the identification of the appropriate plant that for cultivation. The relationship of the field properties and the plant genes associated with it will be identified. The genes that are involved in drought tolerance, disease resistance, and higher yield will be identified using machine learning techniques using data from thousands of genomes from various plants. The fast and efficient method of genome analysis could be achieved by studying the transcriptomes which are the sequences of coding portions of the genome. The strength of transcriptomics is that it could identify genes which are highly expressed in plants that tolerate high stress levels and produce better yields. Comparative study of transcriptomes using AI will help in pinpointing the right genes that need to be targeted in precision agriculture. These data can be used to determine the necessary genome modifications on the plants for better utilization of land and water resources in precision agriculture. The identified modifications can be made into the plant genome using genome editing techniques like CRISPR/Cas9 and TALEN. The modified plant genome structure can be used to breed new varieties of the plant. The same thing can be used to make modifications in the existing plant varieties and to improve their suitability of the field conditions. Thus, the improved crop can be created to improve the productivity.

There are many techniques for sequencing the plant genome as well as their assembly. Further, the genome map will be constructed using de novo or reference-based methods. The identification of plant genome structure helps to identify the plant genes with traits that can be improved for the adaptability of the crop. The modification in the plant genome structure helps to reduce the usage of fertilizer level in their cultivation.

Deep learning techniques are used to generate new DNA sequences and protein coding regions with the desirable properties; these properties are identified from the different data analysis intelligent techniques. There are many genome editing technologies to edit the plant genome structure and to modify the genome structure of a living plant. The use of these technologies requires proper data analysis using artificial intelligence techniques, that enable faster and efficient use of genome editing for the improvements in the plant structure.

Artificial intelligence can be used in the process to identify the changes required in the gene faster. The use of artificial intelligence in prediction of climatic changes and soil characteristics are available. The artificial intelligence techniques are not implemented with respect to the genome data. The implementation of artificial intelligence in the genome data makes the analysis process faster and efficient. The work regarding the genome data analysis using artificial intelligence is lacking its growth. The integration of artificial intelligence in the genome data analysis will result in the huge breakthrough in the bioinformatics.

4 Conclusion

There are many studies concentrating in the area of data analysis and prediction for precision agriculture with genome editing of the plants. Many plants have genetically modified for better yield and performance and adaptability. But effective methods with genome editing for precision agriculture is lacking. The study of the data analysis using artificial intelligence can support genome editing technology to edit the genome of plants to adapt them to the soil and climate conditions. The modification of plant genomic structure helps the farmers to reduce the usage of fertilizers for crop production. Reducing the usage of fertilizers in cultivation of plants helps to reduce environmental pollution and associate health hazards. Thus, genome editing for crops in precision agriculture needs to be an important area of research, which can be optimized using AI data mining techniques. The survey could identify the huge potential of genome analysis and editing for improving the precision farming measures.

References

1. Margaret Rouse (2016) precision agriculture. https://whatis.techtarget.com/definition/precision-agriculture-precision-farming
2. Leonard EC (2015) Precision agriculture. https://doi.org/10.1016/B978-0-08-100596-5.00203-1
3. Liu Y, Tang Q, Cheng P, Zhu M, Zhang H, Liu J, Zuo M, Huang C, Wu C, Sun Z, Liu Z (2019) Whole-genome sequencing and analysis of the Chinese herbal plant gelsemium elegans, 16 August 2019. https://doi.org/10.1016/j.apsb.2019.08.004
4. Harris K, Genome analysis. Nat Gen 51:1306-1307. https://www.nature.com/subjects/genomic-analysis
5. Haleem A, Dr. MohdJavaid, Haleem Khan I (2019) Current status and applications of artificial intelligence (AI) in medical field: An overview, 12 November 2019. https://doi.org/10.1016/j.cmrp.2019.11.005
6. Margaret Rouse (2018) Artificial intelligence, August 2018. https://searchenterpriseai.techtarget.com/definition/AI-Artificial-Intelligence
7. D'Agaro E (2018) Artificial intelligence used in genome analysis studies. The Euro Biotech J
8. Pudumalar S, Ramanujam E, Harine Rajashree R, Kavya C, Kiruthika T, Nisha J (2106) Crop recommendation system for precision agriculture. In: IEEE eighth international conference on advanced computing (ICoAC), 978-1-5090-5888-4/16/$31.00@2016 IEEE
9. Schuster EW, Kumar S, Sanjay E, Jeffrey S, Willers L, Milliken GA (2011) Infrastructure for data-driven agriculture: identifying management zones for cotton using statistical modeling and machine learning techniques, 978-1-4577-1591-4/11/$26.00 ©2011 IEEE
10. Ezziane Z (2006) Applications of artificial intelligence in bioinformatics: a review. Exp Sys App 30(1):2–10
11. Zhang Y, Massel K, Godwin ID, Gao C (2018) Applications and potential of genome editing in crop improvement. Genome Biology. https://doi.org/10.1186/s13059-018-1586-y
12. Andolfo G, Iovieno P, Frusciante L, Ercolano MR (2016) Genome-editing technologies for enhancing plant disease resistance. Frontiers in Plant Science
13. Bagchi A (2000) Artificial intelligence in agriculture. Millennium Development Goal, UN Summit

Efficient Handwritten Character Recognition of MODI Script Using Wavelet Transform and SVD

Solley Joseph and Jossy George

Abstract MODI script has historical importance as it was used for writing the Marathi language, until 1950. Due to the complex nature of the script, the character recognition of MODI script is still in infancy. The implementation of more efficient methods at the various stages of the character recognition process will increase the accuracy of the process. In this paper, we present a hybrid method called WT-SVD (Wavelet Transform-Singular Value Decomposition), for the character recognition of MODI script. The WT-SVD method is a combination of singular value decomposition and wavelet transform, which is used for the feature extraction. Euclidean distance method is used for the classification. The experiment is conducted using Symlets and Biorthogonal wavelets, and the results are compared. The method using Biorthogonal wavelet feature extraction achieved the highest accuracy

Keywords MODI script · Handwritten character recognition · Wavelet transform · Feature extraction method · Image processing

1 Introduction

Character recognition is the process of recognizing characters from a document image and converting the same into the machine-readable form [1]. Unlike non-Indic scripts which are already matured in the field of character recognition, the majority of Indian language scripts still need attention. MODI script is one such script in which the character recognition still needs attention. MODI script was used for writing Marathi until 1950 and has an important role in historical research. It has been observed that various libraries and temples in India have preserved a large number of MODI documents [2].

S. Joseph (✉) · J. George
CHRIST (Deemed to Be University), Bengaluru, India
e-mail: solley.joseph@res.christuniversity.in

S. Joseph
Carmel College of Arts Science and Commerce for Women, Nuvem, Goa, India

D. S. Jat et al. (eds.), *Data Science and Security*, Lecture Notes in Networks and Systems 132, https://doi.org/10.1007/978-981-15-5309-7_24

The basic character set of MODI script consists of 46 characters. It is a Brahmi-based script used mainly for writing Marathi. MODI script was used as a shorthand form and it does not have a termination symbol for sentences or words [3]. The variations in the writing style of individuals, shape similarity of characters, and the absence of word stopping symbol in the text are some of the factors which make MODI character recognition a complex task.

The character recognition process comprises of five major stages such as Pre-processing, Segmentation, Feature Extraction, Classification, and Post-processing. The focus of our study is the feature extraction stage. We have designed a hybrid model called WT-SVD for the feature extraction of MODI script, by combining Wavelet Transform (WT) and Singular Value Decomposition (SVD) techniques.

2 Review of Literature

KNN and Backpropagation Neural Network (BPNN) have been used for the classification of MODI characters, in combination with structural similarity-based feature extraction. This has resulted in a 91–97% recognition rate [4]. Chain code-based method was used by Chandure & Inamdar [5] for MODI script as well as for Devanagari script, in combination with BPNN, KNN, and SVM. The accuracy achieved was 37.5, 60, and 65%, respectively. A hybrid approach for feature extraction, which combines two techniques: Moment invariant and affine moment invariant [6], reported to have achieved an average recognition rate of 89.72%, for the recognition of MODI script. Zernike's complex moments combined with the Euclidian distance classifier were used for MODI script recognition [7] and have achieved 94.78% accuracy. Restricted Boltzmann Machine (RBM) was used as a feature extractor in an experiment for the recognition of a handwritten multilingual data set [8] and it was reported that time complexity reduced using this method.

Wavelet energy-based feature extraction has experimented for Malayalam character recognition [9]. The study was conducted using Daubechies and Biorthogonal wavelets and achieved an accuracy of 95.59% with Daubechies. An experiment using WT methods for the character recognition reported that the Biorthogonal wavelet family gives better results than the Daubechies wavelet family [10].

3 Methodology

The feature extraction in the proposed method is performed using a hybrid combination of WT and SVD. The extracted feature of the image is then subjected to classification using Euclidean distance. WT has diverse applications.

3.1 Feature Extraction

The extraction of distinguishing features is one of the most important stages of the character recognition process. Wavelet functions defined around a certain point are concentrated both in time and frequency. It is a well-suited function in the time-frequency domain analysis for a non-stationary signal. WT can be used to separate the input signal into detail (horizontal, vertical, and diagonal) and approximate sub-signals as shown in Fig. 1. Given 'n' decomposition wavelet coefficients from the three sub-signals are used for formulating wavelet energy [11].

The wavelet energy of horizontal, e_n^H (x, y), vertical e_n^V (x, y), and diagonal e_n^D (x, y) detail sub-signals, at the position (x, y) at each of the level are as follows

$$e_n^H(x, y) = \sum_{r,s \in Z} (D_H)^2 \, G(x-r, \; y-s)$$
$$e_n^V(x, y) = \sum_{r,s \in Z} (D_V)^2 \, G(x-r, \; y-s) \quad (1)$$
$$e_n^D(x, y) = \sum_{r,s \in Z} (D_D)^2 \, G(x-r, \; y-s)$$

where D_H, D_V and D_D are the wavelet coefficients at the position (x, y) of the three sub-signals. Z refers to the set of integers, with r, s is the x-coordinate, and y is the co-ordinate of the neighborhood pixel, respectively. $G(x, y)$ is the Gaussian kernel function given by

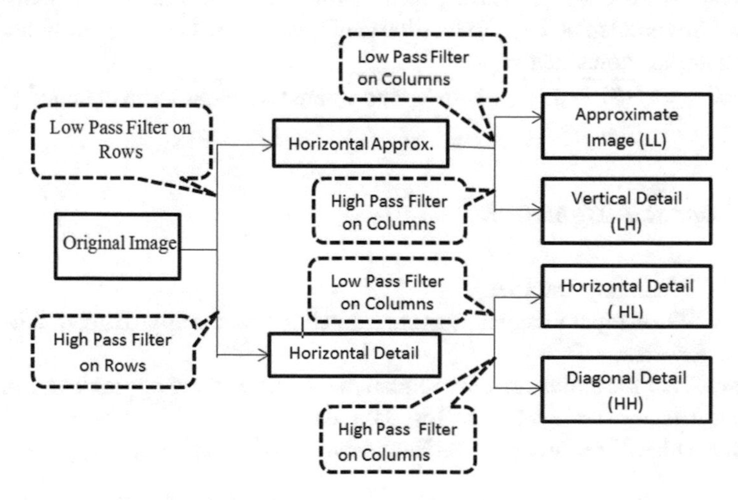

Fig. 1 Discrete wavelet transform

$$G(x, y) = \frac{1}{2\pi\sigma^2} e^{\left(\frac{-(x^2+y^2)}{2\sigma^2}\right)} \quad (2)$$

where σ is the standard deviation of the kernel.

3.2 Singular Value Decomposition

Singular value decomposition (SVD) is a factorization of a matrix to a matrix of lower dimensions. It states that a real matrix 'A' of size m x n can be decomposed into a product of 3 matrices U, Σ, and V such that:-

$$\mathrm{A}_{\mathrm{m}\times\mathrm{m}} = \mathrm{U}_{\mathrm{m}\times\mathrm{n}} \sum_{\mathrm{m}\times\mathrm{n}} \mathrm{V}^{\mathrm{T}}_{\mathrm{m}\times\mathrm{n}} \quad (3)$$

where U and V are the orthogonal matrices and $\sum$ is a diagonal matrix of dimension, same as matrix A. SVD is a powerful technique for dimensionality reduction and has several applications.

3.3 Classification

After the feature extraction stage, the classification by the Euclidean classifier is used for matching the test data samples against the train samples. The distance between the two templates is calculated to produce the similarity score. This distance is calculated by the equation shown below

Distance $= \sqrt{(T_1 - T_2)^2}$, where T_1 and T_2 are templates to be matched (7).

3.4 Proposed Algorithm

1. Read the Training Data set.
2. Apply WT on input image for feature extraction (use the approximate sub-signal) as the features.
3. Apply SVD for dimensionality reduction on the resultant approximate image.
4. Repeat the above steps for the Test data set.
5. Perform classification using the Euclidean Distance method.

As shown in Fig. 2, each input image is vectorized and a matrix is created with each column corresponds to an image vector. The images are then subjected to WT, where the features are extracted from the original image of size 60*60. The extracted feature size of Symlet and Biorthogonal wavelets are 17×17(sym2) and 18×18 (bio1.3),

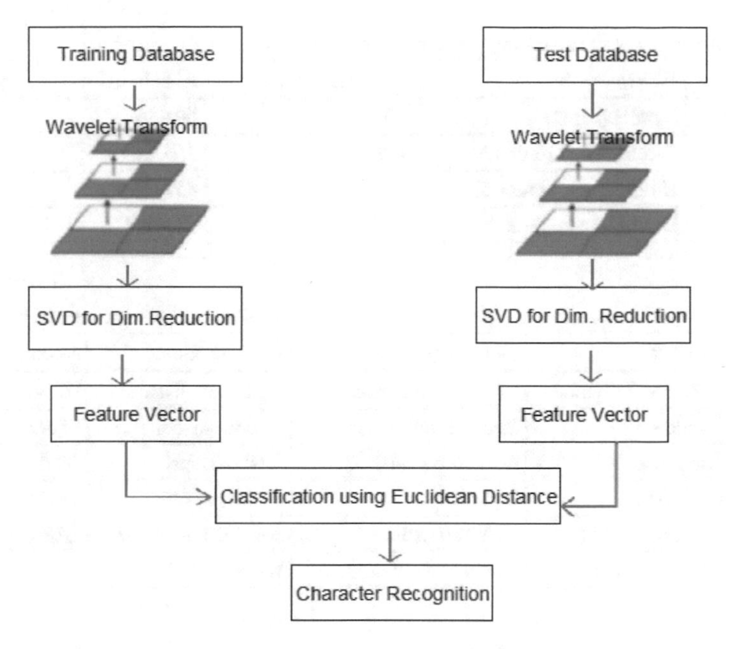

Fig. 2 Proposed WT-SVD model

respectively. SVD is applied to extracted features for dimensionality reduction. The matching and recognition process is performed using the Euclidean Distance method.

4 Experimental Results

The experimental data set consists of 4600 MODI characters (3220 train samples and 1380 test samples). Handwritten characters written by different persons were used for the study. After pre-processing, the grayscale image was size-normalized to fit in a 60 × 60 pixel box. Feature extraction is performed using a single-level discrete wavelet transform. Wavelet families include Haar, Daubechies, Symtlets, Biorthogonal, etc. In our experiment, we have used Biorthogonal and Symlet wavelets. It is reported that the Biorthogonal Wavelet family gives better results for character recognition, and therefore this wavelet is chosen for our experiment.

From the input image (60 × 60), the feature vector is extracted using the WT function 'dwt.' SVD is applied to the extracted features and later classification is performed on it. In the classification process, test data samples are matched against the train samples using the Euclidian distance classifier, and the accuracy is computed. The accuracy achieved using each of the methods (as well as the absolute time) is recorded. The experiment is conducted in a MATLAB environment.

The consolidated report of the experiment is illustrated in Table 1.

Table 1 Accuracy of character recognition using different wavelets

Sr. no	Wavelet filter	Accuracy (%)	Absolute time (in Sec.)
1	Symlet (sym2)	99.49	603.56
2	Biorthogonal (bior 1.3)	99.56	759.45
3	Biorthogonal (bior1.5)	99.42	859.03
4	Biorthogonal (bior3.5)	98.62	759.91

Table 2 Comparison of results, with other methods used in MODI script character recognition

Sr. no	Author	Feature extraction	Classification	Accuracy (%)
1	WT-SVD Method	Wavelet transform	Euclidian distance	99.56
2	Kulkarni [7]	Zernike moments	Euclidean distance	94.92
3	Ramteke, [4]	Structure similarity	KNN, BPNN	93.5
4	Gharde [6]	Hybrid	SVM classifier	89.72
5	Sadanand [12]	Hu's moments	Euclidian distance	82.61
6	Chandure [5]	Chain code histogram	KNN, and SVM	60 & 65

Accuracy is calculated as the number of all correct predictions divided by the total number of data in the test dataset. The highest accuracy achieved is 99.56% using Biorthogonal Wavelet (Bior 1.3). The results of our study are compared with the reported work in MODI character recognition and the outcome of the comparative study is shown in Table 2.

The proposed WT-SVD method has achieved the highest accuracy among all the reported work of MODI character recognition. The WT is an efficient method for feature extraction. By combining it with SVD, we are able to reduce the computational complexity of the system as well. SVD is an effective method for dimensionality reduction and that results in reducing the computational complexity.

5 Conclusion

For MODI script recognition, we have proposed a novel method called WT-SVD. Compared to the reported highest accuracy of MODI Script character recognition, which is 94.92%, the proposed model achieved a much better accuracy of 99.56%. The method is implemented using a hybrid combination of two techniques such as WT and SVD. Dimensionality reduction of the extracted features is achieved using SVD and that resulted in reducing the computational complexity. Euclidean distance classifier is used for classification. It is observed that there is a need for more research work in MODI script and as future work we will be concentrating on the segmentation of unconstrained MODI script documents.

References

1. Pal U, Chaudhuri BB (2004) Indian script character recognition: a survey. Patt Rec 37(9):1887–1899
2. Joseph S, George J (2019) Feature extraction and classification techniques of MODI script character recognition. Pertanika J Sci Technol 27(4):1649–1669
3. Joseph S, George JP, Gaikwad S (2020) Character recognition of MODI script using distance classifier algorithms. Fong S, Dey N, Joshi A. ICT Anal Appl Lect Notes Networks Syst, vol 93, Springer, Singapore
4. Ramteke AS, Katkar GS (2013) Recognition of off-line Modi script: A structure similarity approach. Proce Int J ICT Manag 1(1):12–15
5. Chandure SL, Inamdar V (2017) Performance analysis of handwritten Devnagari and MODI character recognition system, Int Conf Comput Anal Secur Trends, CAST 2016, pp 513–516
6. Gharde SS, Ramteke RJ (2016) Recognition of characters in Indian MODI script. Proc - Int Conf Glob Trends Signal Process Inf Comput Commun. ICGTSPICC 2016, pp 236–240
7. Sadanand K, Borde PL, Ramesh M, Pravin Y (2015) Offline MODI character recognition using complex moments. Proce Comput Sci 58:516–523
8. Solley T (2018) A study of representation learning for handwritten numeral recognition of multilingual data set. Springer, Lect Notes Networks Syst, vol 10, pp 475–482
9. Chacko BP, Krishnan VV, Raju G, Anto BP (2012) Handwritten character recognition using wavelet energy and extreme learning machine. Int J Mach Learn Cybern 3(2):149–161
10. Shelke S (2016) Handwritten character recognition using wavelet transform for feature extraction, no. March 2014, pp 3–7
11. Achuthan A, Rajeswari M, Ramachandram D, Aziz ME, Shuaib IL (2010) Wavelet energy-guided level set-based active contour: a segmentation method to segment highly similar regions. Comput Biol Med 40(7):608–620
12. Sadanand K, Borde P, Ramesh M, Pravin Y (2015) Impact of zoning on Zernike moments for handwritten MODI character recognition. In: IEEE Int Conf Comput Commun Control IC4 2015

An Advanced Machine Learning Framework for Cybersecurity

Angel P. Joshy, K. Natarajan, and Praveen Naik

Abstract The world is turning out to be progressively digitalized raising security concerns and the urgent requirement for strong and propelled security innovations and procedures to battle the expanding complex nature of digital assaults. This paper examines how AI is being utilized in digital security in both resistance and offense exercises, remembering exchanges for digital attacks focused on AI models. Digital security is the assortment of approaches, systems, advancements, and procedures that work together to ensure the confidentiality, trustworthiness, and accessibility of processing assets, systems, programming projects, and information from attacks. Machine learning-based examination for cybersecurity is the following rising pattern in digital security, planned for mining security information to reveal progressed focused on digital threats and limiting the operational overheads of keeping up static relationship rules. In this paper, we are mainly focusing on the detection and diagnosis of various cyber threats based on machine learning.

Keywords Cybersecurity · Machine learning · Threat · GAN · Anomaly detection

1 Introduction

Security assaults are getting progressively pervasive as cyberattackers abuse framework vulnerabilities for monetary profit. The subsequent loss of income can have malicious consequences for governments and organizations the same. Machine learning-based recognition is the most well-known security discovery system being

A. P. Joshy (✉) · K. Natarajan · P. Naik
Department of Computer Science and Engineering, CHRIST (Deemed to Be University), Bangalore, India
e-mail: angel.p@mtech.christuniversity.in

K. Natarajan
e-mail: natarajan.k@christuniversity.in

P. Naik
e-mail: praveen.naik@christuniversity.in

D. S. Jat et al. (eds.), *Data Science and Security*, Lecture Notes in Networks and Systems 132, https://doi.org/10.1007/978-981-15-5309-7_25

used today. The objective of this paper is to build up a repeatable procedure to distinguish cyber threats that is quick, precise, complete, and versatile. This model uses a security investigation to supplement existing security controls to distinguish suspicious client movement happening progressively by applying AI calculations. Cyberattacks have significant impacts on our day by day lives, where they focus from critical infrastructures down to small home networks. Cyberattacks are becoming more complex which makes it harder to protect our arranged frameworks against them. To completely comprehend the security of arranged frameworks inside and out, one must consider not just the effects of individual vulnerabilities when they are exploited, but also the effects of digital threats misusing vulnerabilities in various blends. Thus, it is of vital significance to accomplish three security objectives, which are Confidentiality, Integrity, and Availability (otherwise called the CIA groups of three) [3] AI calculations can be sorted as Predictive (Supervised Learning) or Pattern Discovery (Unsupervised Learning). In predictive learning, there is constantly an objective variable, the estimation of which the AI model figures out how to anticipate utilizing diverse learning calculations. An assortment of machine learning calculations falls under administered picking up, including Linear and Logistic Regression, Decision Tree, and Support Vector Machine (SVM) [2]. Then again, in unsupervised learning, there is no forecast of an objective variable, rather, unsupervised calculations figure out how to discover fascinating affiliations or examples with regards to datasets.

2 Related Work

Most of the model uses the idea of a time schedule. A schedule opening speaks to a window in time which contains total component means that time period [5]. The time schedule is represented by TMS slides over a fixed window of time tm. The initial phase in the process, data collection, includes recognizing and extracting log records [1]. Information pre-processing is required to change the information into a configuration usable by machine learning calculations [4]. In the supervised learning step, the model is prepared and assessed utilizing an order procedure utilizing the marked dataset from the past advance. Grouping, a kind of learning is the procedure of collecting comparable information components into classes or bunches. Euclidean, Manhattan, Minkowski are basic similitude estimates utilized by grouping calculations [3]. There is a wide range of sorts of grouping methods, including yet not restricted to dividing, various leveled, thickness based, and matrix-based strategies [7]. Exception identification is a typical use of grouping. Exceptions are information components that are a long way from every other component and fall outside of any group. Sometimes, the exception may give more understanding into an issue than the typical things [6]. Utilizations of anomaly location incorporate card extortion discovery and observation of electronic trade for crimes. Grouping might be utilized in manual characterization when working with huge datasets which could be very tedious and inclined to human mistakes [8].

K-implies is a typical dividing calculation which ascertains the focal point of each group utilizing the mean estimation of the considerable number of articles in the group [9]. K-medoids is comparative, yet rather than utilizing the mean for the focal point of the group, it utilizes objects situated close to the focal point of the bunch. Dividing based strategies must be expanded when working with exceptionally huge datasets [3]. Chi-square is a typical measurable strategy used to recognize excess. There are other element assessment measures, for example, Information Gain, Gain proportion, and the Gini file [7]. Old style factual techniques which use relationship coefficients, for example, the T-test, F-test, and chi-square, are kinds of channel techniques used to survey variable freedom [2].

3 Proposed Methodology

Nowadays generative adversarial network (GAN) is predominant for its solid generative capacity. In its traditional model, a generator and a discriminator are prepared to produce brings about an antagonistic way where the generator attempts to produce tests that can fit the dispersion of genuine examples, and the discriminator attempts to recognize the produced tests from the genuine examples [5]. Since GAN can become familiar with the circulation of information, it can normally be utilized to gain proficiency with the conveyance of ordinary information, particularly where abnormalities are rare in the preparation set. In testing, we can locate the most comparable example with the testing test and through the characterized inconsistency score dependent on the power of the disparity between the testing and the discovered examples we can know how irregular the samples are [10]. In our model, we make use of this GAN structure. The generator G is the essential model we used on learning. The model effectively produces tests that are proposed to take after those that are in the first preparing appropriation. The generator work is a differentiable capacity that has parameters that can be got of utilizing inclination fair. As a profound deconvolutional neural system, yet on a basic level, it very well maybe some other model that has the differentiable property. The discriminator D is not generally vital once the generator is effectively ready to produce reasonable information. The essential job of the discriminator is to investigate an example to see whether the example looks genuine or not. Like the generator, the discriminator in any sort of differentiable capacity that has parameters we can pick up utilizing angle. The discriminator will likely yield an incentive almost 1 for pictures from the first preparing set and qualities near 0 for pictures from the generator. The main objectives of this model are:

1. Create security displaying and examination strategies to improve versatility and flexibility. The results of this objective are another versatile and security model dependent on the chain of importance created, and a proper meaning of its structures and functionalities.
2. Look at the versatility and flexibility of security displaying and examination for present-day organized frameworks. The results of accomplishing this objective

comprise of a similar examination of security models, considering the complexities and exhibitions of them in an enormous measured and dynamic organized framework. The examinations are made as far as stages in the lifecycle of security models.

3. Create proficient and powerful security appraisal strategies to upgrade the capacities of figuring countermeasures. The subsequent results of this objective are the improvement of new significance measure-based security appraisal strategies and calculations, and assessment of them considering distinctive assault situations brought about by digital assailants found either outside or within the organized framework.
4. Breakdown the consolidated impacts of obscure assaults. This objective will result in the consolidation of obscure assaults into the system of the progressive security model, growing new calculations to recognize critical hosts and vulnerabilities in the arranged framework to figure powerful alleviation methodologies, and assessment of these calculations under different assault situations.

4 Results and Discussion

Our model will have the capability to defend against attacks. At induction time, it finds a nearby yield to a given picture which does not contain the antagonistic changes. This yield is at that point encouraged to the classifier. Our proposed strategy can be utilized with any characterization model and does not adjust the classifier structure or preparing technique. It can likewise be utilized as a protection against any attack as it does not expect information on the procedure for creating the antagonistic models. We exactly show that GAN is reliably powerful against various assault techniques and enhances existing barrier systems.

Figure 1 represents a GAN structure. Gen(x) and z are in real space x and En(z) are in latent space. When the process starts, x and z are changed over into Gen(x) and

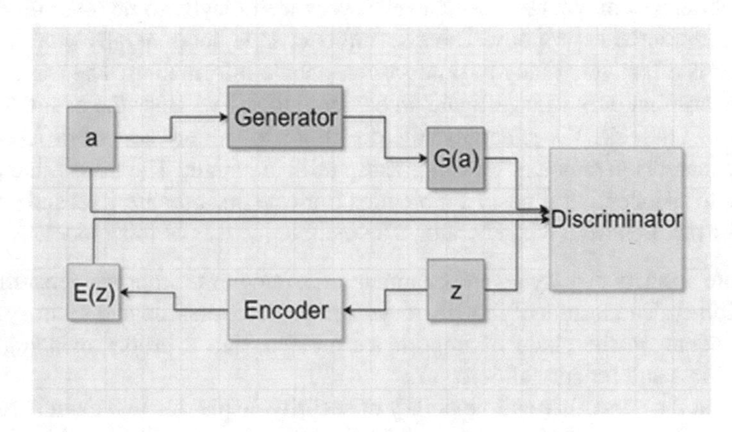

Fig. 1 GAN structure

En(z) separately, at that point the two sets (Gen(x), x) and (z, En(z)) will be given into discriminator, at last, the inclination update will be sent back to optimize the generator and the encoder, where Dis, Gen separately represent discriminator and generator, pZ is the distribution of normal samples z, pX is the distribution over the latent space. Jensen–Shannon (JS) divergence is designed to satisfy the symmetry required by distance [3]. In the objective above where cross-entropy loss function utilizing log is a good strategy, cross-entropy can be utilized to gauge the Shannon Entropy required to take out the vulnerability between two appropriations, so it ought to normally be the estimation of the difference between two conveyances. JS divergence can be seen as a kind of separation, in reality, JS disparity is the establishment most GANs take a shot at, and this uniqueness without a doubt helps a great deal in picture age, in which field the measurement development is not so extreme. For information with discrete highlights in digital interruption identification, for example, the 0–1 portrayal of rationale entryway and non-numeric worth that depends on One-Hot Encoding or Dummy Encoding, its measurement extension can be extreme so that there will be hardly any covering between genuine examples and produced tests. For instance, when an example from idle space is mapped into genuine example space with higher measurement through generator, every one of the assortments in higher measurement space is really be obliged by the example from lower measurement space. So, the component of the help of the higher measurement space is the component of the dormant space in certainty. Under the influence of measurement extension, two conveyances will have scarcely any covering definitely; however, discrete component will bother it due to One-Hot Encoding or Dummy Encoding. Also, at the point when two dispersions have scarcely any covering, JS uniqueness will unavoidably merge to a consistent, at that point prompts the occurrence of disappearing angle.

$$\min_{Gen.En} \max_{Dis} V(Dis,\ En, Gen)$$

$$\begin{aligned} V(Dis,\ En,\ Gen) &= En_{z\sim pZ}\big[En_{x\sim pEn(\cdot|Z)} \mid\mid Dis(Z, X) \mid\mid W\big] \\ &\quad + En_{x\sim pX}\left[En_{z\sim pGen(\cdot|X)}[1- \mid\mid Dis(z, x) \mid\mid w]\right] \end{aligned}$$

where Dis, En, Gen independently speak to discriminator, encoder, and generator, pZ is the circulation of ordinary examples z, pX is the dissemination over the idle space, and pEn(x|z), pGen(z|x) is the conveyance learned by encoder and generator individually. w speaks to the Wasserstein separation, it additionally encourages the segregation procedure contrasted and cross-entropy estimation and enhances the aging procedure to create progressively steady and increasingly premium outcomes. This method is to utilize a standard GAN, prepared uniquely on positive examples, to take in a mapping from the dormant space portrayal to the practical example and utilize this educated portrayal to outline, concealed, examples back to the latent space. Training a GAN on normal samples just, causes the generator to become familiar with the manifold of typical examples. Given that the generator figures out how to produce typical examples, when an irregular picture is encoded its remaking will be non-peculiar; consequently the distinction between the input and the reconstructed image will highlight in the result. The two stages of preparing and distinguishing irregularities help in detection of anomalies. A commonality with all accessible log source types is essential for the reasons for identifying digital assaults [10]. Talking security experts to recognize a rundown of accessible source types is the first step in quite a while assortment. The accessible sources normally vary among associations relying upon their system design. Be that as it may, conceivable source types may incorporate email use action, firewall information, remote passage information, program action [4]. Web application log records are additionally prime contenders for utilization. Coordinating these sources into a solitary storehouse enables us to manufacture a far-reaching image of client movement over different frameworks. Such a storehouse will enable us to pick up understanding into client movement that might be generally missed if analyzing the sources separately. Seeing how any type of assault could show itself in every one of the source types is fundamental for distinguishing potential traits for highlight extraction. We have exhibited GAN-based model for utilizing machine learning examination to improve digital security checking alongside investigation on ideal calculations for cyber threat cases. ML-based investigation is an incredible apparatus. The model we structured is helpful in cyber intrusion identification task. In this model, it strikingly shortens the time cost of preparing and testing processes.

5 Conclusion and Future Work

In this, we explained our GAN-based machine learning technique that can be used for cyber intrusion detection which helps in detecting anomalies using unsupervised training. The proposed idea presented here demonstrates that GAN is a powerful method which can be used in cybersecurity and are relatively easier to train, more cost-effective. Developing better anomaly detection or prevention method for complex and high-dimensional data remain a challenge. So, in that case, we can use this GAN technology where its strong generative ability will learn the distribution of normal status, and identify the abnormal status through the gap between it and the learned distribution.

References

1. Creswell A, White T, Dumoulin V, Arulkumaran K, Sengupta B, Bharath AA (2018) Generative adversarial networks: an overview. IEEE Sig Proc Magaz 35(1):53–65, January
2. Radford A, Metz L, Chintala S (2016) Unsupervised representation learning with deep convolutional generative adversarial networks. In: International Conference on Learning Representations, Workshop Track
3. Vasileios M, Bromander S (2017) Cyber threat intelligence model: an evaluation of taxonomies, sharing standards, and ontologies within cyber threat intelligence. In: Proceedings of the IEEE

4. Miyato T, Kataoka T, Koyama M, Yoshida Y (2018) Spectral normalization for generative adversarial networks. In: International Conference on Learning Representations
5. Donahue J, Krähenbühl P, Darrell T (2017) Adversarial feature learning. In: International Conference on Learning Representations
6. Zenati H, Foo CS, Lecouat B, Manek G, Chandrasekhar VR (2018) Efficient GAN-based anomaly detection
7. Joshy AP, Natarajan K, Pani AK, An innovative approach for risk identification and management in software projects. Int J Comput Sci Eng 7(2)
8. Isola P, Zhu J-Y, Zhou T, Efros AA (2016) Imageto-image translation with conditional adversarial networks
9. Jaikumar P, Gacic A, Andrews B, Dambier M (2011) Detection of anomalous events from unlabeled sensor data in smart building environments. In: 2011 IEEE International Conference on Acoustics, Speech and Signal Processing (ICASSP). IEEE
10. Chhetri SR, Lopez AB, Wan J, Al Faruque MA (2019) GAN-Sec: generative adversarial network modeling for the security analysis of cyber-physical production systems, 2019 Design, Automation & Test in Europe Conference & Exhibition (DATE), Florence, Italy, pp 770–775

Development of Kid-Friendly YouTube Access Model Using Deep Learning

Sanjana Reddy, Nikitha Srikanth, and G. S. Sharvani

Abstract YouTube is a video-sharing platform, used by everyone. It can be accessed by anyone, anywhere, anytime. It enables collaboration, connections among people and gives simple tools for sharing content among everyone. This abundance of expression, content, interactive tools, games and videos may hamper the brain development and safety of the younger audience. There is a lot of content on the internet. The control of this content is a definite challenge leading to legislative concerns. Many threats to security and privacy emerge. In this paper, we propose a simple and interesting way to handle situations where a child might accidentally open a video which includes content that they are not supposed to watch.

Keywords Youtube · Age detection · Sentiment analysis

1 Introduction

In a world where the internet is expanding at a tremendous rate, it brings with it a plethora of opportunities. But it also brings a lot of harm. The minds of young children are shaped by the things they see. The internet influences a lot of their ideas, choices and what they become. Inappropriate and obscene content is not only traumatizing for kids but can also implant wrong ideas into their minds [1, 2].

Our work tries to provide age detection to restrict access to Youtube videos. It is widely motivated by kids accidentally gaining access to explicit content on their parents' phones. There are a lot of attempts being made to detect harmful content. Sometimes, videos are appropriate for adults but not for kids. Videos of this kind

S. Reddy · N. Srikanth (✉) · G. S. Sharvani
Department of CSE, R.V. College of Engineering, Bengaluru, India
e-mail: nikithasrikanth.cs18@rvce.edu.in

S. Reddy
e-mail: sanjanasr.cs18@rvce.edu.in

G. S. Sharvani
e-mail: sharvanigs@rvce.edu.in

D. S. Jat et al. (eds.), *Data Science and Security*, Lecture Notes in Networks and Systems 132, https://doi.org/10.1007/978-981-15-5309-7_26

will not be detected by traditional models. A good example is videos depicting war or violence and this is highly disturbing for a 3-year old who accidentally clicks on it.

Often, the extremity of a video can be gauged by the kind of comments it gets. Advanced computing methods have brought new and better techniques to analyse and classify text into specific categories. Hence, we use these techniques to develop better content classification models that enable us to detect violence, hateful and inappropriate web content through the comments a video gets. If a video has inappropriate parts, it will be spoken about in the comments of that video. Hence, sentiment analysis is performed on these comments. The overall result is a negative sentiment for comments on these inappropriate videos [3, 4].

This enables us to provide an in-built check mechanism for every video that is accessed on YouTube and safe viewing for kids.

In this paper, the problems in the current system of Youtube and what current safety measures look like is briefly talked about, followed by related work in the field, a methodology of the model on which the paper is written and then a conclusion about where this model stands as a contribution to YouTube. The references for ideas used in writing the paper and developing the model are present at the end of this paper.

2 Problem of Current System

YouTube offers amazingly easy graphic user interfaces so even very young children are able to access the next video with ease. Studies show that not many parents are aware of the working of YouTube user interface for filtering inappropriate content. And also, not many parents check the video category of the videos. Many parents also admit that their young children might have accessed inappropriate content by mistake. Though many parents assume they can monitor their children sufficiently and also complete other jobs at the same time, there are situations where children accidentally watch videos which are not meant for their age [5].

YouTube has a ‘Safety Mode’ which allows automatic filtering of content which is inappropriate at the browser level. However, it is not totally accurate. With small children more attention is required. The parent should keep an eye on their child’s actions by being physically present there or in some kind of automated way. However, automatic techniques do not guarantee complete accuracy; some questionable and inappropriate content may be delivered accidently [5].

YouTube Kids is a standalone app that makes watching videos simpler and safer for kids below the age of 13. On this platform, parents can handpick which channels or playlists their child’s viewing can be restricted to. Often we find that parents in today’s world do not have the time or the patience to go through the tutorial that YouTube kids provides to set up these parental controls. It also requires a lot of customized settings to be toggled for parents to be able to achieve the level of restricted access they want to provide to their kids. More often than not, parents end up not having the time to set up these restrictions or even install a separate kids’ version of YouTube for that matter. Parents can also make use of the inbuilt timer that automatically shuts down the app after a certain limit of viewing is reached. Other limits that can be placed include disabling the 'Allow searching' option which will prevent kids from searching for videos on their own. The content level can also be set by parents ranging from 'younger' to 'older' audiences, if they do not wish to choose curated videos themselves. 'Pause watch history' can be turned on so that kids do not get recommended videos based on what they have watched before and they will not be able to choose from these videos.

In this paper, we describe the possibility of restricting access to a video by granting access to them only after detecting their age, which is an interesting and useful feature for improving the safety level of YouTube if accessed by children.

3 Related Work

Detection of age before providing access to videos is a great way to provide access control at the user level. There are many aspects which can be considered to detect age which can be incorporated in access control systems. A method for access control is proposed which is based on age estimation, where the relation between the human and the human auditory perception is considered. Response to auditory perception changes with age. This is a great factor that is used to check the age before giving access. Access is not granted if the person's age is not appropriate for the given content. Biometric approaches like these provide greater security [6].

Another factor which is proposed to be considered is the shape of the ear. An ear has a stable structure and remains the same over time. Its features satisfy the conditions of a good biometric trait. This system helps in controlling access to the contents of social media by detecting the age of the viewer by analysing the shape of the ear [7].

An alert system is proposed wherein it alerts the viewers about the nature of the content. One can enable a warning mode wherein if the kid accidently opens an inappropriate video a loud warning sound is produced which warns their parents about the fact that their kids are trying to access something that is not safe for their age. An automatic video removal system is effective wherein after video content analysis and interpretation of user comments the video can be declared as inappropriate and be removed immediately [3].

4 Methodology

In this paper, a modification on the current Youtube restriction system is proposed, wherein the age of the user is detected every time they access a video and based on the age it is decided whether the user should watch the video or not [8].

We thus provide an alternate system that incorporates a face unlock system for every video that is clicked on YouTube that is based on age detection.

Our model contains two parts:

1. Age detection
2. Sentiment analysis.

Once the age detection process is completed, the detected age is passed to the sentiment analysis model. The sentiment analysis model also calculates the percentage of positive and negative sentiments of the requested video. With a simple if–else mapping it is decided whether the given video with given Negative sentiment should be viewed by the user or not. If the viewer is decided to be underage for the requested video, the access for the given video is denied. If the viewer is allowed to watch it, the Youtube video automatically opens. Below is the algorithm for the entire model.

Step 1: User clicks on the video.
Step 2: Web camera opens and scans user's face.
Step 3: Age is detected.
Step 4: Video ID is extracted from URL and comments are extracted.
Step 5: Comments are preprocessed.
Step 6: Sentiment Analysis is performed on these preprocessed comments.
Step 7: Percentage positive sentiment is obtained as output.
Step 8: If detected age falls under age group mapped with the detected positive sentiment-allow access, Else-deny access.

4.1 Age Detection

The model will take a live video stream from the WebCam. This process involves the following stages: *Image capture, preprocessing and face detection, prediction.*

Image capture. First, the photo is taken from the webcam live by the cv2 module.

Preprocessing and face detection. The image is turned to greyscale. cv2 module's Cascade Classifier class is used to detect faces in the image. OpenCV provides many classifiers which already trained for facial features. Those XML files are available in a directory called haar cascades. After knowing the faces' coordinates, those faces are cropped before feeding to the neural network model.

Prediction. The images are passed to the model by calling the predict method. For the age prediction, the output of the model is a list of 101 values associated with age probabilities ranging from 0 to 100, and all the 101 values add up to 1. Softmax is used for this purpose. Softmax function's output is a vector that shows the probability distributions of all possible outcomes. So, we multiply each value with its related age and sum them resulting in the final predicted age. The age detection is performed for 10 s where the model detects an age for the image of the viewer captured every second [8, 11]. The average of these ages is calculated which is passed to the sentiment analysis model.

Softmax function as shown in Eq. 1:

$$\sigma i(z) := \exp(\lambda zi) \Big/ \sum (\exp(\lambda zj))\ 1 \leq i \leq n. \tag{1}$$

Trained model. Various techniques like batch normalization, dropout and data augmentation can be applied to convolutional neural networks. More accurate systems can be built by simply stacking more and more layers of convolution-batch normalization-relu layers. To some point, accuracy would improve, but beyond about 25 layers, accuracy drops. Accuracy diminishes over many layers due to vanishing gradients and with more layers gradients get small. Hence residual connections are used. It is a simple term to describe connecting the output of previous layers to the output of new layers as seen in Eq. 2. This is shown in Fig. 1 [11].

Consider a layer is $f(x)$. In a standard network, $y = f(x)$ and in a residual network,

$$y = f(x) + x. \tag{2}$$

The feature extraction part of the neural network uses the wide residual Network architecture. It leverages the power of convolutional neural networks to learn the features of the face. Additionally, there is a decrease in depth and increase in the width of residual networks [9].

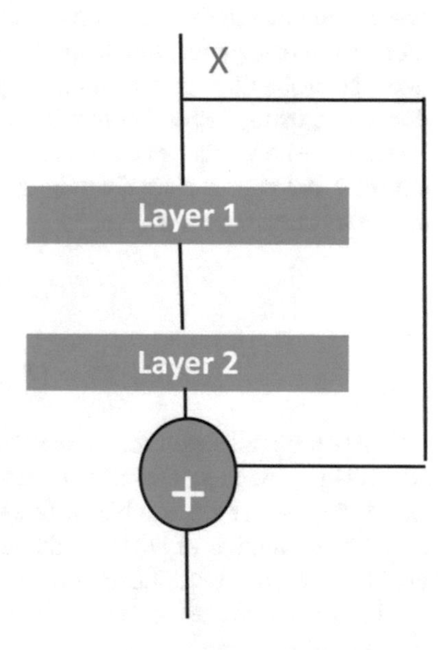

Fig. 1 Residual networks

4.2 Sentiment Analysis

One of the ways we can identify whether a video has explicit or inappropriate content, without having to process every frame of the video, can be through the comments. More likely than not, an explicit video is likely to have explicit comments. Along the same line of reasoning, we can conclude that a completely harmless video will have comments that are completely on the positive side of the spectrum.

We can thus perform text classification algorithms on these comments along the lines of sentiment analysis. The output will be the probability that the comment set leans towards being positive or negative. If the output leans towards a positive sentiment, we can conclude that the video is completely harmless. Similarly, for negative sentiment, we deny access. For intermediate probabilities, we map them with age groups that can view the videos [10].

Training. We perform bigram collocations that pair up words that frequently co-occur in the training data set. These bigrams are then given the tag of being positive or negative, depending on which text file they belong to. We employ NLTK Bigram Assoc Measures and Bigram Collocation Finder and other text preprocessing NLTK modules for this purpose to make it ready for probabilistic analysis. These words are then fed into the Naive Bayes classifier where the model learns how to classify these bigrams using probabilistic measures [10, 12].

Naive Bayes classification provides a way to probabilistically determine if an observation belongs to a certain category. Given two documents, one positive document and one negative, the model learns to classify data into these classes. It calculates the probability that the data are positive, given that it is present in the positive document, using Bayes theorem. Similarly, it does the same for negative bigrams.

Let there be '*p*' classes $C1, C2, \ldots, Cp$. According to this Naive Bayesian classifier, a tuple T belongs to class Cx only when it has a higher conditional probability than any other class Cy, $\boldsymbol{x \neq y}$.

$$P(Cx|T) > P(Cy|T). \tag{3}$$

$$P(Cx|T) = (P(T|Cx * P(Cx)))/P(T). \tag{4}$$

These probabilities are calculated by getting word counts. They are then multiplied according to the Bayes theorem to get the predicted classification as shown in Eqs. 3 and 4. So, we get the number of times a given bigram occurs in the text and divide by the total number of words in the text and add certain constants for smoothing to avoid division or multiplication by zero.

It is particularly useful for sentiment analysis because it doesn't use all the words in the training data as features, but just those words that help express any particular property of input text that we provide. In sentiment analysis, the property of input text that we wish to probabilistically extract is how positive or negative a word is.

It makes a bag of words assumption, which essentially means that we do not care about in what context an offensive word appears in the comments. If it appears, we will classify it as negative, irrespective of its position. So, the words appear as if they are unordered, in a bag.

It also makes a conditional independence assumption that the words are independent of each other, given the class they are a part of, i.e. positive and negative. These assumptions do not affect the accuracy as much but do make the classification faster. We import Python NLTK module with its Naive Bayes classifier for this purpose.

Testing. This trained model can now make predictions on previously unseen data, i.e. comments of any video. There are two ways we can get comments from a video ideally Web scraping and YouTube API itself. We have decided to use YouTube API that requests for comments using video id that we can extract from the URL of the video. We make requests by specifying an API key which is available in the Developer Console's API Access pane for the project. For example, consider a video URL as shown below.

https://www.youtube.com/watch?v=1qw23e4r5tt.

Here, the portion of the URL that says *v=1qw23e4r5tt* represents the video id which help fetch comments of this video. These comments are then given as input after preprocessing is done on the text using Python NLTK. We tokenize the comments to remove stop words, contractions, punctuation. This makes our dataset ready for applying our analysis model on it. A limitation of extracting comments this way is that one API key is given only a limited number of requests [10].

5 Conclusion

It might be unsafe to let children remain even a few minutes alone in front of YouTube without parental guidance since they can easily navigate the playlist and accidently access inappropriate content. Instead of totally relying on automated ways or the presence of a parent to make sure kids are not accidently accessing inappropriate content, this proposed system is reliable and safe. Once it is detected that the content of the video is not appropriate for a kid who is trying to access it, the access to the video is instantly restricted. This is a completely new approach for access control systems since the decision to provide access to the video is done at that instant without any need of previous settings. This system is fast and easy to set up as well. There is no extra effort like setting up of safety mode, installation of Youtube kids or disabling search options to ensure a safe viewing experience for the younger audience. Incorporating this model in Youtube definitely makes it kid-friendly.

References

1. Singh R, Kaushal R, Buduru AB, Kumaraguru P (2019) Fine grained approach for children unsafe video representation and detection
2. Papadamou K, Papasavva A, Zannettou S, Blackburny J, Kourtellisz N, Leon-tiadisz I, Stringhini G, Sirivianos M (2019) Disturbed YouTube for kids: characterizing and detecting inappropriate videos targeting young children. arXiv:1901.07046v2
3. Buzzi M (2011) Children and YouTube: access to safe content: CH Italy
4. Kaushal R, Saha S, Bajaj P, Kumaraguru P (2016) KidsTube detection, characterization and analysis of child unsafe content and promoters on YouTube. arXiv:1608.05966v1
5. Siersdorfer S, Chelaru S, Nejdl W (2010) How useful are your comments? Analyzing and predicting YouTube comments and comment ratings. In: International world wide web conference committee (IW3C2), ACM. 978-1-60558-799
6. Ilyas M, Fournier R, Othmani A, Nait-Ali A (2020) Université Paris Est, LISSI, UPEC, France: BiometricAccessFilter: a web control access system based on human auditory perception for children protection
7. Alghieth M, Alhuthail J, Aldhubiay K, Alshowaye R (2019) Information technology, Qassim University, Qassim, Saudi Arabia: smart age detection for social media using deep learning techniques via ear shape. (IJACSA) Int J Adv Comput Sci Appl 10(11)
8. Tander B, Özmen A, Başkan M (2012) Detection and classification of viewer age range smart signs at tv broadcast. Sig Image Process: Int J (SIPIJ)
9. Zagoruyko S, Komodakis N (2017) Wide residual networks. Université Paris-Est, École des Ponts ParisTech, France. arXiv:1605.07146
10. Parabhoi L, Saha P (2018) Sentiment analysis of YouTube comments on Koha open source software videos. Int J Libr Inf Stud 2231–4911
11. https://github.com/Tony607/Keras_age_gender.git(Easy Real time gender age prediction from webcam video with Keras)
12. https://github.com/sachin-bisht/YouTube-Sentiment-Analysis.git:Scrape all the YouTube comments using api (scraping YouTube comments and identifying the sentiment of comments)

Toxic Text Classification

Sreyan Ghosh, Sonal Kumar, Samden Lepcha, and Suraj S. Jain

Abstract The users of the Internet increase every moment with increasing population and accessibility of the Internet. With the increase in the number of users of the Internet, the number of controversies, arguments and abuses of all kinds increases. It becomes necessary for social media and other sites to identify toxic content amongst a large number of content being posted by the users of the sites every second. The traditional algorithms that depend on users reporting toxic content for it to be deleted and necessary actions to be taken against the users posting the content would take a long time, within which it would have gained media attention and would have lead to huge fights over the content. Thus, it becomes important for the content to be evaluated for toxicity at the time it is posted in order to stop it from being posted. Therefore, we have designed and trained a deep learning model that can be read through the textual content given through it and determine if it is toxic or not.

Keywords Classification · Deep learning · Social media · Internet content · Content moderation

1 Introduction

With the increasing use of social media portals over time, users of these portals, such as news portals, blogs, Question and Answer forums and other websites like

S. Ghosh · S. Kumar · S. Lepcha (✉) · S. S. Jain
Department of Computer Science and Engineering, Christ (Deemed to be University), Bangalore, India
e-mail: sam.lepcha@outlook.com

S. Ghosh
e-mail: gsreyan@gmail.com

S. Kumar
e-mail: skbrahee@gmail.com

S. S. Jain
e-mail: surajsjain@hotmail.com

D. S. Jat et al. (eds.), *Data Science and Security*, Lecture Notes in Networks and Systems 132, https://doi.org/10.1007/978-981-15-5309-7_27

Instagram and Facebook, often fall prey to cyberbullying and end up inviting content that is nasty, toxic, obscene, sexually implicit or threatening. An online survey carried out by the Pew Research Centre in 2017 states that 4 in 10 Americans have personally experienced online harassment. Strikingly, 1 in 5 Americans has witnessed severe forms of online harassment like physical threats, stalking and sexual harassment. The study which was conducted amongst over 1000 adults, which aims to understand the country's exposure to online harassment reported that eight out of 10 people in India have experienced some form of online harassment, with most common forms being abuse and insults.

Keeping online conversations effective and positive should be an important responsibility for all of the social network platform providers. Social media companies (e.g. Twitter, Facebook, etc.) have come under pressure to address this issue, and it has been estimated that hundreds of millions of euros are invested every year. Automatic classification of toxic comments, such as hate speech, threats and insults, can help in keeping discussions fruitful. In addition, the European Commission has taken measures such as the Code of Conduct on countering online hate speech. The INHOPE system which has been funded by the European Commission has reported removal efficiency of illegal content of up to 91% within 72 h where 1 out of 3 content items were removed within 24 h.

The major challenge in solving a problem of this kind is a class imbalance. Since comments of this type are sparse in nature, primitive machine learning models often fail to generalize well which results in overfitting and thus low performance over the unseen data. Thus, in this paper, we propose a novel deep learning model architecture consisting of state-of-the-art capsule network and bidirectional Long Short-Term Memory (LSTM) and Gated Recurrent Unit (GRU) layers. Additionally, we combine the outputs of all these layers to an auxiliary input which consists of sentence meta-features such as length of the sentence, number of uppercase letters, etc. This auxiliary input leads to significant improvement in performance over other networks by a factor of 4% and has not been proposed earlier for sentiment or toxicity classification.

2 Different Approaches

2.1 Primitive Machine Learning Algorithms

Most existing methods frame the problem as a supervised classification problem. This includes manual feature engineering and then running classical machine learning algorithms such as Support Vector Machine (SVM), Naive Bayes and Logistic Regression to the processed data [1–6]. A very common method used to process the data is using bag of words to vectorize the document and then applying Term Frequency–Inverse Document Frequency (TF-IDF) for scoring the words [1, 4, 6, 7].

2.2 *Naive Bayes*

In [8], the author has used the Naive Bayes classifier to train and predict hate comments from racist tweets on Twitter. They achieved an average accuracy of 76% by using a 10-fold cross-validation method. In [1], the authors used bag-of-words representation of the text followed by experimentation with various kernel functions of SVM to achieve 92.78% precision and 90% recall on the PRINCIP project database. In [9], the author used machine learning techniques to classify racist comments from Twitter.

2.3 *Deep-Learning-Based Methods*

The most common type of neural networks used for text classification is convolutional neural networks and long short-term memory networks. In the context of hate speech classification, intuitively, CNN extracts word or character combinations [10] and LSTM learns word or character dependencies in tweets. In [11], the authors have investigated three neural network architectures including CNN and LSTM networks with Glove and FastText embeddings for the detection of hate speech in tweets. The highest $F1$ score they could achieve was 0.93 with an ensemble of LSTM and Gradient Boosted Decision Trees (GBDT) models on a Twitter dataset. In practice, CNN + LSTM structures have outperformed all other structures consisting of just CNN or LSTM networks in tasks such as gesture, activity recognition and named entity recognition [12–14].

2.4 *Ensemble Learning*

Burnap and Williams [2] studied the advantages of ensembles of different classifiers. They combined results from three feature-based classifiers. Further, the combination of results from logistic regression and a neural network has also been studied by researchers [6, 7]. Zimmerman et al. [15] investigated in ensembling models with different hyperparameters.

3 Data

The dataset was taken from Quora Insincere Classification competition on Kaggle. The training set has 13,06,122 samples and the test set has 56,370 samples. It is due to the imbalanced nature of the dataset that makes this problem more challenging.

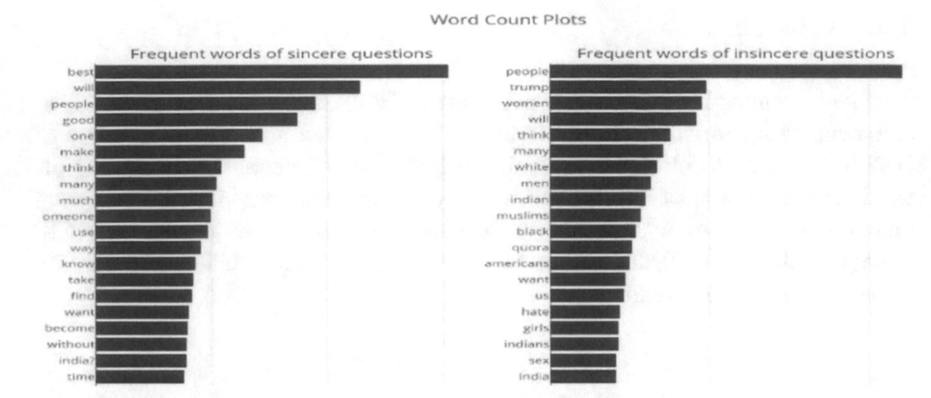

Fig. 1 Word count plot for the most frequently occurring words in sincere (non-toxic) and insincere (toxic) comments

The number of non-toxic comments in the dataset is 12,25,312 and the number of toxic comments is 80,810 (Fig. 1).

4 Cleaning the Data

For cleaning the data, we have taken the help of pre-trained GloVe embeddings. First, we build our vocabulary from all the texts in the corpus train and test combined and then compared this vocabulary with the words in the GloVe embedding. If words in the embedding did not match words in the embedding, it meant it requires cleaning or correction. For our experiment, we just cleaned the top 200 mismatched words with the highest frequency of occurrence in the text corpus. The goal was to match as many words as possible from the corpus to the embeddings for better classification.

4.1 Cleaning Contractions

A contraction is the shortened form of commonly spoken or written words, syllables or group of words generally formed by missing a letter or two from the original text. Examples of such contractions in the corpus are ‘aren’t’ (originally are not) and ‘can’t’ (originally cannot).

4.2 Cleaning Misspelled Words

The most commonly misspelled words in the corpus were ‘favorite’ (originally ‘favourite’) and ‘canceled’ (originally ‘cancelled’). A common analysis showed that people often misspelled words which consisted of multiple occurrences of the same adjacent letters like ‘ss’ or ‘ll’.

4.3 Cleaning Numbers and Punctuation

GloVe embeddings do not contain embeddings for any punctuations. Thus, all the punctuations were cleaned except ‘?’ and ‘,’. Keeping these showed a significant improvement in the performance of our model. All numbers were also removed from the corpus.

4.4 Cleaning Special Characters

The corpus had words from other foreign languages such as Hindi, Sanskrit and French. All such special characters were removed. In some special cases, the word was replaced by its corresponding word in English if the meaning of the word was known.

5 Extra Features

As we have mentioned earlier, our model has an auxiliary input which contains features related to the comment which we are trying to classify. The features used for our model are listed below:

1. Number of unique words in a sentence.
2. The ratio of unique to the total number of words in a sentence.
3. The total length of the sentence (in terms of words and letters).
4. Total number of uppercase letters in a sentence.
5. The ratio of uppercase to the total number of letters in a sentence.
6. Number of most prominent monograms, bigrams and trigrams in the sentence (two features, one each for toxic and non-toxic words).

6 The Model Architecture

6.1 Bi-LSTM

LSTM, commonly known as long short-term memory networks learn long-term dependencies in the data. A bidirectional LSTM learns bidirectional long-term dependencies between each time step in time-series or sequential data. The learning algorithm is fed once from the beginning to the end of the layer and once from the end to the beginning of the layer. This way the LSTM learns faster and better.

6.2 Bi-GRU

GRU, more widely known as gated recurrent unit, is an alteration over LSTM which has only two gates, in particular, the reset gate and update gate. Bi-GRU like Bi-LSTM learns through both forward and backward propagations.

6.3 Capsule Network

A capsule layer helps to preserve hierarchical pose relationships between object parts for better classification and object detection. It builds relative relationships between objects and is represented numerically as a 4D pose matrix. A CNN cannot do this because it does not understand 3D spaces. One of the biggest benefits of capsule is that it achieves high performance with just a fraction of the data when compared to CNN (Fig. 2).

7 Formulas

The metric that was primarily used for the evaluation of the proposed model is $F1$ score also known as F-score or F-measure. It is a metric that has been widely used in natural language processing tasks such as named entity recognition. In statistics, it is a measure of the test's accuracy and takes into account both precision and recall to compute its final score. The equations are defined below:

$$\text{Precision} = \frac{TP}{TP + FP} \tag{1}$$

Precision is the proportion of number of correct positive outcomes in the dataset divided by the all-out number of positive outcomes returned by the model.

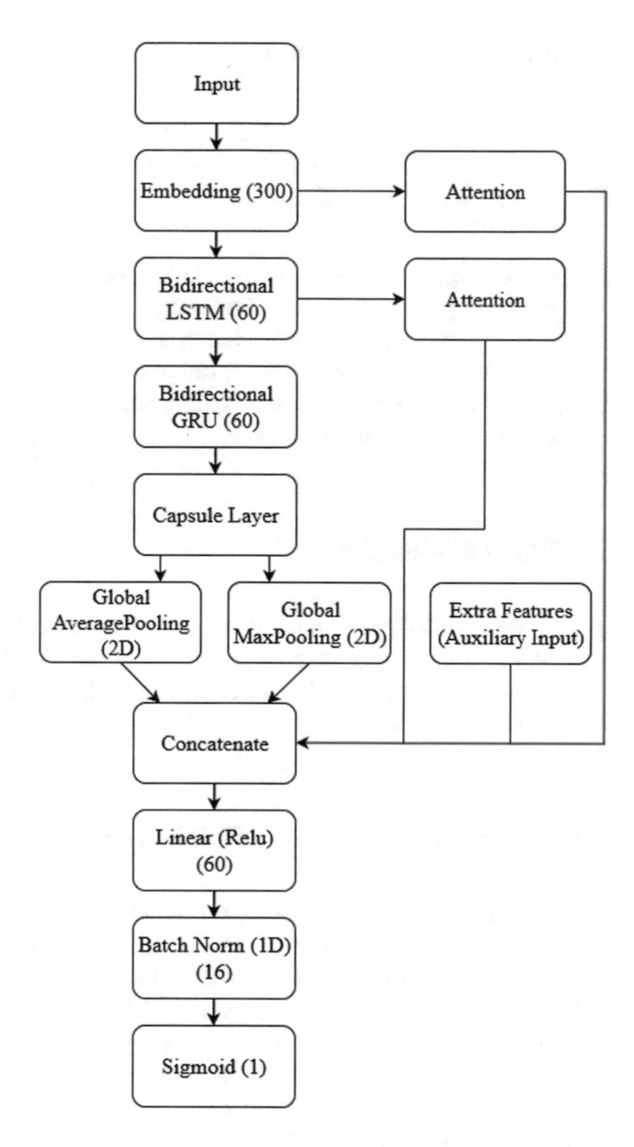

Fig. 2 Architecture overview

$$\text{Recall} = \frac{TP}{TP + FN} \tag{2}$$

Recall is the proportion of number of correct positive outcomes divided by the all-out number of samples that should have been distinguished as positive.

$$F - \text{score} = \frac{\text{Precision} \cdot \text{Recall}}{\text{Precision} + \text{Recall}} \tag{3}$$

Finally, the $F1$ score is commonly defined as the harmonic mean of precision and recall.

8 Cyclic Learning Rate

Cyclic learning rate is used to train a neural network with a learning rate that changes in a cyclic way for each batch of training instead of setting it to a constant value. The learning rate schedule varies between two preset bounds. In our experiments, the upper and lower bounds of the learning rate were set to 0.001 and 0.002, respectively. This helps to come out of a saddle point or local minima faster and leads to convergence at much lesser epochs.

9 Word Embedding

A 300-dimensional pre-trained word embedding was used in the model. This word embedding was an ensemble of paragram, glove and Wikinews pre-trained word embedding. An ensemble of the three resulted in more word coverage and performed better than any individual.

10 Experimentation and Final Results

We have used $F1$ score to assess the performance of our model on the test set. The model was trained on binary cross-entropy loss with cyclic learning rate. The model was trained for five epochs and for validation purposes fivefold K-fold cross-validation was used to validate the model. Thus, in each cycle, the model was trained on 4/5th of the data for five epochs and validated on 1/5th. After each training cycle, the model was made to run on the test set and the final scores were achieved after ensembling all these five predictions.

Final Training Loss = 49.4498.

Average validation loss (across folds for 5 epochs) = 51.8612.

Additionally, thresholds between 0.1 and 0.501 with increments of 0.01 were tested on the data with $F1$ score as the validation measure on the training set itself. The best threshold obtained was 0.2900 with an $F1$ score on the train with 0.6820.

The final $F1$ score obtained after ensembling the predictions across folds and setting the optimal threshold was 0.70235.

11 Conclusion

This work proposes an efficient sequence model architecture that can be used for many applications that would require determination of the toxicity of a statement.

In this way, time and work required for determining toxicity of any statement in applications of any kind would reduce to a great extent, allowing conflicts to be prevented from happening and maintaining a peaceful environment amongst the users on Internet platforms so that the platforms can serve their purpose without being accused of providing a way for toxic content to be on their platform for which they could be accused of.

References

1. Greevy E, Smeaton AF (2004) Classifying racist texts using a support vector machine. In: Proceedings of the 27th annual international conference on research and development in information retrieval—SIGIR ′04, 2004. https://doi.org/10.1145/1008992.1009074
2. Burnap P, Williams ML (2015) Cyber hate speech on Twitter: an application of machine classification and statistical modeling for policy and decision making. Policy Internet 7(2):223–242. https://doi.org/10.1002/poi3.85
3. Davidson T, Warmsley D, Macy M, Weber I (2017) Automated hate speech detection and the problem of offensive language. In: Eleventh international AAAI conference on web and social media, ′05, 2017
4. Djuric N, Zhou J, Morris R, Grbovic M, Radosavljevic V, Bhamidipati N (2015) hate speech detection with comment embeddings. In: Proceedings of the 24th international conference on world wide web—WWW ′15 Companion, 2015
5. Li Q, Shah S, Liu X, Nourbakhsh A (2017) Data sets: "Word embeddings learned from tweets and general data". In: Eleventh international AAAI conference on web and social media, ′05, 2017
6. Gao L, Huang R (2017) Detecting online hate speech using context aware models. In: RANLP 2017—recent advances in natural language processing meet deep learning. https://doi.org/10.26615/978-954-452-049-6_036
7. Risch J, Krestel R (2018) Aggression identification using deep learning and data augmentation. In: Proceedings of the first workshop on trolling, aggression and cyberbullying (TRAC—′08, 2018), pp. 150–158
8. Kwok I, Wang Y (2013) Locate the hate: detecting tweets against blacks. In: Twenty-seventh AAAI conference on artificial intelligence—′06, 2013. https://doi.org/10.5555/2891460.2891697
9. Waseem Z (2016) Are you a racist or am I seeing things? Annotator influence on hate speech detection on Twitter. In: Proceedings of the first workshop on NLP and computational social science. https://doi.org/10.18653/v1/w16-5618
10. Gambäck B, Sikdar UK (2017) Using convolutional neural networks to classify hate-speech. In: Proceedings of the first workshop on abusive language online. https://doi.org/10.18653/v1/w17-3013
11. Badjatiya P, Gupta S, Gupta M, Varma V (2017) Deep learning for hate speech detection in tweets. In: Proceedings of the 26th international conference on world wide web companion—WWW ′17 companion
12. Tsironi E, Barros P, Weber C, Wermter S (2017) An analysis of convolutional long short-term memory recurrent neural networks for gesture recognition. Neurocomputing 268:76–86. https://doi.org/10.1016/j.neucom.2016.12.088

13. Ordóñez F, Roggen D (2016) Deep convolutional and LSTM recurrent neural networks for multimodal wearable activity recognition. Sensors 16(1):115. https://doi.org/10.3390/s16010115
14. Chiu JPC, Nichols E (2016) Named entity recognition with bidirectional LSTM-CNNs. Trans Assoc Comput Linguist 4:357–370. https://doi.org/10.1162/tacl_a_00104
15. Zimmerman S, Kruschwitz U, Fox C (2018) Improving hate speech detection with deep learning ensembles. In: Proceedings of the eleventh international conference on language resources and evaluation—LREC, ′05, 2018

Tag Indicator: A New Predictive Tool for Stock Trading

Tessy Tom, Ginish Cheruparambil, and Antony Puthussery

Abstract In this paper, TAG—an indicator for stock market prediction in which volume-based means for measuring potential trading and investing decision-making is introduced. This task has been in correlation of the changes in the volume with the changes in the actual trade volume. Using this, a concise trading strategy is formulated. Hoping to outperform the market and analyze the results by back testing across intraday, price data for the last 1 year, 2019, is performed. It was discovered that about 48.9% of the time, the volume-based trading strategy outperformed and the returns from market are also healthy enough to support the claim. Statistical methods like linear regression, mean square error in prediction and stochastic gradient descent are applied. Furthermore, while the scope of the study was limited to a few stocks in Nifty in order to mitigate selection bias, nonetheless, we hypothesize that numerous other assets that similarly possess a predictable correlation to volumes based on daily high and low are likely to exist.

Keywords TAG indicator · Predictive stock market tool · Stochastic gradient descent method · Mean square error (MSE)

1 Introduction

The curiosity of many economists, statisticians, and teachers of finance has been in developing and testing models of stock price behavior. Early part of 1960s, Eugene Fama wrote a dissertation which later turned out to be the EMH—Efficient Market Hypothesis—considered as the fundamental law of the stock exchange [2]. Fama later wrote that efficiency can be defined as the price must represent all the available information in the market. This theory has brought out the fact that there could be three iterations which range from weak, semi-strong, and strong. Weak, semi-strong, and strong iterations follow these expectations in the market like: weak suggests all the available information are already accounted in the price of the asset, semi-strong

T. Tom · G. Cheruparambil · A. Puthussery (✉)
CHRIST (Deemed to be University), Bengaluru, India
e-mail: frantony@christuniversity.in

D. S. Jat et al. (eds.), *Data Science and Security*, Lecture Notes in Networks and Systems 132, https://doi.org/10.1007/978-981-15-5309-7_28

suggests that both technical and fundamental cannot help you to beat market, and strong suggests that all these are being priced in already [4]. Though EMH sounded great in the beginning it has got its own short comings.

Outperforming the market is a very difficult task to achieve. This outlook emerged from EMH which is a most commonly accepted theory. This theory postulates that the winning strategies are attributed to chance rather than the efficient skill of the trader. There is no reliable software yet for stock prediction. Considering the fluctuation of a stock index, forecasting is a difficult task, i.e., whether a stock index will rise or fall. A few scholars noted the fact, but their systems do not work yet well since different periods of a stock have different inherent laws [1, 3, 5]. Thus, one should not depend on a single model or a set of parameters to solve the problem. In this paper, a new indicator, namely, TAG, in which volume is used as a major means for identifying trading and investing decisions has been introduced. Here intraday trading strategy is followed.

2 TAG: An Indicator

With the resources at hand, a way to predict market for a shorter period of time is executed. It can also be elaborated for a larger time frame. For testing the strategy in an unbiased manner, all range of stocks for analysis in the beginning are considered. For verification, we took Axis Bank due to its accumulation and distribution nature. Any company which operated in a price range of 500–1000 suits well for the specific strategy. Apart from this, the success of the strategy was tested with few of other already existing strategies too. It gave us a better result, to refine it we started to calculate it from different points and time periods. An accumulation and distribution calculated from specific nodal points give us the edge to understand the whole aspect of doing this.

The TAG indicator can be used at different time frames as a trader wishes, even if a trader wants to keep the chart time frame constant, TAG indicator can be operated at different time frames within that constant chart time frame. In this indicator, we consider the volume which carriers forward to the next candle which can be a game changer. This specific volume is taken into account for the development of this indicator. The extraction of the volume which carry forward is done with the help of candlestick charts.

For refining the TAG indicator and to back test the trading strategy utilized by the indicator, trade volumes, closing price, and the buy and sell price provided by our system for Axis bank over a period of 1 year were collected. The data collected were sorted out on the basis of the formula which is created. The whole volume is split in terms of buy and sell volume. Then price got plotted as buy price, sell price, short price, and cover price.

The sample size used for back testing and refining consisted of 340 pairs of closing and opening price along with buy and sell price with specific volume at these periods. Once these datasets were arranged, then the portioning of the data subsets on the basis of buy, sell, short, and cover began. As the time framed used for testing the strategy was small as equivalent to 5 minutes, the data set for 1 year turned out to be a huge set. AmiBroker software for running the strategy and retrieving data is utilized. With this tool the fitness and relevancy of the strategy are analyzed.

3 Methodology

The data collected using the strategy is being used for the analysis. If the volume is higher according to the strategy, the entry price is taken as the close price of the asset at 5 minutes interval for the day trading. The parameter settings also need to be changed to the hourly for getting a better signal in the 5 minutes candle. The same way for sell also, when the signal emerges the same candle close is being chosen as the entry price. The data along with volume is collected from the trading platform. Also, for the validation of the newly introduced indicator by minimizing the error in the prediction, stochastic gradient descent approach is used in the formulated linear regression model. This implementation is done in the Python platform. In order to fit the regression line, we tune the parameters slope (m) and intercept (c). The optimality for parameters is achieved using mean squared error (MSE).

4 Mathematical Formulation

The mean squared error function for the problem is derived as follows:

$$\text{MSE} = \frac{1}{n}\sum_{i=1}^{n}(y_i - \hat{y}_i)^2$$

where $\hat{y}_i = mx_i + b$

And MSE with input parameters can be taken as

$$f(m, b) = \frac{1}{n}\sum_{i=1}^{n}(y_i - (mx_i + b))^2$$

$$\frac{\partial f}{dm} = \frac{1}{n}\sum_{i=1}^{n} -2x_i(y_i - (mx_i + b))^2$$

$$\frac{\partial f}{db} = \frac{1}{n}\sum_{i=1}^{n} -2(y_i - (mx_i + b))$$

By analyzing these derived functions, we can understand the way we should tune parameters and how much and update the parameters accordingly by iterating through our derived functions and gradually minimizing MSE. The additional parameter learning rate helps us in defining the step we take toward updating parameters with each iteration. Here, the learning rate is set as a very small quantity, 0.0001, to make sure that the model would not jump over a minimum point of MSE and converge nicely. With the optimized error we modified the tool and the TAG indicator is developed. Following analysis and predictions are made using this indicator.

5 Analysis

The table given below shows last 1-year trading diary. The risk or the exposure percentage of total capital and the returns in percentage and the ending capital as of December 31, 2019. The initial capital for trading used is one lakh and the ending capital is one lakh thirty-one thousand and six hundred.

	All trades	Long trades	Short trades
Initial capital	100000.00	100000.00	100000.00
Ending capital	131605.71	124568.87	107036.84
Net profit	31605.71	24568.87	7036.84
Net profit (%)	31.61	24.57	7.04
Exposure (%)	98.75	50.21	48.54
Net risk adjusted return (%)	32.01	48.93	14.50
Annual return (%)	125.92	91.92	22.36
Risk adjusted return (%)	127.51	183.07	46.07

Gradient descent process is performed in Python environment, taking the input feature X as a NumPy vector, a list of values and start the implementation by defining m and b with random values. Optimal values of parameters are obtained by iterating through the derived functions using a *for* loop. Here each step is controlled by a learning rate and progresses in every step and being tracked. To make the stochastic gradient descent faster, we use only one sample at a time to update parameters. Convergence to the optimum error is made faster by increasing the epochs and learning rate parameters (Figs. 1 and 2).

A drawdown is a dimension of decline from an asset's high value to its lowest value over a period of time. The drawdown is usually expressed as a percentage from high to low. It can be measured on any asset including individual stocks or sectors. However, it is most valuable as a measurement of portfolio risk. The above figure shows that the maximum and minimum losses you can incur by adopting this strategy are 8.3 and 0.8%. We have brought down this risk by applying the stop-loss or in other words reduced the risk which is much lesser.

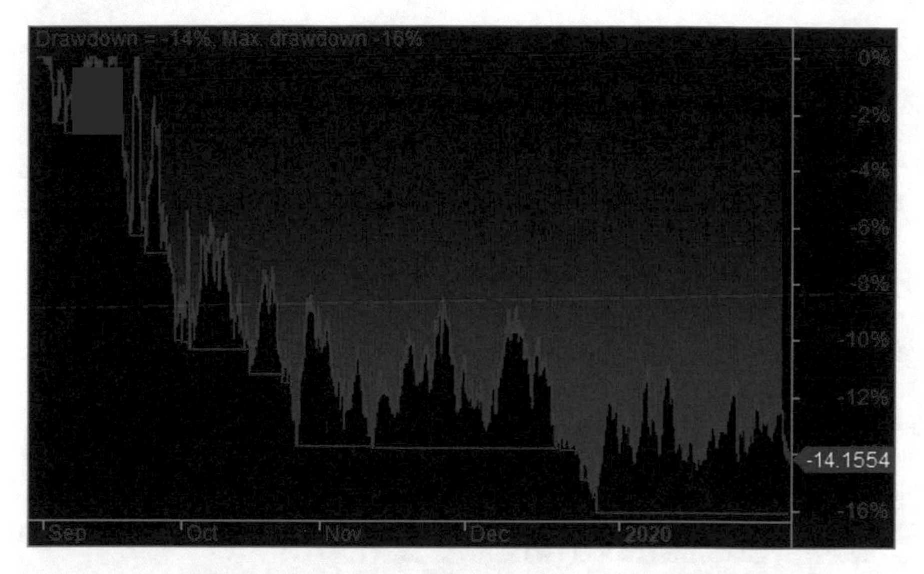

Fig. 1 Risk factor pattern before mathematical treatment

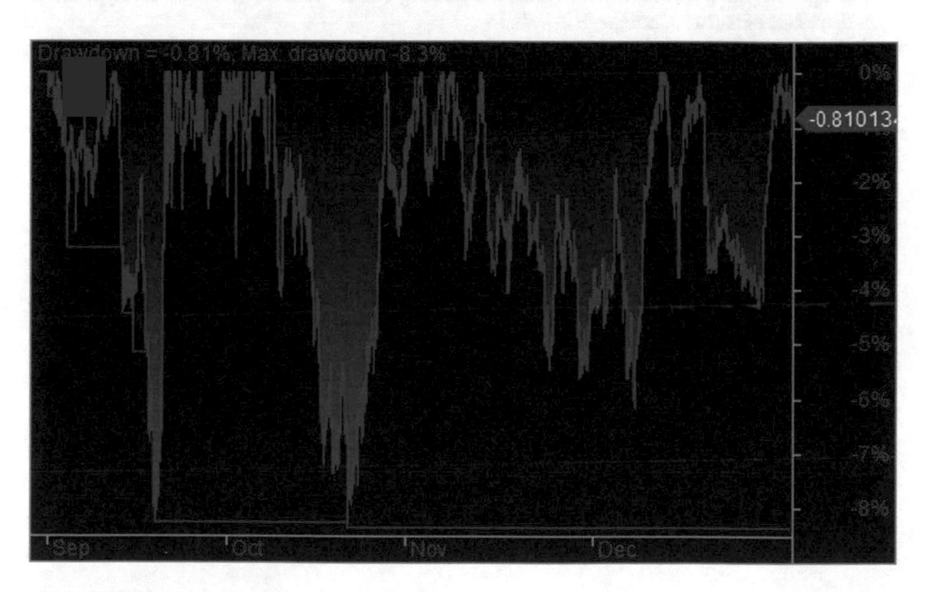

Fig. 2 Risk factor pattern after error optimization

6 Conclusion

TAG indicator is a volume-based indicator developed for stock market prediction, in which volume is used as an important means for measuring potential trading and investing decision-making. During the course of development of this new indicator, a rigorous mathematical treatment such as stochastic gradient descent method is applied to optimize the error factor in the prediction, which in turn reduced the risk factor. This tool gives a far better result in predicting future movements of assets in financial market especially stocks. We have used and proved this strategy which is well applicable with stocks having price between Rs. 1000 and Rs. 500. This can be seen as a drawback of the strategy. This has led us into the exploration of further scrutinizing the strategy for all stocks.

References

1. Abhyankar A, Ghosh D, Levin E, Limmack R (1997) Bid-ask spreads, trading volume and volatility: intra-day evidence from the London stock exchange. J Bus Finance Acc 24:343–362. https://doi.org/10.1111/1468-5957.00108
2. Eugene F (1965) The behavior of stock-market prices. J Bus, XXXVIII 34–105
3. Hussain SM (2011) The intraday behaviour of bid-ask spreads, trading volume and return volatility: evidence from DAX30. Int J Econ Finance 3(1)
4. Nath T (2015) Investing basics: what is the efficient market hypothesis, and what are its shortcomings? https://www.nasdaq.com/articles/investing-basics-what-efficient-market-hypothesis-and-what-are-its-shortcomings-2015-10-15
5. Sarkissian J (2016) Quantum theory of securities price formation in financial markets, arXiv: 1605.04948v2 [q-fin.TR]

A Predictive Model on Post-earthquake Infrastructure Damage

Aishwaria Paul and Jossy P. George

Abstract Disaster management initiatives are employed to mitigate the effects of catastrophic events such as earthquakes. However, post-disaster expenses raise concern for both the government and the insurance companies. The paper provides insights about the key factors that add to the building damage such as the structural and building usage properties. It also sheds light on the best model that can be adopted in terms of both accuracy and ethical principles such as transparency and accountability. From the performance perspective, random forest model has been suggested. From the perspective of models with ethical principles, the decision tree model has been highlighted. Thus, the paper fulfills to propose the best predictive model to accurately predict the building damage caused by earthquake for incorporation by the insurance companies or government agency to minimize the post-disaster expenses involved in such catastrophic event.

Keywords Predictive model · Earthquake · Random forest · Decision tree · Disaster management

1 Introduction

Earthquakes are considered as one of the most destructive natural hazards among the other catastrophic perils in the world. Over the past 10 years, the world has witnessed probably the biggest earthquakes by far in terms of the intensity and magnitude, with far-reaching death toll and property damage leading to rebuilding and associated costs of unprecedented scale. One of the primary outcomes of an earthquake is the shaking of the building or infrastructure and this results in the loosening of building components that leads to subsequent damage or collapse. A key realization is that

A. Paul · J. P. George (✉)
CHRIST (Deemed to be University), Bengaluru, India
e-mail: frjossy@christuniversity.in

A. Paul
e-mail: aishwaria.paul@mba.christuniversity.in

D. S. Jat et al. (eds.), *Data Science and Security*, Lecture Notes in Networks and Systems 132, https://doi.org/10.1007/978-981-15-5309-7_29

in the recent times insurance companies as well as government agencies have paid more attention to build a comprehensive view of risks and insurance covers related to the events of earthquake. In order to restore the earthquake-damaged community swiftly, a well-prepared damage assessment technique is essential. Presently, the research in the field of earthquake risk assessment is directed with the thought that earthquakes though cannot be prevented could be assessed through the knowledge gained through the past events to build an earthquake-resilient society for the future. It gives attention to taking actions that ensure the risk preparedness with respect to the structural aspects of a property and in doing so, the focus is oriented toward becoming more resilient to cost disruptions that result from such catastrophic events.

This study on post-earthquake building damage prediction models undertaken by the researcher can be attributed to the passion of the researcher toward developing solutions to solve real-time problems for the betterment of the society and the planet. Through this paper the researcher obtains a platform to pursue the passion and aspires to tackle the complex aspects of uncertainty and massive destruction caused by earthquakes and thereby propose a solution that improves the certainty and minimize the impact caused by the catastrophic destructions of earthquakes.

This article is presented in following sections. In Sect. 2, several studies on structural damages are discussed. Sections 3 and 4 present problem statements and describe a methodology applied in this work. Section 5 analyzes the experiments performed and discusses the results. Lastly, in Sect. 7, the conclusion is given by providing main aspects of this research work.

2 Literature Review

At an International level, over the years, most of the researches on earthquake and earthquake damage assessment and recovery have revolved around several theories that attempt to disentangle the complex, uncertain nature of the catastrophic events of earthquake and have witnessed an increasing number of risk assessment and mitigation alternatives being explored. The USGS Earthquake Hazards Program reports that in the past decade from 2000 to 2015, the earthquakes resulted in 801,629 death tolls across the world [1]. Based on the past data, the studies conducted by GeoNet estimate that in New Zealand a high impact earthquake with magnitudes 7 and above in the Richter scale occurs 1 per two and half years and low impact earthquakes of magnitudes 4–5 occur 1 per day [2]. On January 2010, an earthquake of magnitude 7.0 devastated Haiti and is reported as the fifth deadliest earthquake recorded worldwide according to Than et al. due to its chaotic and catastrophic impact of 222,000 deaths [3]. Again in the month of March in 2011, Japan witnessed a massive quake of magnitude 9, the most powerful known seismic event ever reported to hit the country as reported by Kenneth Pletcher, in his article published in Britannica and resulted in death tolls close to 18000 [4].

Several studies have been undertaken to study the structural damages of reinforced concrete (R/C) structures post the earthquake in various regions. Some of them includes the empirical vulnerability assessment of non-engineered R/C structures conducted by Ahmad S., Khan et al., analyzing and predicting the performance of R/C structures under seismic activities as outlined by the study performed by Ahmad et al. [5], Bilgin and Frangu [6] and Masi et al. [7]. Besides analyzing the structural properties of R/C buildings with respect to the damage intensity, several other literature also exist that study the correlation of damage intensity with other structures such as masonry buildings, stone and adobe structures and have been analyzed by the studies given in the literatures of Keshab Sharma [8] and Blondet [9], respectively. In order to identify the contributing as well as discriminating factors that affect the degree of damage in the buildings affected by the earthquake, several research works have been undertaken. According to Jia and Yan [10] and Huang et al. [11], the most contributing factors identified include the roof type, structure type as well as the frame structure. Again by the studies outlined in Goda et al. [12], the number of floors, age of the building, height, and area of the building where also identified as significantly contributing factors besides the ones mentioned earlier. Also, according to the studies of Jurukovski [13], foundation type as well as land surface conditions where also highlighted as yet another important parameter that accounts for the damage intensity. Although these factors were obtained from the analysis of a few researchers, the inferences gathered were found to binding and consistent with other research works such as those mentioned in Saputra et al. [14] and Anbazhagan et al. [15].

3 Problem Statement

The absence of appropriate disaster management initiatives to mitigate the effects of catastrophic events such as earthquakes increases the post-disaster expenses. This in turn raises concern for both the government and the insurance companies. Thus, based on the aspects of building properties, the project aims to build an accurate model to predict the building damage caused by earthquake to develop better preventive disaster management initiatives and thereby minimize the post-disaster expenses.

3.1 Scope of the Study

From the perspective of an insurance company or the government, the post-disaster expense is considered as a major area of concern under the risk assessment and management planning as they are the most affected stakeholders in such events. As a result, the assessment or anticipation of the damage level imposed post the disaster becomes extremely vital to approve/disapprove the insurance claims as well as damage recovery process. Thus, through this paper, the researcher aims to propose an AI-integrated system for the insurance companies and the government to estimate the damage level accurately and draw essential insights for insurance claim approval or damage recovery process, respectively. In case of insurance company, the study helps in developing an AI-integrated insurance claims system that uses an accurate predictive model to look over the traditional time-consuming process of insurance claim approval and focuses on a single index which is the damage grade to decide the severity of damage and thereby fasten the insurance claim process. Again, with respect to the government, the accurate damage level prediction by the integration of AI can create a faster and approximated view of the impact caused by the earthquake and thereby catalyze the damage recovery process.

3.2 Objectives of the Study

1. To identify the key factors that determine the level of damage caused during the earthquake.
2. To predict the damage level caused during an earthquake using multiple prediction models.
3. To identify the best model that predicts the damage level with highest accuracy.

4 Research Methodology

The CRISP-DM methodology has been adopted for the study. It also includes an exploratory analysis and discriminant analysis to explore the characteristics of features. Further, it also focuses on identification of the best model to predict the building damage caused by earthquakes. The various predictive models analyzed for the study include a logistic regression model, classification models as well as black box models.

4.1 Data Understanding

Although the primary objective of the survey was the identification of eligible beneficiaries for the government assistance for house reconstruction, it also collected several socio-economic information. Thus, in addition to the house reconstruction, the data collected through the survey now serves as a valuable resource for researchers, local governments for wide range of uses.

In the study, the dataset used has been collected from the Nepal Open Portal and mainly consists of information on the buildings' structure and their legal ownership. There are 136991 cases measured across 40 features. Each row in the dataset represents a specific building in the region that was hit by the earthquake. The dataset does not have any missing values and most of the data collected are binary or categorical in nature indicating a limitation of availability of amble scale or quantitative information.

4.2 Modeling

In this study, in order to predict the damage level caused by an earthquake given the structural and legal ownership aspects of building, six predictive models have been identified which are listed as multinomial logistic regression, decision tree with tuning, random forest with tuning, KNN classifier, SVM, and neural network.

The selection of these models is based on the nature of the dependent variable (damage_grade) which is a categorical variable with three categories. The above-mentioned models all prove to serve the purpose of classifying the buildings into the three categories and thereby generate efficient and accurate results. Another aspect that is addressed under this phase is the selection of appropriate train test split of the data for the purpose of model building as well as model evaluation. In order to achieve this, a basic classification model was run on three combinations of train test spilt, namely, 75–25, 70–30, and 65–35, respectively. Among the combinations, the one with the least error is selected as the standard train test split throughout the analysis.

From the Table 1, it was observed that when the train test split combination was 75–25 the error in prediction for both the train and test models is the least as compared to the results of other combinations irrespective of the method used. Hence, in this study, in order to build the models, the standard assumed for the train test split combination is 75–25, i.e., 75% of the data is used for training the model and 25% of the data is used for testing the model accuracy.

Table 1 Train test split combinations

Chaid	Data split					
	75–30		75–25		65–35	
	Train	Test	Train	Test	Train	Test
	70	30	75	25	65	35
Accuracy (%)	70.80	69.80	70.30	69.80	70.90	69.30
Error rate (%)	29.20	30.20	29.70	30.20	29.10	30.70
Difference in train and test error rate	**1.00%**		**0.50%**		**1.60%**	
Cart	Data split					
	75–30		75–25		65–35	
	Train	Test	Train	Test	Train	Test
	70	30	75	25	65	35
Accuracy (%)	66.40	66.20	66.50	66.30	66.30	66.50
Error rate (%)	33.60	33.80	33.50	33.70	33.70	33.50
Difference in train and test error rate	**0.20%**		**0.20%**		**−0.20%**	

5 Experimental Analysis

5.1 Bivariate Analysis

Under this analysis, the focus is on identifying the relationship between each variable in the dataset with the dependent variable as well as to explore the hidden relationship among the independent features. The analysis is performed by utilizing a correlation plot to understand the strength and the direction of correlation among the variables under investigation. The correlation plot for the variables has been developed using the Anaconda Spyder software and utilizes the Matplotlib package to plot the correlation plot and the parameter used to measure the correlation among the variables is Spearman's correlation.

From Fig. 1, the following observations are derived:

- The damage grade has negative correlation with almost all the independent features except with buildings with mud mortar stone structures, age and count families where there is a slightly positive correlation between the damage level and the variable.

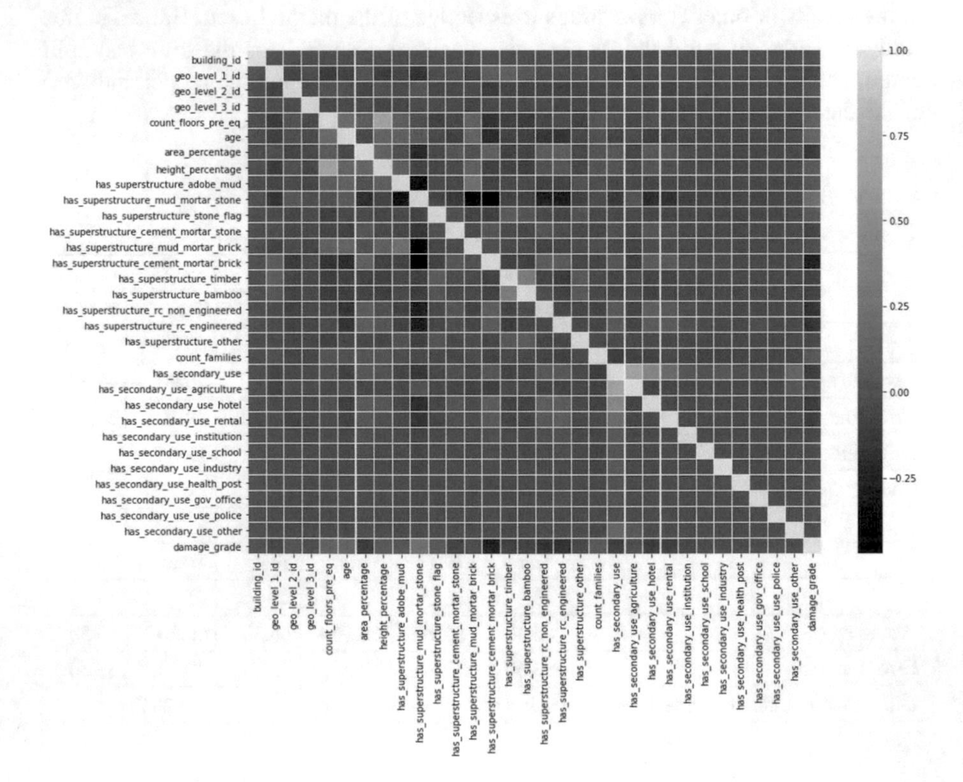

Fig. 1 Correlation plot

Table 2 Wilks' lambda

Test of function(s)	Wilks' lambda	Chi-square	df	Sig.
1 through 2	0.749	38373.163	76	0.000
2	0.975	3316.398	37	0.000

- The variables such as area percentage has structure cement mortar brick, has structure R/C non-engineered, has structured R/C engineered, has secondary use rental, has secondary use hotel, and has secondary use having a strong negative correlation with the damage grade. This indicates that as the area increases and as the structural properties of the buildings become stronger and also as the buildings have a secondary use especially as a rental or a hotel, i.e., better maintenance, the damage level caused by the earthquake reduces.

5.2 *Bivariate Analysis*

Model significance for discriminant analysis is shown in Table 2.

- Null Hypothesis (H0): The model is not significant.
- Alternate Hypothesis (H1): The model is significant.

Rejection Rule: p-value $<= \alpha$. Reject H0.
From the Wilks' lambda table,
the p-value $= 0.000$ when $\alpha = 5\%$ or 0.05.
That is, p-value $< \alpha$.
Therefore, reject H0 and accept H1.

Thus, at 5% level of significance the model is significant for the study.

5.3 *Structure Matrix*

From Fig. 2, it is observed that the following predictors have the highest discriminatory power to discriminate among the damage grade levels caused by an earthquake:

- has superstructure mud mortar stone,
- has superstructure cement mortar brick,
- has super structure R/C engineered,
- roof type and ground floor type,
- foundation type,
- has super structure R/C non-engineered,
- has super structure timber,
- has superstructure mud mortar brick,

Fig. 2 Structure matrix

Structure Matrix

	Function	
	1	2
has_superstructure_mud_mortar_stone	.650*	-.072
has_superstructure_cement_mortar_brick	-.528*	.219
has_superstructure_rc_engineered	-.450*	-.261
roof_type	-.422*	-.225
ground_floor_type	-.385*	.084
has_superstructure_rc_non_engineered	-.352*	-.008
area_percentage	-.258*	.108
count_floors_pre_eq	.253*	-.067
has_secondary_use_hotel	-.197*	.085
has_secondary_use_rental	-.191*	-.031
legal_ownership_status	.183*	-.041
has_superstructure_adobe_mud	.136*	.067
count_families	.104*	-.077
plan_configuration	-.093*	-.017
height_percentage	.093*	-.045
geo_level_2_id	.083*	-.079
age	.074*	.061
other_floor_type	.064*	-.046
has_superstructure_other	-.059*	.020
has_secondary_use_institution	-.057*	.018
has_secondary_use_school	-.024*	-.024
has_secondary_use_industry	-.021*	.004
has_secondary_use_gov_office	-.020*	.005
geo_level_1_id	-.231	-.421*
foundation_type	-.143	.369*
has_superstructure_timber	-.092	.305*
has_superstructure_mud_mortar_brick	.086	.279*
has_superstructure_bamboo	-.086	.277*
has_superstructure_cement_mortar_stone	-.084	.247*
position	.075	-.244*
has_secondary_use	-.141	.217*
has_secondary_use_agriculture	.055	.205*
has_superstructure_stone_flag	.108	-.184*
has_secondary_use_other	-.029	.076*
geo_level_3_id	.004	-.045*
land_surface_condition	-.002	-.024*
has_secondary_use_health_post	-.012	.015*
has_secondary_use_use_police	-.005	-.006*

Pooled within-groups correlations between discriminating variables and standardized canonical discriminant functions

Variables ordered by absolute size of correlation within function.

*. Largest absolute correlation between each variable and any discriminant function

- has superstructure bamboo,
- area percentage, count of floors,
- has superstructure cement mortar stone,
- position and has secondary use,
- has secondary use agriculture,
- has secondary use hotel,
- has secondary use rental,

- has superstructure stone flag,
- legal ownership status,
- has superstructure adobe mud,
- count families, and
- height percentage.

Thus, these predictors are utilized for building the predictive models among the pool of predictors used for investigation.

5.4 Predictive Modeling

a. Multinomial Logistic Regression
   It is observed from Fig. 3, that the accuracy of the model in correctly predicting the damage level is only 58.25% and the micro-averaged $f1$ score is also just 0.58 indicating a good precision and recall rate.
b. Decision Tree
   From the classification report as shown in Fig. 4, it is observed that as compared to the multinomial logistic regression, the accuracy of the decision tree model has improved significantly. For the decision tree, the accuracy obtained is 64.32% and the micro-averaged $F1$ score value is 0.64. This indicates that the model has good accuracy as well as good precision or recall as the value of $F1$ score is comparatively higher than the previous model. However, it may be caused due to overfitting and in such case the model is further trained under the random forest technique which is discussed in the next section.
c. Random Forest
   From Fig. 5, it is observed that the random forest model has indeed improved the model to a greater aspect in terms of the accuracy as well as the micro-averaged

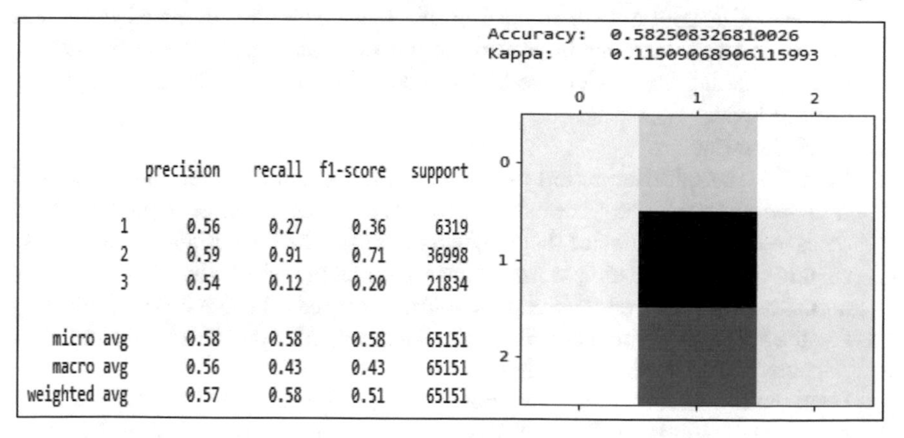

Accuracy: 0.582508326810026
Kappa: 0.11509068906115993

	precision	recall	f1-score	support
1	0.56	0.27	0.36	6319
2	0.59	0.91	0.71	36998
3	0.54	0.12	0.20	21834
micro avg	0.58	0.58	0.58	65151
macro avg	0.56	0.43	0.43	65151
weighted avg	0.57	0.58	0.51	65151

Fig. 3 Structure matrix

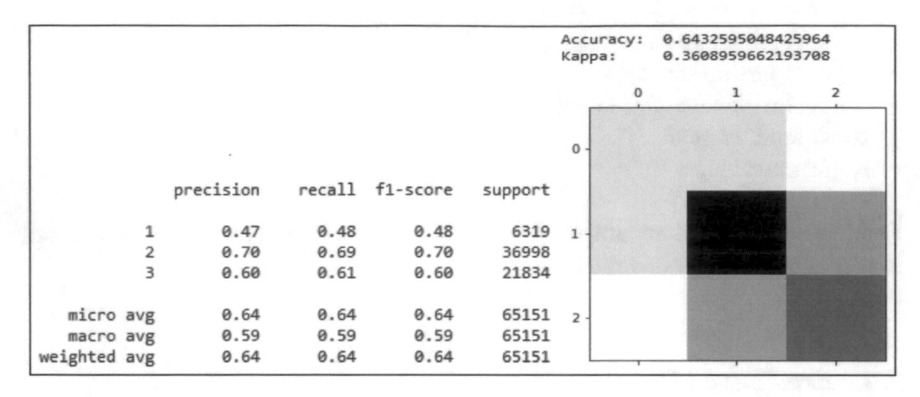

Accuracy: 0.6432595048425964
Kappa: 0.3608959662193708

	precision	recall	f1-score	support
1	0.47	0.48	0.48	6319
2	0.70	0.69	0.70	36998
3	0.60	0.61	0.60	21834
micro avg	0.64	0.64	0.64	65151
macro avg	0.59	0.59	0.59	65151
weighted avg	0.64	0.64	0.64	65151

Fig. 4 Classification report

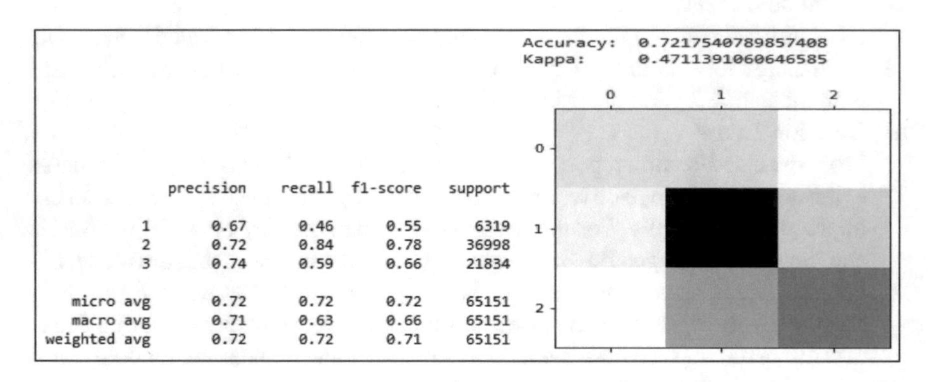

Accuracy: 0.7217540789857408
Kappa: 0.4711391060646585

	precision	recall	f1-score	support
1	0.67	0.46	0.55	6319
2	0.72	0.84	0.78	36998
3	0.74	0.59	0.66	21834
micro avg	0.72	0.72	0.72	65151
macro avg	0.71	0.63	0.66	65151
weighted avg	0.72	0.72	0.71	65151

Fig. 5 Classification report

$F1$ score. It is noted that the accuracy of the model is 72.17% and the micro-averaged $F1$ score is 0.72 indicating that the model has addressed the issues of decision trees and has improved the precision and recall rate of the model, thereby making the model suitable for adoption for the prediction of building damage level.

d. KNN Classifier
From the classification report as shown in Fig. 6, it is observed that as noticed from the literature study, the model confirms to be less accurate in predicting the damage level and indicates that the factors used in the prediction are complex to capture through such simple models based on feature similarity. The prediction accuracy obtained is 64.06% and the micro-averaged $F1$ score is 0.64 indicating that the accuracy of the model is low but has acceptable precision and recall rate.

e. Support Vector Machines (SVM)
From the classification report as shown in Fig. 7, it is observed that the accuracy of the prediction is 63.97% and the micro-averaged $F1$ score is 0.64 implying

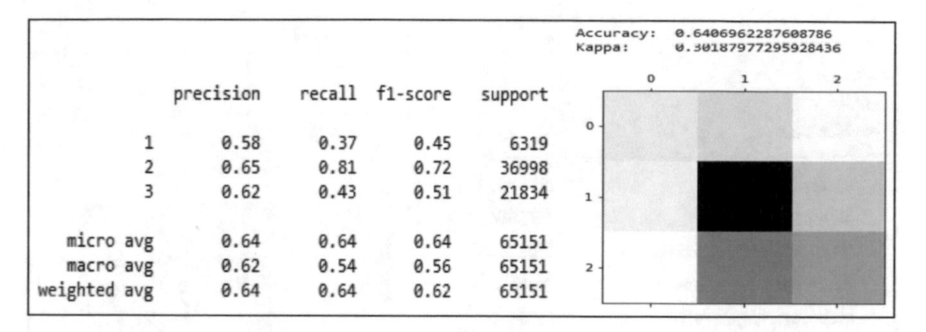

Accuracy: 0.6406962287608786
Kappa: 0.30187977295928436

	precision	recall	f1-score	support
1	0.58	0.37	0.45	6319
2	0.65	0.81	0.72	36998
3	0.62	0.43	0.51	21834
micro avg	0.64	0.64	0.64	65151
macro avg	0.62	0.54	0.56	65151
weighted avg	0.64	0.64	0.62	65151

Fig. 6 Classification report

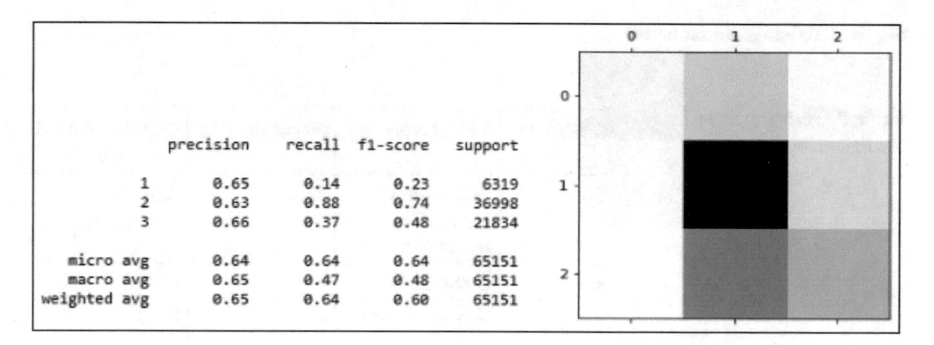

	precision	recall	f1-score	support
1	0.65	0.14	0.23	6319
2	0.63	0.88	0.74	36998
3	0.66	0.37	0.48	21834
micro avg	0.64	0.64	0.64	65151
macro avg	0.65	0.47	0.48	65151
weighted avg	0.65	0.64	0.60	65151

Fig. 7 Classification report

that the SVM model although a good classification model may not be suitable for the prediction of the building damage level.

f. Multiclass Neural Network

From the above report as shown in Fig. 8, it is observed that accuracy of the model is 63.7% and the micro-averaged precision and micro-averaged recall values indicate that that the model has comparatively high precision and recall rate.

6 Performance Analysis

From Table 3, it is clearly highlighted that the best model that can be utilized for the prediction of the building damage level caused by earthquake is the random forest model due to its high accuracy and high precision or recall rate.

However, when we consider the ethical aspects of AI technologies which are given prime importance in all the upcoming studies the selection of the random forest model for the prediction is debatable as random forest models are still considered as a black box model due to the issues of transparency as well as accountability.

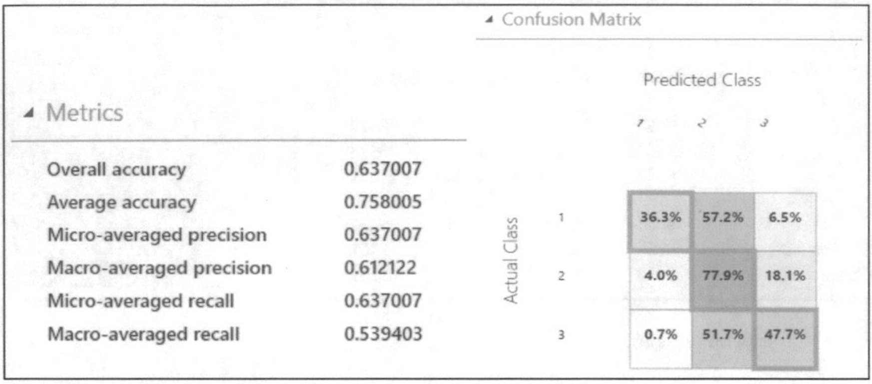

Metrics

Overall accuracy	0.637007
Average accuracy	0.758005
Micro-averaged precision	0.637007
Macro-averaged precision	0.612122
Micro-averaged recall	0.637007
Macro-averaged recall	0.539403

Confusion Matrix

Actual Class \ Predicted Class	1	2	3
1	36.3%	57.2%	6.5%
2	4.0%	77.9%	18.1%
3	0.7%	51.7%	47.7%

Fig. 8 Multiclass neural network

Table 3 Performance of predictive models

Model no:	Model type	Accuracy (%)	Micro-averaged $F1$ score
1	Multinomial logistic regression	58.25	0.58
2	Decision tree	64.32	0.64
3	Random forest	72.17	0.72
4	K-nearest neighbor classifier	64.06	0.64
5	Support vector machine (SVM)	63.97	0.64
6	Multiclass neural network	63.70	0.63

Thus, from the perspective of adopting models abiding the ethical principles, a better choice would be to adopt the decision tree model and work toward improving its prediction accuracy.

6.1 Findings

The findings obtained from the study for the respective objective are presented below. First and foremost, the objective was to identify the significant predictors that contribute toward the damage level to infrastructure at the event of an earthquake. This aspect of the study was effectively outlined via the discriminant analysis. Based on the analysis it was observed that

(a) The structural aspects such as the material used or not used for the construction like R/C engineered, R/C non-engineered, mud mortar stone, cement mortar brick, etc. play a critical role in damage level of infrastructure post-earthquake.
(b) Besides the material, the other structural aspects that contribute significantly include the roof type, foundation type, ground floor type, area and height percentage, and the number of floors.
(c) Another interesting aspect, besides the structure that contributes to the damage, is the usage of the infrastructure. It is observed that there is a correlation between the secondary usage nature of a building and the damage level, especially when the secondary usage was hotel, rental, or agriculture, the damage induced was minimum.

For the next two objective, to predict the damage level of infrastructure using multiple prediction models and thereby identifying the best predictive model, the combined results are utilized to arrive at the following important findings:

(a) The relationship between the predictors and the damage level is not entirely linear. This is observed from the inability of the multinomial logistic regression model to capture the results accurately irrespective of tuning.
(b) The relation of the predictors and the damage level is also slightly complex in nature to be captured by simple classification models such as KNN classifier. This is indicated by the accuracy value. However, the precision and the recall rates of the model are comparable.
(c) The results obtained by using SVM and multiclass neural network models are very close in terms of the accuracy levels. However, both the models are inefficient as compared to the hierarchical models in capturing the relation between the predictors and the target variable.
(d) Among all the models analyzed in the study, it is observed that the decision tree and the random forest model were the most successful at predicting the results accurately. This indicates that the hierarchical models are better at capturing the relation between the predictors and the damage level.

Under this analysis, the focus is on identifying the relationship between each variable in the dataset with the dependent variable as well as to explore the hidden relationship among the independent.

7 Conclusion

Prediction of infrastructure damage in devastating events of earthquake is a highly sensitive and crucial area of research that demands high level of accuracy for insightful decisions. Through the current study undertaken by the researcher, two models have surfaced that may be taken up by either the local government agencies or the insurance companies to be incorporated in the present systems to improve and fasten the decision-making process. The models identified are random forest model and the decision tree model. The random forest model has high accuracy levels, whereas the decision tree model has comparatively lower accuracy level but higher interpretability and accountability.

References

1. USGS Earthquake Hazards Program (2017). https://earthquake.usgs.gov/earthquakes/browse/stats.php
2. GeoNet (2016). http://info.geonet.org.nz/display/quake/Earthquake+Facts+and+Statistics
3. Than K (2010) Haiti Earthquake "Strange", strongest in 200 years. https://www.nationalgeographic.com/news/2010/1/100113-haiti-earthquake-red-cross/
4. Kenneth Pletcher JP (2020) Japan earthquake and tsunami of 2011. Britannica.com: https://www.britannica.com/event/Japan-earthquake-and-tsunami-of-2011/Aftermath-of-the-disaster
5. Ahmad S, Khan SA, Pilakoutas K, Khan QUZ (2015) Empirical vulnerability assessment of the non-engineered reinforced concrete structures using the Kashmir earthquake damage data. Bull Earthq Eng 13(9):2611–2628. https://doi.org/10.1007/s10518-015-9735-0
6. Bilgin H, Frangu I (2017) Predicting the seismic performance of typical R/C healthcare facilities: Emphasis on hospitals. Int J Adv Struct Eng 9(3):277–292. https://doi.org/10.1007/s40091-017-0164-y
7. Masi A, Chiauzzi L, Santarsiero G, Manfredi V, Biondi S, Spacone E, Del Gaudio C, Ricci P, Manfredi G, Verderame GM (2017) Seismic response of RC buildings during the Mw 6.0 August 24, 2016 Central Italy earthquake: the Amatrice case study. Bull Earthq Eng 1–24. https://doi.org/10.1007/s10518-017-0277-5
8. Keshab Sharma MS (2017) Geotechnical and structural aspect of 2015 Gorkha Nepal Earthquake and lesson learnt. J Inst Eng 20–36
9. Blondet M, Serrano M (2017) Training in earthquake resistent adobe brick construction in the Peruvian Andes. In: 16th world conference on earthquake engineering
10. Jia J, Yan J (2015) Analysis about factors affecting the degree of damage of buildings in earthquake. J Phys: Conf Ser 628(1). https://doi.org/10.1088/1742-6596/628/1/012062
11. Huang M, Zhao L, Li P (2013) Earthquake damage prediction system of buildings group based on damage factors method. Appl Mech Mater 341–342:1496–1499. https://doi.org/10.4028/www.scientific.net/AMM.341-342.1496
12. Goda K, Kiyota T, Pokhrel RM, Chiaro G, Katagiri T, Sharma K, Wilkinson S (2015) The 2015 Gorkha Nepal earthquake: insights from earthquake damage survey. Front Built Environ 1(June):1–15. https://doi.org/10.3389/fbuil.2015.00008
13. Jurukovski D (1997) Earthquake consequences and measures for reduction of seismic risk. Ren Fail 19(5):621–632. https://doi.org/10.3109/08860229709109028
14. Saputra A, Rahardianto T, Revindo MD, Delikostidis I, Hadmoko DS, Sartohadi J, Gomez C (2017) Seismic vulnerability assessment of residential buildings using logistic regression and geographic information system (GIS) in Pleret Sub District (Yogyakarta, Indonesia). Geoenvironmental Disasters 4(1). https://doi.org/10.1186/s40677-017-0075-z
15. Anbazhagan P, Ramyasri S, Moustafa SSR, Al-Arifi NSN (2017) Intensity based building damage level prediction model from past Earthquakes for risk assessment. Disaster Adv 10(3):1–15

On Some Classes of Equitable Irregular Graphs

Antony Puthussery and I. Sahul Hamid

Abstract Graph labeling techniques are used by data scientists to represent data points and their relationships with each other. The segregation/sorting of similar datasets/points are easily done using labeling of vertices or edges in a graph. An equitable irregular edge labeling is a function $f : E(G) \to N$ (not necessarily be injective) such that the vertex sums of any two adjacent vertices of G differ by at most one, where vertex sum of a vertex is the sum of the labels under f of the edges incident with that vertex. A graph admitting an equitable irregular edge labeling is called an equitable irregular graph (EIG). In this paper, more classes of equitable irregular graphs are presented. We further generalize the concept of equitable irregular edge labeling to k-equitable irregular edge labeling by demanding the difference of the vertex sum of adjacent vertices to be $k \geq 1$.

Keywords Edge labeling · Datasets · Equitable irregular graph · k-equitable irregular edge labeling

By a graph $G = (V, E)$, we mean a finite, undirected graph with neither loops nor multiple edges. For graph theoretic terminologies, we refer to the book by Chartrand and Lesniak [1]. All graphs in this paper are assumed to be connected and non-trivial. Throughout the paper the order and size of G are denoted by n and m, respectively.

The study of graph labeling is one of the fastest growing areas. Graph labeling techniques are used by data scientists to represent data points and their relationships with each other. The segregation/sorting of similar datasets/points are easily done using labeling of vertices or edges in a graph. A graph labeling is an assignment of integers to the vertices or edges or both subject to certain conditions. The interest in graph labelings can trace its roots back to a paper [5] by Alex Rosa in the late 1960s. A detailed survey of graph labeling by Gallian is given in [2].

A. Puthussery (✉)
CHRIST Deemed to be University, Bengaluru, India
e-mail: frantony@christuniversity.in

I. S. Hamid
The Madura College, Madurai 11, India
e-mail: sahulmat@yahoo.co.in

D. S. Jat et al. (eds.), *Data Science and Security*, Lecture Notes in Networks and Systems 132, https://doi.org/10.1007/978-981-15-5309-7_30

An equitable irregular edge labeling is a function $f : E(G) \to N$ (not necessarily be injective) such that the vertex sums of any two adjacent vertices of G differ by at most one, where vertex sum of a vertex is the sum of the labels under f of the edges incident with that vertex. A graph admitting an equitable irregular edge labeling is called an equitable irregular graph (EIG). This concept was introduced in [3] and further studied in [4] wherein several families of equitable irregular graphs and some properties of EIG have been presented. In this paper, we present more classes of equitable irregular graphs. We further generalize the concept of equitable irregular edge labeling to k-equitable irregular edge labeling by demanding the difference of the vertex sum of adjacent vertices to be $k \geq 1$.

1 More Classes of Equitable Irregular Graphs

In this section, we obtain some classes of equitable irregular graphs such as gear graphs; total graph of paths, wheels and complete bipartite graphs; and Mycielskian graph of cycles. Recall that the gear graph G_k is obtained from the wheel W_k by subdividing each of its rim edges exactly once. So the gear graph G_k has $2k-1$ vertices and $3k-3$ edges.

Theorem 1.1 *Gear graph G_k where $k \geq 3$ is equitable irregular.*

Proof Let $V(G_k) = \{v_0, v_1, v_2, \ldots, v_{k-1}, u_1, u_2, \ldots, u_{k-1}\}$ and $E(G_k) = \{v_0u_i : 1 \leq i \leq k-1\} \cup \{u_iv_i : 1 \leq i \leq k-1\} \cup \{v_iv_{i+1} : 1 \leq i \leq k-2\} \cup \{v_1v_{k-1}\}$. Define an edge labeling f as follows. $f(v_0v_i) = 1;\ 1 \leq i \leq k-1$; $f(u_iv_i) = \frac{k-1}{2};\ 1 \leq i \leq k-1$; $f(v_iv_{i+1}) = f(v_1v_{k-1}) = 1;\ 1 \leq i \leq k-2$. And, $S_f(v_0) = \left(\frac{k-1}{2}\right)(f(v_0u_i)) = \frac{k-1}{2}$; $S_f(u_i) = f(v_0u_i) + f(u_iv_i) = 1 + \frac{k-1}{2}$; $S_f(v_i) = f(u_iv_i) + f(v_iv_{i+1}) + f(v_{i-1}v_i) = \frac{k-1}{2} + 2$. Hence, f is an equitable irregular edge labeling of G_k so that the gear graph G_k) is 1-equitable irregular.

The *total graph* $T(G)$ of a graph G is the graph whose vertex set is $V(G) \cup E(G)$ and two vertices are adjacent whenever they are either adjacent or incident in G.

Theorem 1.2 *The total graph of path, $T(P_n)$, is equitable irregular.*

Proof Let $V(T(P_n)) = \{v_1,\ v_2,\ \ldots,\ v_n,\ u_1,\ u_2,\ \ldots,\ u_{n-1}\}$ and $E(T(P_n)) = \{v_iv_{i+1}\} \cup \{v_iu_i\} \cup \{v_{i+1}u_i\} : 1 \leq i \leq n-1\} \cup \{u_iu_{i+1} : 1 \leq i \leq n-2\}$. Certainly, the order and size of $T(P_n)$ are $2n-1$ and $4n-5$, respectively. Consider the edge labeling f of $T(P_n)$ that assigns the label 1 to the edges u_1v_1 and u_nv_{n-1}; and the label 1 to all the remaining edges. Then, for each $i \in \{2, 3, \ldots, n-1\}$, we have $S_f(v_i) = f(v_iv_{i+1}) + f(v_{i-1}v_i) + f(u_{i-1}v_i) + f(u_iv_i) = 4$; $S_f(u_i) = f(v_iu_i) + f(u_{i-1}u_i) + f(u_iu_{i+1}) + f(u_iv_{i+1}) = 4$; $S_f(v_1) = f(u_1v_1) + f(v_1v_2) = 3$; $S_f(v_n) = f(u_{n-1}v_n) + f(v_{n-1}v_n) = 3$. Hence, f is an equitable irregular edge labeling of $T(P_n)$ so that the total graph of path, $T(P_n)$, is an equitable irregular.

Remark 1 The total graph of a cycle is a four-regular graph and thus $T(C_n)$, for all n, is an equitable irregular graph. In general, the total graphs of regular graphs are equitable irregular.

Theorem 1.3 *The total graph of wheel, $T(W_n)$, is equitable irregular.*

Proof As W_4 is regular, $T(W_4)$ is equitable irregular. Let $n \geq 5$. Let $V(T(W_n)) = \{v_0, v_1, v_2, \ldots, v_{n-1}, u_1, u_2, \ldots, u_{n-1}, u'_1, u'_2, \ldots, u'_{n-1}\}$ and $E(T(W_n)) = \{v_0v_i : 1 \leq i \leq n-1\} \cup \{v_0u_i : 1 \leq i \leq n-1\} \cup \{u_iu_j : i \neq j \text{ and } 1 \leq i, j \leq n-1\} \cup \{u_iv_i : 1 \leq i \leq n-1\} \cup \{u_iu'_i : 1 \leq i \leq n-1\} \cup \{u_{i+1}u'_i : 1 \leq i \leq n-2\} \cup \{u_1u'_{n-1}\} \cup \{v_iv_{i+1} : 1 \leq i \leq n-2\} \cup \{v_1v_{n-1}\} \cup \{v_iu'_i : 1 \leq i \leq n-1\} \cup \{v_{i+1}u'_i : 1 \leq i \leq n-2\} \cup \{v_1u'_{n-1}\} \cup \{u'_iu'_{i+1} : 1 \leq i \leq n-2\} \cup \{u'_1u'_{n-1}\}$. Define an edge labeling f of $T(W_n)$ as follows. $f(v_0v_i) = f(v_0u_i) = 1$, for $1 \leq i \leq n-1$; for $1 \leq i \leq n-2$, $f(v_iv_{i+1}) = f(v_{i+1}u'_i) = f(u_{i+1}u'_i) = 1$.

And, $f(v_1v_{n-1}) = f(u_1u'_{n-1}) = f(u'_{n-1}v_1) = 1$; $f(u_iu_j)=1$, for $i \neq j$ and $1 \leq i, j \leq n-1$; $f(u_iu'_i) = 1$, $1 \leq i \leq n-1$; $f(u'_iu'_{i+1}) = \lceil \frac{n}{2} - 1 \rceil$, $1 \leq i \leq n-2$; $f(u'_1u'_{n-1}) = \lceil \frac{n}{2} - 1 \rceil$; $f(u_iv_i) = f(v_iu'_i) = n-3$, $1 \leq i \leq n-1$.

$S_f(v_0) = \sum_{n-1}^{i=1} f(v_0v_i) + \sum_{n-1}^{i=1} f(v_0u_i) = n-1+n-1 = 2n-2$; $S_f(u_i) = f(v_0u_i) + f(u_iv_i) + f(u_iu_j) + f(u_iu'_i) + f(u'_{i-1}u_i) = 1 + (n-3) + (n-2) + 1 + 1 = 2n-2$; $S_f(v_i)$=$f(v_0v_i) + f(u_iv_i) + f(v_iv_{i+1}) + f(v_{i-1}v_i) + f(v_iu'_i) + f(u'_{i+1}v_i) = 1 + (n-3) + 1 + 1 + (n-3) + 1 = 2n-2$; $S_f(u'_i) = f(u_iu'_i) + f(u_{i+1}u'_i) + f(u'_iu'_{i+1}) + f(u'_{i-1}u'_i) + f(v_iu'_i) + f(v_{i-1}u'_i) = (n-3) + 1 + \lceil \frac{n}{2} - 1 \rceil + \lceil \frac{n}{2} - 1 \rceil + 1 + 1 = 2\left(\lceil \frac{n}{2} - 1 \rceil\right) + n = 2n-2$ or $2n-1$.

Hence, f is an equitable irregular edge labeling of $T(W_n)$ so that the total graph of wheel, $T(W_n)$, is an equitable irregular.

Theorem 1.4 *The total graph $T(K_{2,s})$ is equitable irregular.*

Proof Let X and Y be the parts of $K_{2,s}$ with $X = \{x_1, x_2\}$ and $Y = \{y_1, y_2, \ldots, y_s\}$. Let $V(T(K_{2,s})) = \{x_1, x_2, y_1, y_2, \ldots, y_s, u_1, u_2, \ldots, u_{2s}\}$ and $E(T(K_{2,s})) = \{x_iy_j : 1 \leq i \leq 2 \text{ and } 1 \leq j \leq s\} \cup \{x_iu_j : 1 \leq i \leq 2 \text{ and } 1 \leq j \leq s\} \cup \{u_iy_i : 1 \leq i \leq s\} \cup \{u_{s+i}y_i : 1 \leq i \leq s\} \cup \{u_iu_j : i \neq j \text{ and } 1 \leq i, j \leq s\} \cup \{u_iu_{s+1} : 1 \leq i \leq s\} \cup \{u_iu_j : i \neq j \text{ and } s \leq i, j \leq 2s\}$. Define an edge labeling f as follows. $f(x_iy_j) = f(x_iu_j) = 1$, $1 \leq i \leq 2$ and $1 \leq j \leq s$; $f(u_iu_j) = 1$, $i \neq j$ and $1 \leq i, j \leq s$; $f(u_iu_j) = 1$, $i \neq j$ and $s \leq i, j \leq 2s$; $f(u_iy_i) = f(u_{s+i}y_i) = s-1$, $1 \leq i \leq s$; $f(u_iu_{s+1}) = 1$, $1 \leq i \leq s$.

$S_f(x_i) = f(x_iy_j) + f(x_iu_j) = s + s = 2s$; $S_f(y_i) = 2(f(x_iy_j)) + f(u_iy_i) + f(u_{s+i}y_i) = 2 + (s-1) + (s-1) = 2s$; $S_f(u_i) = f(x_iu_j) + f(u_iu_j) + f(u_i u_{s+1}) + f(u_iy_i) = 1 + (s-1) + 1(s-1) + 1 = 2s$.

Hence, f is an equitable irregular edge labeling of $T(K_{2,s})$, so that the total graph of complete bipartite graph, $T(K_{2,s})$, is an equitable irregular.

For a given graph G with $V = V(G) = \{v_1, v_2, \ldots, v_n\}$, denote $V' = \{v'_1, v'_2, \ldots, v'_n\}$ to be the corresponding set of V, the Mycielskian graph $\mu(G)$ of G is the graph with vertex set $V(\mu(G)) = V \cup V' \cup \{u\}$ and edge set $E(\mu(G)) = E(G) \cup \{v'_iv_j : v_iv_j \in E(G)\} \cup \{uv'_i : 1 \leq i \leq n\}$.

Theorem 1.5 *The Mycielskian graph $\mu(C_n)$ of the cycle C_n is equitable irregular.*

Proof Let $V(\mu(C_n)) = \{v_1, v_2, \ldots, v_n, v'_1, v'_2, \ldots, v'_n, u\}$ and $E(\mu(C_n)) = \{uv'_i : 1 \le i \le n\} \cup \{v_i v_{i+1} : 1 \le i \le n-1\} \cup \{v_1 v_n\} \cup \{v'_i v_{i+1} : 1 \le i \le n-1\} \cup \{v'_n v_1\} \cup \{v_i v'_{i+1} : 1 \le i \le n-1\} \cup \{v'_1 v_n\}$. Define a labeling f as follows. $f(uv'_i) = 1$; $1 \le i \le n$; $f(v'_i v_{i+1}) = f(v_i v_{i+1}) = 1$; $1 \le i \le n-1$; $f(v_i v'_{i+1}) = n-2$; $1 \le i \le n-1$; $f(v_n v_1) = f(v'_n v_1) = 1$; $f(v_n v'_1) = n-2$.

$S_f(u) = \sum_{i=1}^{n} f(uv'_i) = \sum_{i=1}^{n} (f(uv'_i) + f(v'_i v_{i+1})) + f(v_n v'_1) = n$; $S_f(v_i) = n+1$.
Hence, f is an equitable irregular edge labeling of $\mu(C_n)$, so that the Mycielskian graph $\mu(C_n)$ is an equitable irregular.

In the following theorem, we obtain some properties of trees which admit equitable irregular labeling.

Theorem 1.6 *If a tree is an equitable irregular graph on $n \ge 4$ vertices, then the following conditions hold:*

(i) If v is a support vertex, then $deg\ v = 2$ and the label of the non-pendant edge incident with v under any EIL is 1.
(ii) The number of pendant vertices of T is at most $\lfloor \frac{n}{2} \rfloor$.
(iii) $\Delta(T) \le \lfloor \frac{n}{2} \rfloor$ with equitability if and only if T is either P_4 or isomorphic to the graph in Fig. 1.

Proof Suppose T is an EIG with an EIL f.

(i) Let v be the support vertex of T and let v' be the pendant neighbor of v. If $f(vv') = a$ then $S_f(v') = a$ and $S_f(v) > a$ so that the equitability condition implies that $S_f(v) = a + 1$. This is possible only when $deg\ v = 2$ and $f(vw) = 1$, where w is the non-pendant neighbor of v. Hence, (i) is proved.

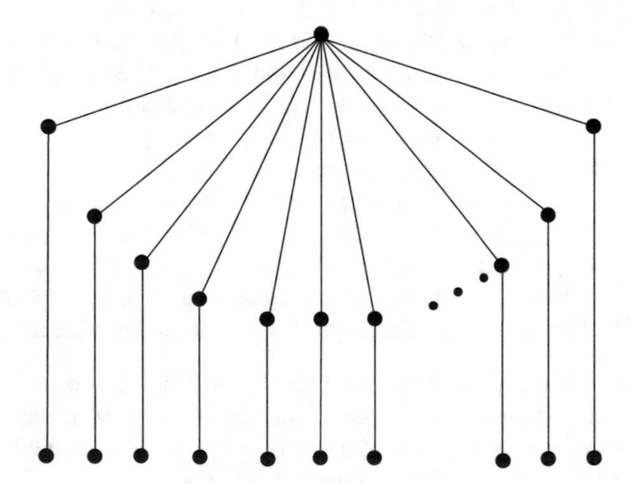

Fig. 1 The graph given in Theorem 1.6

(ii) Let l be the number of pendant vertices of T. As support vertices are of degree 2, it follows that the number of support vertices of T is also exactly l. This implies that T has at least $2l$ vertices so that $l \leq \lfloor \frac{n}{2} \rfloor$; proving (ii).

(iii) As the number of pendant vertices in a tree is at least $\Delta(T)$ it follows from (ii) that $\Delta(T) \leq \lfloor \frac{n}{2} \rfloor$. Now, let $\Delta(T) = \lfloor \frac{n}{2} \rfloor$ and let v be a maximum degree vertex. Suppose n is even. Then $n = 2\Delta(T)$. Since $n = 2\Delta(T)$ at least one neighbor of v must be a pendant vertex and so it follows from (ii) that $deg\ v = 2$. That is $\Delta(T) = 2$ which implies that $n = 4$ and so $T = P_4$.
On the other hand, suppose n is odd. Then $n = 2\Delta(T) + 1$. If $\Delta(T) = 2$, then $T = P_3$. Assume that $\Delta(T) \geq 3$. By (ii), v is not a support vertex. Therefore, every neighbor of v is adjacent to one of the remaining $\Delta(T)$ vertices of T. This implies that T is isomorphic to the graph G in Fig. 1.

2 K-Equitable Irregular Graphs

In this section, we define the notion of k-equitable irregular edge labeling of a graph and obtain some basic results.

Definition 2.1 For a positive integer k, k-equitable irregular edge labeling of a graph G is a function $f : E(G) \rightarrow N$ such that $|S_f(u) - S_f(v)| \leq k$ for any two adjacent vertices u and v of G, where $S_f(u)$ is the sum of the labels of the edges incident with the vertex u. A graph G admitting an k-equitable irregular edge labeling is said to be k-equitable irregular graph $(k - EIG)$.

Example 2.2 Consider the graph G of Fig. 2. As G has a support vertex of degree 3, the difference of the vertex sums of the support vertex and its pendant neighbor would be at least 2 under any edge labeling of G so that G cannot be 1-EIG. However, the graph G is 2-EIG as it admits an 2-EIL as shown in Fig. 2.

Remark 2 (i) Suppose an k-EIG G has a support vertex v with deg $v = r$. Then under any edge labeling f of G, if $f(v'v) = a$, where v' is pendant vertex of v, then $S_f(v') = a$ whereas $S_f(v') = a + r - 1$ so that $|S_f(v) - S_f(v')| = r - 1 \leq k$ and hence $r \leq k + 1$. Thus, the degree of a support vertex (if any) in a k-EIG is at most $k + 1$.

(ii) If G is k-EIG, then G is l-EIG for every $l \geq k$. But not the converse, as seen in Example 2.2.

Lemma 2.3 *Every graph G is* $(\Delta(G) - \delta(G)) - EIG$.

Proof Consider the constant function $f : E(G) \rightarrow N$ defined by $f(e) = 1$, for all $e \in E(G)$. Certainly, $S_f(u) = \deg u$, for every $u \in V(G)$. So for any two vertices u and v with e with deg $v \geq$ deg u, we have $|S_f(v) - S_f(u)| = |\deg v - \deg u| \leq \Delta - \delta$. This completes the proof.

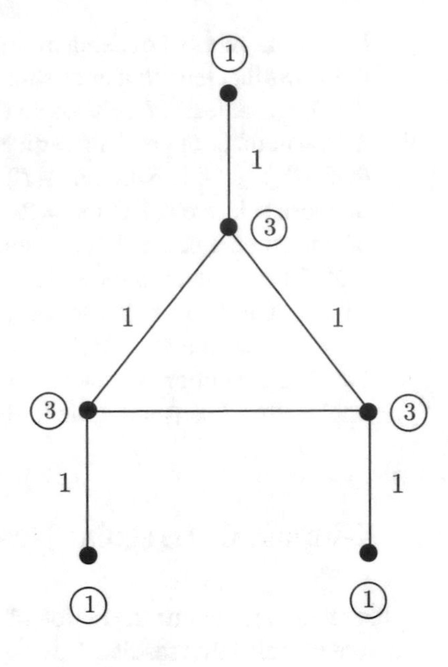

Fig. 2 A graph which is 2-EIG but not 1-EIG

We now present some classes of graphs which are k-equitable irregular, where $k \geq 2$.

Theorem 2.4 *$\mu(W_n)$ is 2-equitable irregular.*

Proof Let $V(\mu(W_n)) = \{v_0\} \cup \{v_i : 1 \leq i \leq n-1\} \cup \{v'_0\} \cup \{v'_i : 1 \leq i \leq n-1\} \cup \{u\}$ and $E(\mu(W_n)) = \{v_0v_i : 1 \leq i \leq n-1\} \cup \{v_iv_{i+1} : 1 \leq i \leq n-1\} \cup \{v_{n-1}v_1\} \cup \{v'_0v_i : 1 \leq i \leq n-1\} \cup \{v'_iv_0 : 1 \leq i \leq n-1\} \cup \{v'_iv_{i+1} : 1 \leq i \leq n-2\} \cup \{v'_iv_{i-1} : 2 \leq i \leq n-1\} \cup \{v'_{n-1}v_1\} \cup \{v'_1v_{n-1}\} \cup \{uv'_i : 1 \leq i \leq n-1\}$. Define an edge labeling f of $\mu(W_n)$ as follows. $f(v_0v_i) = f(v_iv_{i+1}) = f(v_{n-1}v_1) = 1;\ 1 \leq i \leq n-1;\ f(v'_iv_0) = 1;\ 1 \leq i \leq n-1;\ f(v'_0v_i) = f(uv'_i) = 2;\ 1 \leq i \leq n-1;\ f(v'_iv_{i+1}) = f(v'_iv_{i-1}) = n-2; 1 \leq i \leq n-1;\ f(v'_{n-1}v_1) = f(v'_1v_{n-1}) = n-2;\ f(uv'_0)=1$.

$S_f(v_0) = \sum_{n-1}^{i=1}\big(f(v_0v_i) + f(v'_iv_0)\big) = (n-1) + (n-1) = 2n-2;\ S_f(v'_0) = 2(f(v'_0v_i)) + f(uv'_0) = 2(n-1) + 1 = 2n-1;\ S_f(u) = f(uv'_i) + f(uv'_0) = 2(n-1) + 1 = 2n-1;\ S_f(v_i) = f(v'_0v_i) + f(v_0v_i) + f(v_iv_{i+1}) + f(v_{i-1}v_i) + f(v'_{i+1}v_i) + f(v'_{i-1}v_i) = 2 + 1 + 1 + 1 + (n-2) + (n-2) = 2n+1;\ S_f(v'_i) = f(v'_iv_{i+1}) + f(v'_iv_{i-1}) + f(v'_0v_i) + f(uv'_0) = (n-2) + (n-2) + 2 + 1 = 2n-1$. Hence, f is an 2-equitable irregular edge labeling of $\mu(W_n)$, so that the Mycielskian graph of wheel is 2-equitable irregular.

Theorem 2.5 *$\mu(G_k)$, where G_k is the gear graph and is 3-equitable irregular.*

Proof Let $V(\mu(G_k)) = \{v_0\} \cup \{v_i : 1 \leq i \leq k\} \cup \{u_i : 1 \leq i \leq k\} \cup \{v'_0\} \cup \{v'_i : 1 \leq i \leq k\} \cup \{u'_i : 1 \leq i \leq k\} \cup \{u\}$ and $E(\mu(G_k)) = \{v_0u_i : 1 \leq i \leq k\} \cup \{u_iv_i : 1 \leq i \leq k\} \cup \{v_iv_{i+1} : 1 \leq i \leq k-1\} \cup \{v_kv_1\} \cup \{v'_0u_i : 1 \leq i \leq k\} \cup \{u'_iv_0 :$

$1 \leq i \leq k\} \cup \{u'_i v_i : \quad 1 \leq i \leq k\} \cup \{v'_i u_i : \quad 1 \leq i \leq k\} \cup \{v'_i v_{i-1} : \quad 2 \leq i \leq k\}$ $\cup \{v'_i v_{i+1} : 1 \leq i \leq k-1\} \cup \{v'_k v_1\} \cup \{v'_1 v_k\} \cup \{uv'_i : 1 \leq i \leq k\} \cup \{uu'_i : 1 \leq i \leq k\} \cup \{uv'_0\}$. Define an edge labeling f of $\mu(G_k)$ as follows. Let $f(v_0 u_i) = f(u_i v_i) = f(u'_i v_0) = 1, \quad 1 \leq i \leq k$; $f(v_i v_{i+1}) = f(v'_i v_{i+1}) = 1, \quad 1 \leq i \leq k-1$; $f(v_k v_1) = 1$; $f(v_i v_{i-1}) = f(v'_i v_{i-1}) = 1, \quad 2 \leq i \leq k$; $f(v'_0 u_i) = 2 \quad 1 \leq i \leq k$; $f(u'_i v_i) = 2k-3, 1 \leq i \leq k$; $f(v'_i u_i) = 2k-2, 1 \leq i \leq k$; $f(v'_k v_1) = f(v'_1 v_k) = 1$; $f(uv'_i) = f(uu'_i) = f(uv'_0) = 1, 1 \leq i \leq k$. $S_f(v_0) = \sum_k^{i=1} \left(f(v_0 u_i) + f(u'_i v_0)\right) = k + k = 2k$; $S_f(u_i) = f(v_0 u_i) + f(u_i v_i) + f(u_i v''_0) + f(u_i v'_i) = 1 + 1 + 2 + (2k-2) = 2k+2$; $S_f v_i = f(u_i v_i) + f(v_i v_{i+1}) + f(v_i v_{i-1}) + f(v_i u'_i) + f(v_i v'_{i+1}) + f(v_i$ $v'_{i-1}) = 1 + 1 + 1 + (2k-3) + 1 + 1 = 2k+2$; $S_f(v'_0) = (f(uv'_0)) + 2\sum^{i=1}$ ${}_k f(v'_0 u_i) = 1 + 2k = 2k+1$; $S_f(u'_i) = f(u'_i v_i) + f(u'_i v_0) + f(u'_i u) = (2k-3) + 1 + 1 = 2k-1$; $S_f(v'_i) = f(v'_i u_i) + f(v'_i v_{i+1}) + f(v'_i v_{i-1}) + f(v'_i u) = (2k-2)$ $+ 1 + 1 + 1 = 2k+1$; $S_f(u) = \sum_k^{i=1} \left(f(uv'_i + f(uv'_i)\right) + f(uv'_0) = k + k + 1 = 2k+1$.

Hence, f is an 3-equitable irregular edge labeling of $\mu(G_n)$, so that Mycielskian graph $\mu(G_n)$ of gear graph is 3-equitable irregular.

3 Conclusion and Scope

This paper extends the study of the equitable irregular edge labeling of graphs. The study includes the presentation of more classes of equitable irregular graphs such as gear graphs, the total graphs of paths; wheels; and the complete bipartite graph $K_{r,s}$. It is also proved that the Mycielskian graph of cycles is equitable irregular. Further the concept of k-equitable irregular graphs was introduced and a study has been initiated. The study also throws some light for the data scientists in understanding the proper nodes/points for the classification as the need arises. The following are some interesting problems for further investigation:

(i) We have proved in Lemma 2.3 that every graph G is $\Delta(G) - \delta(G) - EIG$. So, for every graph G, there exists a least positive integer k such that G is k-EIG. Let us say this least k to be the *equitable irregular index*, and let us denote it by $\sigma(G)$. Note that $\sigma(G) \leq \Delta(G) - \delta(G)$. One may now initiate a study on this parameter.

(ii) More classes of graphs that are k-EIG may be found.

References

1. Chartrand G, Lesniak L (2005) Graphs and digraphs. CRC Press, Boca Raton
2. Gallian JA (2017) A dynamic survey of graph labeling. Electron J Combin #DS6
3. Hamid IS, Kumar SA (2010) Equitable irregular edge-weighting of graphs. SUT J Math 46 79–91
4. Rajan M, Abraham VM (2017) A study on equitable irregular edge-weighting of graphs. J Comp-Math 8:442–451
5. Rosa A (1966) On certain valuations of the vertices of a graph. In: International symposium on theory of graphs, p p 349–355

Therapy Recommendation Based on Level of Depression Using Social Media Data

Tanvi Rath, Yash Purohit, Rishab Rajput, Arpita Ghosh, and Samiksha Shukla

Abstract Social media is a massive platform with currently over 100 million registered users. It is a platform where individuals express themselves along with their interests. These expressions of individual can be used to identify their mental status. That being said, depression and anxiety are the dominant cause for illness and ill-health across the world. Studies show that user's mental health can be predicted by their everyday use of language. This paper examines the tweets for analyzing the linguistic and behavioral features for classifying the levels of depression among the users. In order to classify the levels of depression, a knowledge base of the words that are associated with depression/anxiety has been created. The model evaluated this using simple text mining techniques to measure the mental health status of the users and provide appropriate recommendations.

Keywords Depression · Anxiety · Social media · Language · Mental health · Knowledge base · Recommendation system

1 Introduction

Behavioral psychopathology relates tension and despair closely. Anxious depression is described as a state for the individuals who are diagnosed with depression, present in a way that is more regular with feeling anxious in place of sad. It is a Major Depressive Disorder (MDD) with a co-morbid tension sickness. Symptoms of tension and depression are of identical depth and none of them genuinely predominates, characterizing a mixed tension-depressive sickness.

These findings highlight the importance of issues concerning anxious depression as an extreme case of psychological disorder. Social media is omnipresent and allows people to self-express, live connectedly, and keep in touch with buddies and acquaintances throughout the globe. Social media and intellectual fitness of users can be associated in three exceptional methods such as **social media anxiety disorder**,

---

T. Rath · Y. Purohit · R. Rajput · A. Ghosh (✉) · S. Shukla
CHRIST (Deemed to be University), Lavasa Pune, India
e-mail: arpita.ghosh@christuniversity.in

D. S. Jat et al. (eds.), *Data Science and Security*, Lecture Notes in Networks and Systems 132, https://doi.org/10.1007/978-981-15-5309-7_31

an active social media user, who is distressed by way of bad interactions and social comparisons on social networking sites affecting their self-esteem and mental wellness. **Anxious depression** social media verbalization, an active social media person who makes use of social media postings as an outlet to share feelings in a non-threatening atmosphere. **Social anxiety**, a passive social media consumer who is comfortable with digital connections as a substitute to real interactions.

Though feelings are tough to articulate, however, online self-expression presents a means to bring a mental condition into a physical form. Social media can facilitate pre-analysis of a clinical intellectual health situation associated with anxiety, depression, or anxious depression in active extroverts who verbalize and share their inner anxiety. The proposed model explores opportunities which use written social media textual content that people are already producing in abundance. Social media text consists of implicit information which can be thus used for detection of depression.

This paper is conferred in following sections: Sect. 2 discusses previously proposed approaches. Section 3 presents the proposed framework and algorithm and describes methodology that is applied. Section 4 presents the therapy recommendation system based on the result. Section 5 provides the experiments performed and discusses the results. Lastly, Sect. 6 presents the conclusion and future scope.

2 Related Work

Various studies have been conducted to investigate social media's capacity for predicting multiple forms of depression in social media users. In [1], authors have created a lexicon of 60 words related to depression, and have used tweet timing, tweet frequency, negative polarity tweets, and polarity contrast as features to train their predictive model. To detect anxious depression, three multinomial classifiers are used, namely, multinomial Naïve Bayes, gradient boosting, and random forest and to generate final prediction, an ensemble vote classifier is used.

In [2], authors conducted survey for candidates to identify participants with depression using DAAS-21 standards and their levels of depression are characterized. They collected letters written by participants that were used to analyze depression levels. The depressive and non-depressive groups were compared using Mann–Whitney U test and were conducted separately for men and women, and then the ranks were compared.

Thorstad et al. [3] took into account the tweets from clinical subreddit and non-clinical subreddit and features were identified with the help of binary logistic regression, were classified with different illness, and were used for training.

Similarly, in [4], essays were collected accompanied by meta-data for its author including big five personality traits, calculated Linguistic Enquiry and Word Count (LIWC) features and Latent Dirichlet Allocation (LDA) features for each document, and used linear regression to predict neuroticism score for each test document. In [5], Choudhury et al. employed crowd sourcing methodology to collect information on the set of depressed users and their tweets, and features like post-centric and user-centric are identified to characterize the postings. The post features include emotion, time, and linguistic style. User-centric features include engagement and ego-network. Principle Component Analysis (PCA) is employed with standard SVM classifier with Radial Basis Function (RBF) kernel and used fivefold cross-validation with 100 randomized experiments run.

In another study in [6], the authors propose multi-task learning for detecting suicide risk and mental health conditions, and their model includes neuroatypicality, suicide attempt, as well as the related mental health conditions of anxiety, depression, eating disorder, panic attacks, schizophrenia, bipolar disorder, Post-traumatic Stress Disorder (PTSD), and effect of modeling gender to increase accuracy.

Twitter is a widespread social networking site which allows people to put forth their opinions on to a wide-reaching platform [7]. The popularity of Twitter is strengthened by the statistic that 500 million tweets are posted daily. This in turn offers huge volume of data accessible for any topic with people casting their observations on it.

The study in [8] shows another dimension of data fetching by means of live tweets, Tweepy, and Twitter API.

3 Proposed Model

Depression is one of the major concerns in this new age of social media and technology. Mental health is as important as keeping check on one's physical health. This model helps to identify the early onset of depression and classifies the severity of depression detected from the tweet. For this model, the study focused on the users' everyday use of language in social media sites, such as Twitter. Figure 1 presents a framework which consists a database of dictionary of words with different severity level related to anxiety/depression to help classify the tweets based on severity level.

3.1 Data Collection

Social media is a great means of motivating people to reach out. Whether it is in the form of a Facebook status, writing on walls, updating Twitter, sharing images, all these are aspects of individual personality which helps to represent individual values, point of view, and to convince people what they are. They give the individual a voice to be themselves and be imaginative. Hence, to collect an individual's everyday use of language and how they express themselves, Twitter dataset "Sentiment140" containing 1,600,000 tweets was downloaded from Kaggle, which contains the following fields: *Target: the polarity of the tweet, Id: the id of the tweet, Date: the date of the tweet, Flag: the query, User: the user that tweeted, and Text: the text of the tweet.*

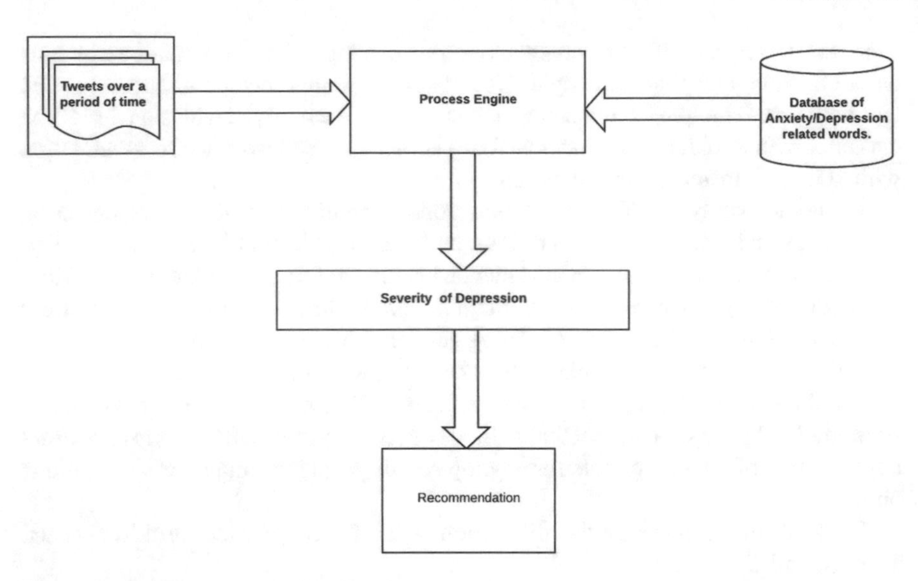

Fig. 1 Proposed recommendation framework

3.2 Database Generation

A knowledge base containing a dictionary of words was created in consultation with psychotherapist for the detection of severity levels. In this dictionary, Level 1 is inclusive of all the words contributing to low levels of depression, Level 2 contains all the words contributing to moderate levels of depression, and Level 3 indicates high level of depression.

3.3 Feature Engineering

According to previous studies [3], it is said that mental health can be detected by an individual's way of communication and use of everyday language. With the rise in popularity of social media, it gives us a back door entry for data on how individuals express themselves and their choice of words for expression. In this model, these tweets help to approximately detect the severity of depression of the user.

3.4 Proposed Methodology

Proposed Classification Methodology for identification of level of Depression using Social Media Data

Input: Test Set *T*, Knowledgebase Dictionary *words*
Output: Level of Depression and Therapy Recommendation

n: Number of tweets in test set.
df: Data frame list [total number of tweets, high level of depression, moderate level of depression, mild level of depression].
T_i: Tweet *i* in the test set (*i*=1, 2… n).
words: Dictionary of words for three level of Depression.
level1, *level2*, *level* 3: Level of depression based on the word used in social media post.

While *i* in Test Set do
high=0; *moderate*=0; *mild*=0;
Split the *tweet* T_i
for *j* in tweet
if *j* in *words* then
if (value for word ==1) then
mild = *mild*+1;
else if (value for word ==2) then
moderate = *moderate*+1;
else if (value for word ==3) then
high = *high*+1;
if (*high*>0) then
level3=*level3*+1
else if (*moderate*>0)
level2=*level2*+1
else if (*low*>0)
level1=*level1*+1
df ⟵ append depression levels (Number of tweets *n*, *level3*, *level2*, *level1*)

4 Therapy Recommendation System

Depression has crept into our society stealthily and is spreading among most individuals, though in different stages. Largely, therapy for depression includes more of lifestyle changes. It is only in extreme cases that clinical help and medicines are recommended.

The ones with mild depression (Level 1) tend to be moody, insecure, and afraid. These individuals can help themselves by building their confidence, maintaining a regular routine that includes some physical activity, and having a good support group. It is equally important for those around to be aware that if someone shows signs of mild depression reach out and support or else chances of the individuals getting sucked into depressive states may increase.

Individuals with Level 2 depression are first recommended to undergo digital detox, that is, keep away from smartphone and social media for a few days. They can be reintroduced to the use of social media by using it only during the predetermined scheduled time. Bringing in self-discipline for Level 2 individuals is very important, as they have a tendency to give up and feel disheartened and hurt very easily. Group therapy is recommended for such individuals, where there are various activities and interventions that help them open up and share. In the process of sharing, the individual realizes that they are not alone and boosts their morale and it is also easy to build a support group among them. Also recommended for Level 1 individuals are lifestyle changes in terms of the food they eat, exercise routine, sleeping hours, and reducing stress in life.

In the case of Level 3 depression, it is important to know the duration of previous history of depression and the duration through which it lasted. In critical cases where suicide is on their mind, which can be identified by the use of pronouns, I, Me, and myself and a constant resigned feeling of inadequacy which makes them withdraw from all social interactions in person. Clinical help is highly recommended in such cases. A one-on-one talk therapy to help them pour out their hearts and find a purpose and some meaning in life will give them a sense of belonging and hope. In this case too, lifestyle changes are required to support clinical help as it will motivate individual and instill a feel good factor. It is recommended that Level 3 individuals establish regular sleep and wake times; avoid excess eating, smoking, or drinking alcohol before sleep; create a proper environment for sleep; and take regular physical exercise. Please note that all interventions for depression should be delivered by competent practitioners.

5 Results

Table 1, represent the analysis of severity level by increasing the number of tweets.

Table 1 Number of tweets aanalyzed with severity level

No. of tweets	High	Moderate	Low
16,000	4110	30	31
32,000	8113	78	57
48,000	12,138	106	91
64,000	16,265	133	117

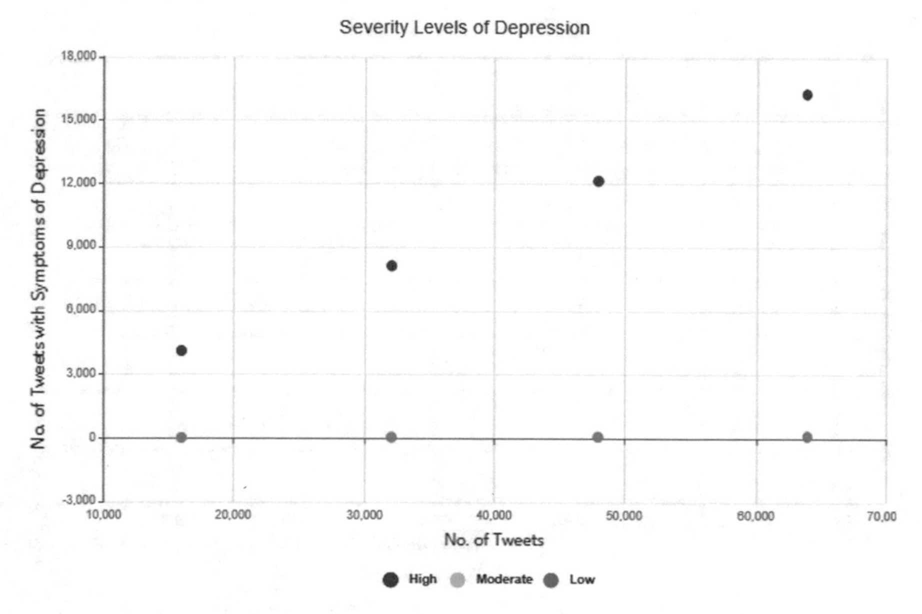

Fig. 2 Severity Level with respect to Tweet

According to the model, with more than 60,000 tweets that are analyzed, 4132 tweets are related to high level of depression, 288 tweets are related to moderate level of depression, and 244 tweets are related to low level of depression.

Figure 2 shows how the levels of severity of depression change with respect to the number of tweets being analyzed. Since the scale of high rates of depression is larger than moderate and low, moderate and low levels fall almost in the same line, hence Fig. 3 represents the level of growth of moderate and low levels of depression. Therefore, the model shows that with increased number of tweets being analyzed the severity levels also increase.

Figure 4 shows the severity levels, when 64,000 tweets were analyzed, 20.2% of tweets have severe level of depression, 0.2% of tweets have moderate level of depression, 0.1% of tweets have low level of depression, and 79.5% had no symptoms.

6 Conclusion

The developed social-media-based depression analyzer and therapy recommender model can help government to identify mental health of different age groups. Considering the analysis of the tweets government can plan policies to implement the recommendations proposed by the system. Uncovering methods that work with such text would help assist people with different types of mental health concerns to identify the appropriate therapy for the individual. The accuracy level of the proposed model

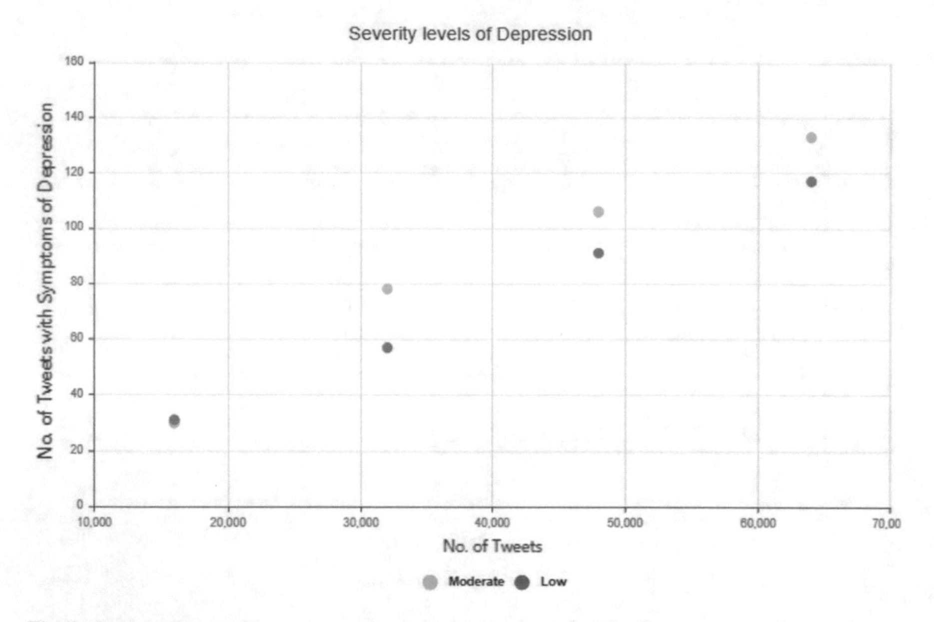

Fig. 3 Severity level with respect to tweet moderate versus low level

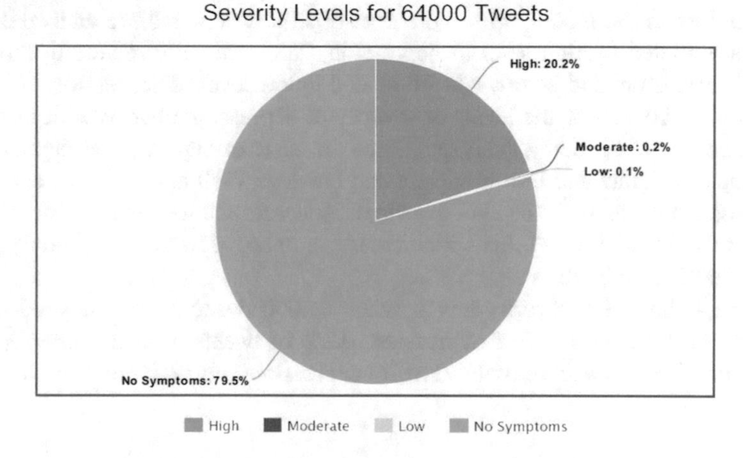

Fig. 4 Percentile of tweets with recorded severity levels

depends on the language, if a tweet contains the word in any other languages other than English it may not be able to identify. The classification output is represented in the form of line graph. In future, the study can be conducted for multilingual dictionary.

References

1. Kumar A, Sharma A, Arora A (2019) Anxious depression prediction in real-time social data
2. Havigerová JM, Haviger J, Kucera D, Hoffmannová P (2019) Text-based detection of the risk of depression
3. Thorstad R, Wolff P (2019) Predicting future mental illness from social media—a big-data approach
4. Resnik P, Garron A, Resnik R (2013) Using topic modeling to improve prediction of neuroticism and depression in college students
5. Choudhury M, Sounts S, Horvitz E (2013) Social media as a measurement tool of depression in populations
6. Benton A, Mitchell M, Hovy D (2017) Multi-task learning for mental health using social media text
7. Phand SA, Phand JA (2017) Twitter sentiment classification using Stanford NLP. In: 2017 1st international conference on intelligent systems and information management (ICISIM)
8. Dhanush M, Ijaz Nizami S, Patra A, Biswas P, Immadi G (2018) Sentiment analysis of a topic on twitter using tweepy. Int Res J Eng Tech 5(5):2881. e-ISSN: 2395–0056

Phoneme and Phone Pre-processing Using CLIR Techniques

Mallamma V. Reddy and Manjunath Sajjan

Abstract Phoneme is a study of the particular units of sound in determined language that recognizes single word from another, while phone is any distinct speech regardless of whether the exact sound is critical to the meaning of word and the pre-processing of phoneme and phone typically involves the removal of unwanted noise such as coughs, sneezes, and any aberrant peaks so that audio files are perceived effortlessly and morphing of prefix and suffix is done to get the root word. In this paper, Python programming has been used to develop the user-friendly graphical user interface which implements the working model of the phoneme and phone processing for cross-language inputs through the microphone and speaker as an output with no distortion. To accomplish this task the Cross-Language Information Retrieval (CLIR) techniques can be used.

Keywords CLIR · IR · Phone · Phoneme · Python

1 Introduction

Phoneme and phone play an important role in communication in our daily life. It may be a face to face, video conference, telephonic conversion, or by any other media. The main motive of communication is to transfer the information without any type of disturbance and one such type of communication should be saved without the background noise, which is an undesirable signal that causes miscommunication. The undesirable signal may be in the form of "background sound," "cough," "sneezes," "hiss sound by the speaker," or any other abnormal peaks.

Good quality of restored audios can be observed by assigning a suppression rule in the frequency spectra. The results found were important in the field of speech

M. V. Reddy (✉) · M. Sajjan
Rani Channamma University Belagavi, Belagavi, Karnataka, India
e-mail: mvreddy@rcub.ac.in

M. Sajjan
e-mail: mssajjan04@gmail.com

D. S. Jat et al. (eds.), *Data Science and Security*, Lecture Notes in Networks and Systems 132, https://doi.org/10.1007/978-981-15-5309-7_32

enhancement. Moreover, this method is a simple implementation to solve the problem of background noise [1]. Based on the signal-to-noise ratio of each segment, short-time spectral attenuation analyzes a signal's characteristic using a short-time Fourier transform segment by segment and sets a suppression level [2]. A research carried out by Pascal and Jozue shows the different suppression rules that can be applied based on the SNR value of a signal. The study emphasizes on the problem of a single microphone frequency-domain noise reduction in the noisy environment [3]. The study done by Pascal and Jozue tabulates different suppression estimates. The Wiener suppression rule is used for suppression value which is proportional to the SNR estimate of the input signal.

These days, the chief uses of acoustic DSP are top notch acoustic sign coding and the advanced age and control of instrumental sound signs, which are regular research themes including perceptual estimation systems and investigation/union techniques. Lesser yet exceptionally huge themes are portable amplifiers utilizing signal handling innovation and equipment structures for advanced sign preparing of acoustic. A short diagram of each segment will show the wide assortment of specialized material to expel the clamor from the sound sign introduced in this paper.

2 Literature Survey

The paper has given sufficient information on how the actual recording with background noise was collected and later the FFT of the signal was analyzed, and the energy of the segment, noise energy, and an estimate of the SNR were calculated per segment [1]. Further, it also attempts to give an insight of using FFT and estimate the SNR appropriately [2]. This book has widely covered the application of DSP algorithms to sound and acoustics. There are entire classes of algorithms that the discourse network which is not keen on seeking after or utilizing [3]. They proposed three new objective measures that can be used for prediction of fluency of speech in noisy conditions. This also explores the possibility of predicting the eloquence of noise-suppressed speech, thus reducing the SNR [4]. In this paper, MATLAB-based GUI for recording the audio and validation is carried out using different filters for audio noise reduction. They have come up with low-pass filter, high-pass filter, and band-pass filter which are efficient to remove noise from the signal thus recorded [5]. This article describes the mechanism of yielding the frequency-dependent change in the I/O transfer function that is defined as the frequency response using wavelet transform alongside modified universal threshold [6]. The authors have used the equiripple method and pass-band ripple method to find the difference between comparing results, showing the amount of SNR.

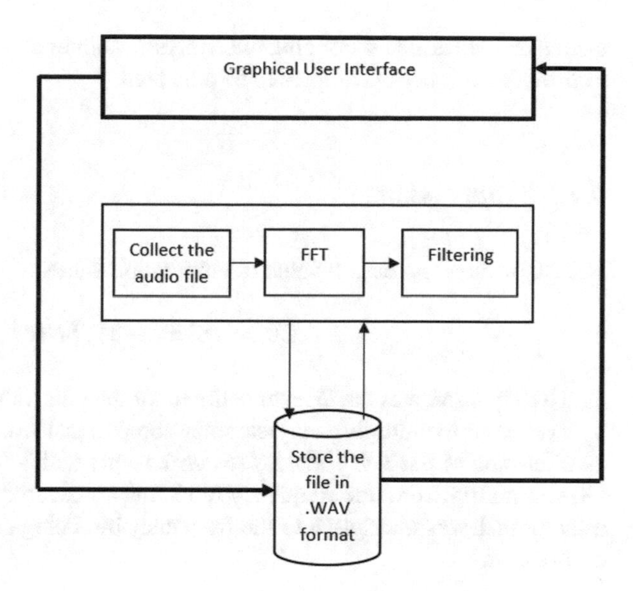

Fig. 1 Proposed system block diagram

3 Proposed Methodology

The proposed methodology is designed with a user-friendly GUI which takes the audio signal from the user and sends it to back to processing where the noise reduction procedure will be done and it will be stored in .wav format in the file. The file is sent for pre-processing, pre-processing consists of three major components: segmentation, suppression, and storing in .wav format. The audio signal received from GUI was segmented into frames. FFT was used to evaluate the spectrum of the signal and to estimate the SNR of the signal and lastly the filtered audio signal is stored in the database and it can be retrieved effortlessly (Fig. 1).

4 Experimental Setup

4.1 Segmentation and SNR Estimation

Let z(t) be the noisy sound signal, x(t) be the clear signal, and y(t) be the background noise added to the signal. Now Eq. 1 shows the actual input signal.

$$z(t) = x(t) + y(t) \tag{1}$$

In the study of the signal, the FFT of the sound input is also equal to the FFT of the signal and the noise component itself. Hence, in order to eradicate the noise, the multiplication of a reduction feature is essential to decrease the level of the signal to a considerable amount. For efficient analysis, segmentation of the audio is necessary to limit the number of samples to be analyzed.

4.2 Suppression

Let G0 be the suppression value or attenuation value.

$$G0 = (SNR - 1)/SNR \tag{2}$$

This equation was the Wiener estimate of the gain value in the suppression rule.

The idea is to multiply each frequency component to a suppression array generated as a function of the SNR. The noise components will be suppressed to a negligible. G0 was multiplied to the frequency bin being considered as a noise component and a factor of 1 was multiplied to the frequency bin being considered as a clear signal component.

4.3 Filters

Filters are used to process signals in a frequency-dependent manner. The principle thought of a channel is to look at the recurrence subordinate nature of the impedance of capacitors and inductors. Consider a voltage divider where the shunt leg is responsive impedance. At the point when the recurrence is changed, the estimation of the receptive impedance changes and the voltage divider proportion changes. This system yields the recurrence subordinate change in the I/O move work that is characterized as the recurrence reaction [5]. Channels have numerous consistent applications. To balance out speakers by moving off the increase at higher frequencies where extreme phase shift may cause motions, a straightforward, single-pole, low-pass channel (the filter) is regularly utilized. To block DC balance in high addition enhancers, a basic, single-pole, high-pass channel can be utilized. Channels can be utilized to isolate signals, passing those of intrigue, and debilitating the undesirable frequencies [4].

5 Implementation and Result

This paper presents a procedure to remove noise from the sound sign. This is done by recording the sound in the .wav format. Playback of the sound will appear to be noisy. It depends on GUI as shown in Fig. 2 (graphical user interface) in Python. In the GUI, we will take the filter button and afterward click on the filtered voice button than the phoneme and phone pre-processing is done to the loud solid to filter sound sign.

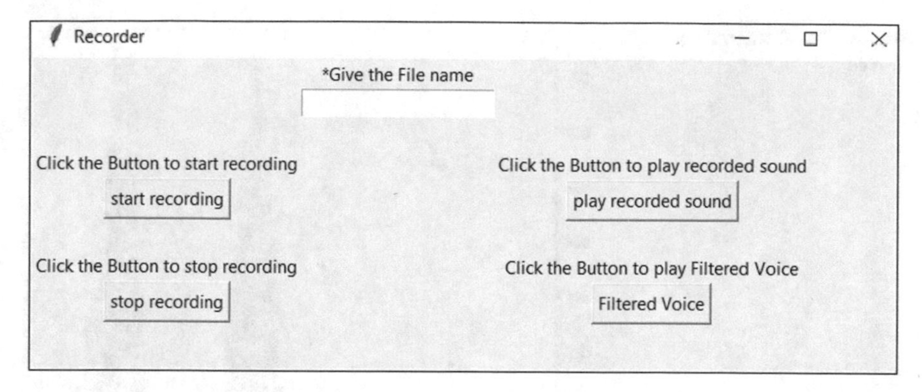

Fig. 2 Graphical user interface

Once recording is done, click on play recorded sound button from the GUI to play the recorded sound and the plot of same audio will be displayed on the screen as shown in Fig. 3 (plot of noisy sound recorded using GUI), which may have background noise. X-axis represents the time in second and Y-axis represents the amplitude of frequency.

To remove the background noise from the recorded file, click on the filtered voice button present in the GUI. It will send the file to remove the noise by applying filtering methods like FFT, low-pass filtering, high-pass filtering, and band-pass filtering and after that it will play the audio and plot the graph as shown in Fig. 4 (plot of filtered sound using FFT and band-pass filters).

Fig. 3 Plot of noisy sound waves

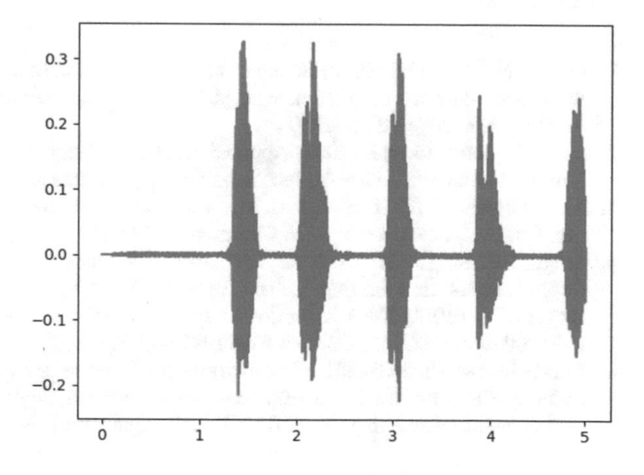

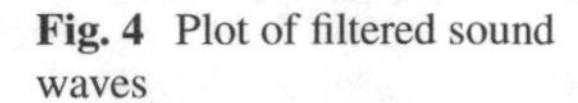

Fig. 4 Plot of filtered sound waves

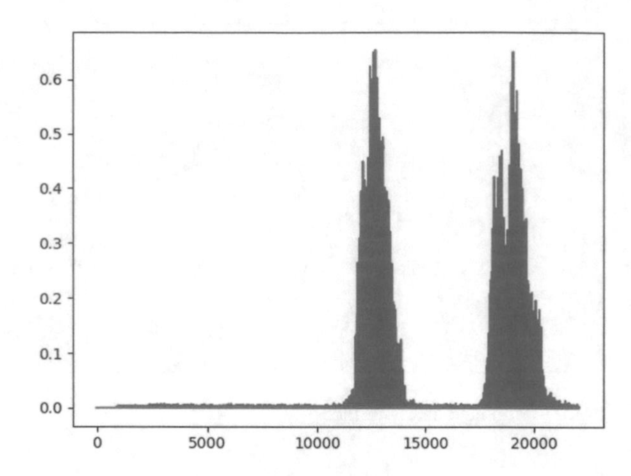

6 Conclusion

The noise reduction system is intensely reliant on the setting and application. In certain conditions, for instance, we need to expand the fluency or improve the general discourse quality or investigation of commotion crossing out techniques. Noise reduction technology is utilized to lessen the undesirable encompassing sound and is applied through various strategies.

References

1. Gapuz JN, Ladion GJ (2018) Background noise reduction based on wiener suppression using a priori signal-to-noise ratio estimate in python. Int J Electr Electron Data Commun 6(7). ISSN(p): 2320-2084, ISSN(e): 2321-2950
2. Kars M, Brandenberg K (2002) Application of digital signal processing to audio and acoustic. Kluwer Academic Publishers New York, Boston, Dordrecht, London, Moscow
3. Ma J, Loizou PC (2011) SNR loss: a new objective measure for predicting the intelligibility of noise-suppressed speech. Speech Commun 53:340–354
4. Singla EM, Singh EH (2015) Review paper on frequency based audio noise reduction using different filters. Int J Sci Eng Technol Res (IJSETR) 4(5)
5. Aggarwal R (2011) Noise reduction of speech signal using wavelet transform with modified universal threshold. Int J Comput Appl (0975–8887) 20(5)
6. Meidani M, Mashoufi B (2015) The importance of precise sizing of passband ripple in specifying order of FIR filter. In: 2nd international conference on electrical, computer, mechanical and mechatronics engineering (ICE2015), 27–28 August 2015, Istanbul, Turkey

Analysis of Various Applications of Blockchain Technology: A Survey

Sheetal and G. Shanmuga Rathinam

Abstract Blockchain is a core technology used as a solution for double spending by Satoshi Nakamoto. This technology is a right solution for problems like forgery and double spending. The technology created many opportunities for the developers and its application not only proved promising solutions for finance industry but many other domain areas such as education, machinery, and commerce industry. This paper covers a survey of using blockchain in different domains and identified successful applications as proof of concept, and it also explores problems with respective domains for future research.

Keywords Blockchain applications · Cryptocurrency · Challenges · Domain

1 Introduction

Blockchain is a decentralized, distributed digital ledger with real-time updating facility that has capacity to store data from financial domain to physical assets connected with peer-to-peer network. The data is stored in the form of transactions and encrypted using cryptographic algorithms and time stamped. A digital asset named cryptocurrency (or crypto currency) is framed for securing financial transactions that involve strong cryptography to work as medium for exchange and verify the transfer of assets. Many researchers are trying to find problems and also try to provide a useful solution in their applications using blockchain technology. However, still there is confusion in adoption of the technology and which domain this application best suited for and what kind of application we need to apply for what reason the solution can be applied. So it is always better to know complete details of their approach toward blockchain application in different domains. This

Sheetal (✉)
Department of Computer Applications, Presidency College, Bengaluru, India
e-mail: sheetal.sunil@gmail.com

G. Shanmuga Rathinam
Computer Science and Engineering, Presidency University, Bengaluru, India
e-mail: shanmugarathinam@presidencyuniversity.in

D. S. Jat et al. (eds.), *Data Science and Security*, Lecture Notes in Networks and Systems 132, https://doi.org/10.1007/978-981-15-5309-7_33

survey also lists the implementation of the technology and represents it as Proof of Concept (POC) for each domain.

The goal of this paper is to conduct a literature review of the different areas for usage of blockchain application those prove their nature as proof of concept for being successful blockchain applications and identify problems.

The structure of the paper is as follows: Sect. 2 contains a review of theoretical background, Sect. 3 presents analysis of blockchain applications in different domains, Sect. 4 provides classification of blockchain applications based on domains with Proof of Concept (POC) and challenges, and Sect. 5 completes with conclusion and references.

2 Related Work

Nakamoto [1] proposed a system which carries out electronic transactions without involving any third party. The concept of cryptography, peer-to-peer network, and proof for validating the transaction to eliminate double spending problem was designed which also rewarded the miners.

Valentina et al. [2] explain how blockchain works and considered insurance domain for blockchain applications and selected few use cases for which some of the prototypes are already defined and have to be analyzed and still few are in the form of theory which can be implemented. SWOT analysis is used to analyze any blockchain application.

Guo et al. [3] say banking sector in China currently faces risks and external and internal issues but blockchain can disrupt and also specified few specific applications such as credit mechanism, payment clearing and distributed clearing mechanism, and say about a transaction done from NAB to Canadian Imperial Bank of Commerce which took very less time and conclude with obstacles in implementing the same.

Dimiter et al. [4] explain blockchain as no single point of failure and use IPFS to access data. The author focuses on health care ecosystem and advice for dynamic consent management and quotes an example of Nebula Genomics Blockchain where the customer can charge for data access and can safe guard the data reaching in wrong hands. Finally concludes the potential benefits of using blockchain to disrupt traditional ways.

Chen et al. [5] consider present educational blockchain applications and say learning is earning. The students can be motivated by earning when learning. Usage of blockchain can reduce degree frauds, fake certificate can be avoided, and we can also use the technology for economic transactions and employment applying with smart contracts.

Yi et al. [6] say that exploring blockchain techniques were suggested in e-voting system concerning security issues. Synchronized, user credential model and withdrawal model were designed and implemented in Linux and the proposal proved successful for voting-based systems.

Wilson et al. [7] proposed a blockchain-based model on two aspects concerning the legal issues of metrology for controlling of measuring instrument and conclude with challenges that have to be addressed.

3 Analysis of Blockchain Applications in Different Domains

See Table 1.

Table 1 Analysis of blockchain applications

Approach	Year	Objective	Method/Tool	Findings and discussion	Domain
Nakamoto et al. [1]	2008	Provide a solution for double spending	Digital signatures, hash-based proof of work	Proposed peer-to-peer mechanism for electronic without involving a financial institution	Finance
Valentina et al. [2]	2018	Study of blockchain in insurance sector	SWOT analysis of blockchain	Few drawbacks were identified in order to implement blockchain in insurance	Insurance
Guo et al. [3]	2016	Analysis of using blockchain in banking	Comparison of pros and cons	Proposed regulatory sandbox and standards	Banking
Dimiter et al. [4]	2019	Review on storing electronic medical records	Pilot study	Potential benefits of blockchain such as decentralized management, data provenance, cryptographic protocols, immutable audit trails, and security	Health care

(continued)

Table 1 (continued)

Approach	Year	Objective	Method/Tool	Findings and discussion	Domain
Chen et al. [5]	2018	Exploring blockchain as solution for problem in education Industry	Survey was conducted by referring many research papers	Proposed innovative applications in education industry and also recognized pros and challenges of the technology	Education
Yi et al. [6]	2019	Providing a security for e-voting system which solves a forgery problem of votes	Voting theory, ECC cryptographic algorithm and theory based on software engineering	E-voting system using blockchain technology was proposed and the technique involved was ECC cryptographic algorithm which is not ready for an attack by quantum computer has to be addressed	Government
Wilson et al. [7]	2017	Control of MI and to solve its threats in legal metrology while designing, deploying, and inspecting	Conceptual model was considered for implementing combining MI and blockchain	They were successful in proving financial blockchain and surveillance activities but came across challenges in measuring big data, privacy, and oracle authentication	Legal

4 Classification of Blockchain Applications Based on Domains with Proof of Concept (POC)

Blockchain in insurance: [8] Blockchain possesses potential features like reducing cost, fraud prevention, and for tracking of asset and assessment, it can be one of the impactful developments for insurers and their customers. 4 to 5 percentage points can be reduced its combined operating ratio for worldwide property and casualty insurance industry.

POC: DYNAMICS Ethereum-based blockchain is used to solve supplemental unemployment insurance.

POC: [9] ICICI bank is the first bank of India to execute banking transactions on blockchain with supporting partner as Emirates NBD Mumbai: The largest private sector ICICI bank by consolidated assets has used blockchain technology for successful executed transactions in international trade finance and remittance with supporting partner Emirates NBD, a leading banking group in the Middle East. The biggest winner at IBA awards 2018 for its reinvestment and being successful for using technology as key of success is ICICI bank. The bank has tested on new technology trends like blockchain, artificial intelligence, cybersecurity, and many more in usage. The technology disrupts the financial messaging technique and makes convenience for heralds of instant cross-border remittances (retail customers).

Blockchain in health care: The aim to improving healthcare application using blockchain technology is to store data and give access to permissioned users. The advantages of sharing data reduce the expenses of the patient and disrupt the fraud documentation.

POC: [10] Medical chain uses blockchain technology to maintain health records as single document of proof with security. The point of access is given to doctors, hospitals, labs, pharmacists, and health insurers and records their transactions on distributed ledger.

Blockchain in education: The blockchain has potential into one's work and educational history. For recruitment and performance process, the blockchain can assist in the candidate sorting process.

POC: University of Nicosia uses blockchain technology in education industry for managing certificates received from MOOC platforms and other education sectors such as Sony Global Education and also Massachusetts Institute of Technology used blockchain technology.

Blockchain in government: Shifting from centralized to decentralization technology can improve the efficiency and also has ability to solve complex problems. One of the following projects developed for Australian government stands as proof for adoption.

POC: [11] A project for Australian water industry was developed by Donaghy Company to increase participation and confidence in water trading. Ethereum public blockchain civic ledger developed water ledger to update and verify all water trades in state registries for proof of water trading along with the location of trade point. Most difficulty was to consider physical asset for tokenization. It proved that blockchain technology could solve complex problems involving 15,000 rules.

Blockchain in legal: Many illegal activities to be disrupted by use of blockchain technology. The land mafia is biggest headache for the government which should be legalized to minimize illegality. The legal documentation of land registry is required around the world. Blockchain technology will be able to develop legally unassailable agreements based on smart contract technology.

POC: [12] Blockchain is helping build Andhra Pradesh's new capital: February 2019 farmer registered his plots at sub-registrar office located at Thullur Town, Guntur District. An appointment was booked by the farmer through an application which is run by government. He used Unique Identity Aadhaar number provided by the Government of India. The registration process was completed by sending details

Table 2 Challenges with respective domains

Domain	Problems
Blockchain in insurance	Scalability, mining cost and time, security, stringent risk prevention and for fraud detection apparatuses
Blockchain in banking	Time delay, complications during fund transfer. Efficiency and security are problems to be addressed. Every month digital clearing to be settled between banks and has to be reviewed which is costly
Blockchain in health care	Efficiency in electronic health records, maintaining updated records and speed up processing
Blockchain in education industry Blockchain in government	Information asymmetry, online fee payment, maintenance of E-Marks cards to avoid fake certificates, fair evaluation and to measuring learning
Blockchain in legal	Accessibility and fraud free with transparency, corruption and dishonesty Oracle authentication, privacy, big data legal issues of the usage of the technology, litigation due to contracts paid and unpaid

to registration office such as map of the plot, unique QR code, and certificate, which was all done using blockchain technology. The entire process was completed in a couple of hours.

There are several issues in implementing blockchain technology with respective domains, so researchers have to address several challenges few are listed above (Table 2).

5 Conclusion

In this paper, we have analyzed briefly the various applications, their technique, and methods used which involved blockchain technology. Proof of concepts in different domains that used blockchain as a solution in their applications is listed and also suggested areas of research in each domain. According to observation, most of the applications are involving blockchain for financial purpose in respective domains. So with interest a financial problem related to banking transaction process is selected and also a real-time problem in banking system is identified which is in concern, and also trying to provide a solution for the same using blockchain technology in our further research.

References

1. Nakamoto S (2008) Bitcoin: a peer-to-peer electronic cash system. Retrieved from https://bitcoin.org/bitcoin.pdf
2. Valentina G, Fabrizio L, Cludio D, Chiara P, Victor S (2018) Blockchain and smart contracts for insurance: is the technology mature enough?: Future Internet
3. Guo, Liang (2016) Blockchain application and outlook in banking industry. Financ Innov 2:24 Springer-Open
4. Dimiter VD (2019) Blockchain applications for healthcare data management. The Korean society of Medical informatics: Healthcare Informatics Research 25.1.51
5. Chen G, Bing X, Manli L (2018) Nian-Shing: exploring blockchain technology and its potential applications for education. Smart Learn Environ 5:1 Springer-Open
6. Yi H (2019) Securing e-voting based on blockchain in P2P network. EURASIP J Wirel Commun Netw 137 (Springer-Open)
7. Melo Jr WS, Alysson B, Carmo LFRC (2017) How Blockchain can help legal Metrology. Association for computing Machinery, Las Vegas, NV, USA
8. disruptor daily (2019) [Online]. https://www.disruptordaily.com/. Accessed March 2019
9. ICICI Bank executes India's first banking transactions on blockchain in partnership with EmiratesNBD (2018) [Online]. https://www.icicibank.com/aboutus/article.page?identifier=news-icici-bank-executes-indias-first-banking-transactions-on-blockchain-in-partnership-with-emirates-nbd-20161210162515562. Accessed 2018
10. Medicalchain (2019) [Online]. https://medicalchain.com/en/. Accessed March 2019
11. K. Donaghy (2018) How governments are using blockchain technology [Online].https://www.intheblack.com/articles/2018/08/22/how-governments-using-blockchain-technology. Accessed March 2019
12. Ananya (2018) Blockchain is helping build Andhra Pradesh's new capital—but can it cut through the red tape? [Online]. https://scroll.in/article/887045/blockchain-is-helping-build-andhra-pradeshs-new-capital-but-can-it-cut-through-the-red-tape. Accessed March 2019

Implementation of Integer Factorization Algorithm with Pisano Period

Tessy Tom, Amrita Tamang, Ankit Singh Garia, Anjana Shaji, and Divyansh Sahu

Abstract The problem of factorization of large integers into the prime factors has always been of mathematical interest for centuries. In this paper, starting with a historical overview of integer factorization algorithms, the study is extended to some recent developments in the prime factorization with Pisano period. To reduce the computational complexity of Fibonacci number modulo operation, the fast Fibonacci modulo algorithm has been used. To find the Pisano periods of large integers, a stochastic algorithm is adopted. The Pisano period factorization method has been proved slightly better than the recently developed algorithms such as quadratic sieve method and the elliptic curve method. This paper ideates new insights in the area of integer factorization problems.

Keywords Cryptography · Network security · Integer factorization algorithm · Pisano period

1 Introduction

With the advancement of technology, most of the communication is done online. It is inevitable to make sure that the messages or data sent reach the legitimate recipient and that there is no space for the eavesdropper to misuse the information. This is where the concept of cryptography comes into relevance. Cryptography is the branch of applied mathematics, which provides the techniques for confidential exchange of information using mathematical concepts and algorithms. Day by day, this process is refined to be more secure and optimized. The basic procedures that are connected with these security features are known as cryptosystem. The two forms of encrypting data are symmetric algorithm and asymmetric algorithm. In symmetric encryption, single key is used to perform both the encryption and decryption. The Advanced Encryption Standard (AES) is used as a prominent tool to encrypt the information. In asymmetric encryption, the public key which is used for encryption can be shared

---

T. Tom (✉) · A. Tamang · A. S. Garia · A. Shaji · D. Sahu
CHRIST (Deemed to be University), Bengaluru, India
e-mail: tessytoms@gmail.com

D. S. Jat et al. (eds.), *Data Science and Security*, Lecture Notes in Networks and Systems 132, https://doi.org/10.1007/978-981-15-5309-7_34

with others, whereas the private key be kept in secret. Integer factorization has been studied extensively, since many years, due to its importance in cryptography and related fields. Its importance depends on the difficulty of factoring the public keys (RSA).

The problem of factorization of integers into the prime factors is fascinating and is an important application of number theory. The problem of integer factorization appears difficult, both in a practical and theoretical sense (because none of the algorithms run in polynomial time). The major factor affecting its performance is due to the largeness of number and the complexity in the process of factorization. In this paper, starting with a broad literature review, the implementation of integer factorization with Pisano period using a modified algorithm is performed. The algorithmic complexities of different methods are compared. It is proved that the adopted algorithm is a slightly better algorithm in terms of time complexity than the recent developed algorithms such as the quadratic sieve method and the elliptic curve method. This paper might throw some light into a completely new idea of integer factorization with the Pisano period.

2 Literature Review

Cryptosystems are mainly based on integer factorization. The complexity of this factorization is giving an advantage for data security. Integer factorization is the process of decomposing any given composite number into a product of smaller integers which are usually prime [3]. In [2], the authors have developed a framework to convert an arbitrary integer factorization problem to an executable model by writing it as an optimization function and then transforming the k-bit coupling ($k \geq 3$) to quadratic terms using ancillary variables.

The public key cryptography is based on modern algorithms [3], which develop integer factoring algorithms such as sieve of Eratosthenes, Fermat algorithm, and quadratic sieve. These algorithms are mainly based on special-purpose factorization algorithm and general-purpose factorization algorithm. Integer factorization is used in the security of IOT devices [4]. Integer factorization of large numbers can be done by Pisano period factorization method [6]. This method takes less time compared to others. The computation speed can also be better described by MFFV3 (modified Fermat factorization version 3) by using DLSDT (difference's least significant digit table) [3].

The integer factorization of very large composite numbers is of a major concern for number theorists, mathematicians, computer scientists, and cryptographers in the recent decades. The role of integer factorization is not just the direct attack means on the asymmetric cryptographic algorithm RSA, but also is the most critical entry for RSA security analysis. Therefore, any optimization in the large integer factorization attracts the keen attention of the cryptography scientists. Though not proven theoretically that the security of RSA is equivalent to integer factorization, it requires large integer factorization.

From the literature, it is clear that there are many ways of cracking RSA which have been proven to be feasible, but not limited to the chosen ciphertext attack, the public module attack, the small exponential attack, the Wiener attack, the Coppersmith theorem attack, and the side channel attack [7]. Integer factorization is the most plausible means of attacking. Common large integer factorization methods include Fermat's factorization method, continued fractional factorization method (CF) [8], the quadratic sieve method (QS), Pollard's p-1 algorithm, the elliptic curve factorization method (EC), and the general number field sieve method (GNFS) [8]. Quadratic sieve factorization algorithm was invented by Carl Pomerance in 1984. Currently, it is considered as the second most efficient algorithm after the number field sieve and fastest method of factoring the numbers up to 110 digits. The number field sieve is the most advanced factorization algorithm discovered by John Pollard and is currently the best known method for factoring large numbers over 100 digits [8]. From the Fibonacci sequence, the series obtained by taking each element of the sequence modulo any modulus is periodic and these periods are said to be Pisano periods [5]. Further extended study using Pisano period factorization could be seen in [1].

3 Methodology

In this paper, the study is extended to some recent developments, namely, integer factorization with the Pisano period. A comparative study on their efficiency with respect to the time span and the number of steps required to factorize a given composite number has been performed. Finally, an improvisation in the integer factorization, including the Pisano period, has been implemented in the selected data. In order to compare the efficiency of basic algorithms, two important factors, namely, execution time and number of steps required in factoring are taken into consideration. In order to test the efficiency of the trial division algorithm, composite digits of varying sizes are taken into consideration and the amount of steps and the time it took to factorize the number are checked. Further, integer factorization is concerned about the Pisano period, which is a period during which the Fibonacci sequence repeats after reducing the original modulo n is considered. All these implementations are made in Python platform.

The four specific factorization algorithms are studied in detail, namely, trial division, Fermat's algorithm, Pollard p-1, and Pollard rho method. Trial division is the simplest method existing to find the factors of a given number. In this method, a sequential approach is followed to find the factors of the number. Fermat factorization approach is used in the problem of factorizing a number N sequentially and is used for finding factors that are suspected to have prime factors of large value.

Pollard rho (ρ) method introduced by J. Pollard is a probabilistic method of factorization and can be considered as one of the Monte Carlo methods. Pollard p-1 factorization method is based on Fermat's theorem and is almost same as pollard *rho* method of factorization. These factorization algorithms are simple in nature but have large number of drawbacks and cannot be implemented for the efficient factorization of composite numbers. In the general-purpose factorization algorithms, the factorization solely depends on the size of the number whose factor has to be found. Two general-purpose integer factorization algorithms, namely, quadratic sieve and number field sieve are also analyzed here.

4 Comparative Study of Existing Methods and Implementation of Pisano Period Factorization

It is found that in the case of special-purpose integer factorization, the factor does not depend on the size of the integer but rather depends on the property of the factors of the number which in turn makes the loop up to infinity. It leads to lag in getting the factors of the number in desired time interval. Pollard *rho* method is the most efficient method of factorization compared to the other special-purpose factoring algorithm. This method is considered the best factorization algorithm when the factors of a 9–15 digit number have to be obtained. Twelve different digit composite numbers were taken into consideration and are factored into prime numbers using different factorization algorithms. The step size and execution time were calculated. The following table and graph give details of this comparative study. Some special-purpose algorithms are also included in the present study (Fig. 1).

From the graph, it is clear that the Pollard *rho* and the Pollard p-1 take very less time to factor the same number compared to Fermat and the trial methods. For the numbers having less than or equal to 16 digits, all the factorization takes negligible time to factor the number. But for the numbers with more than 16 digits, Fermat factorization takes the longest execution time followed by the trial division method. Again, the Pollard p-1 and rho methods have almost same efficiency and far better than the Fermat and the trial factorization methods. In the present work along with the complexity analysis of the existing integer factorization algorithms, recently introduced concept, integer factorization with the Pisano period also has been implemented.

The Pisano Period Searching Algorithm

Input: A (Digit Difference), L (any large Integer), SL (Sorting Length);
Output: PP (Pisano Period);
Step 1: **FUNCTION** Search (A, L, A)
Step 2: Area of Search = [L - $10^{(\theta+A)/2}$, L + $10^{(\theta+A)/2}$];
Step 3: Now sort the first SL; REPEAT AGAIN
Step 4: Then Generate r & Compute F(ModA);
Step 5: **IF** F(ModA) hits the SORTED one THEN;
Step 6: PP = r - Position(Which has been hit);
Step 7: **END IF UNTIL** PP is Confirmed to be a period;
Step 8: **RETURN** PP;
Step 9: **FUNCTION END.**

Numbers	Fermat Factorization	Trial factorization	Pollard p-1	Pollard rho
561	0.000032	0.0007	6.0E-06	6 .5E-06
1003	0.0015	0.008	9.0E-06	0.0003
90005	0.0022	0.0026	6.0E-06	0.0012
875673	0.019	0.008	6.0E-06	6.5E-04
7894537	0.0036	0.019	4.2E-06	6.0E-04
45678943	3.73	0.0017	9.5E-06	1.5E-04
1.53E+10	2.32	1.67	6.0E-06	6.5E-06
4.11E+08	11.46	0.0085	6.0E-06	6.5E-06
4.57E+09	1E+155	0.016	4.2E-06	6.5E-06
5.54E+16	1E+155	0.11	6.0E-06	6.5E-06
5.62E+28	1E+155	0.38	4.2E-06	7.5E-06
5.68E+65	1E+155	1E+155	6.5E-06	7.0E-07

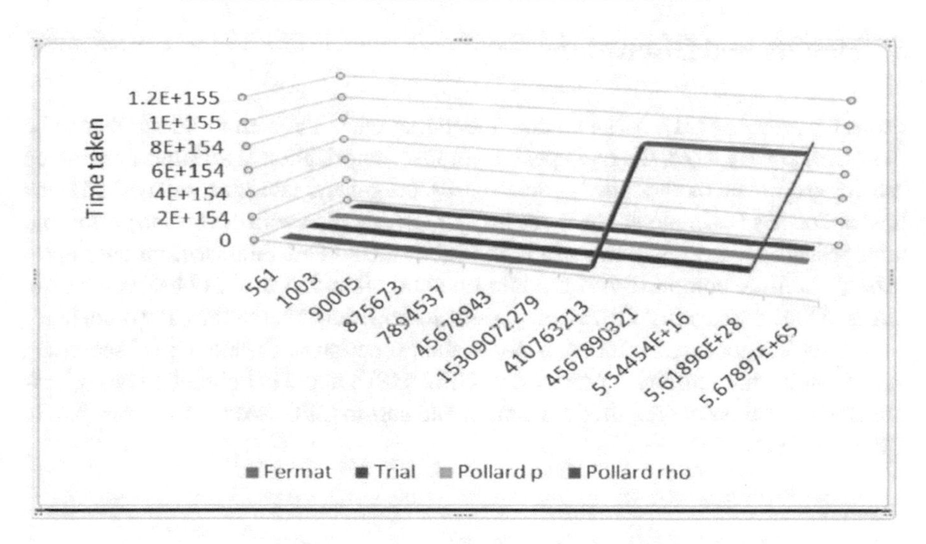

Fig. 1 Comparative study on step size and execution time of various algorithms

The following graph gives the details of time taken for the execution of Pisano period searching algorithm in three different processors, CPU, GPU, and TPU simultaneously (Fig. 2).

It is observed that the algorithm took least time to search and sort in GPU than in CPU and TPU.

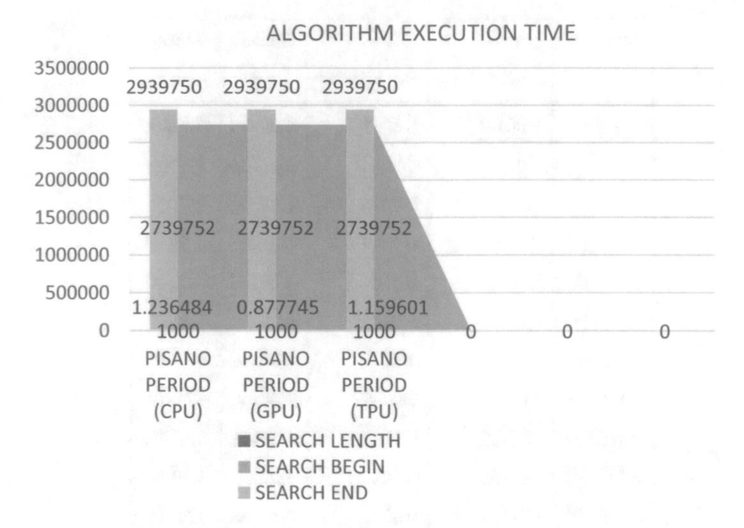

Fig. 2 Integer factorization algorithm with Pisano period execution time

5 Results and Discussions

From the present study, it can be concluded that when the composite number to be factorized is very large, the four special-purpose factorization algorithms considered are not so efficient methods and, in such cases, the general factorization method is the best option. By the implementation of integer factorization with the Pisano period and after reducing the original modulo n, the subsequent "Fast Fibonacci modulo algorithm," the time complexity of modulo operation is reduced. It is observed that in terms of time complexity, the Pisano period factorization is better than any other algorithms for integer factorizations. Also, in the execution of Pisano period searching algorithm in three different processors, CPU, GPU, and TPU simultaneously, it is observed that the time required for search and sort in GPU was much lesser than in CPU and TPU.

6 Conclusion

In this paper, four basic special-purpose factorization algorithms are compared for their efficiency in terms of the time span and the number of steps required to factorize any given composite number. The time taken and the number of steps required to factorize for each of the chosen composite numbers are noted and the graphs are plotted. From the graphs, it is observed that out of the four algorithms, the trial division method is the least efficient even though it gives prime factors within a small time interval and thus not considered. Fermat's factorization algorithm is better than trial division algorithm but it enters into an infinite loop while finding the proper factor for very large composite numbers. Hence, it also cannot be considered as a proper factorization algorithm. Pollard *rho* and Pollard p-1 methods are comparatively more efficient than the former two algorithms but fail to give proper results when the digit exceeds 16. The Pisano period factorization is found better than any other algorithm for integer factorizations. This paper ideates new insights in the area of integer factorization problems. The study can be diversified into many more directions by incorporating various mathematical concepts such as periodicity of prime numbers and more efficient algorithms in integer factorization can be evolved.

References

1. Liangshu W, Cai HJ, Zexi G (2019) The integer factorization algorithm with Pisano period. IEEE Access
2. Shuxian J, Keith A, Alexander JM, Travis SH, Sabre K (2018) Quantum Annealing for prime factorization
3. Josef P (2018) Integer factorization—cryptology meets number theory Scient. J Gdynia Maritime Uni
4. Sitalakshmi V, Anthony O (2019) New method of prime factorization based attacks on IO. J Cryptograp
5. Yuan MH (2007) The periodicity of Fibonacci modulus sequence. Math Pract Theory 3(1):119–121
6. Willrich K (2019) Pisano periods: a comparison study. In: Mathematics Senior Capstone Papers, vol 10
7. Guo M (2009) RSA algorithm and its challenges. J Henan Mech Electr Eng College 17(2):29–31
8. Koblitz N (2009) A course in number theory and cryptography. Springer, Berlin

Correction to: A Survey on Guiding Customer Purchasing Patterns in Modern Consumerism Scenario in India

Sonia Maria D'Souza, K. Satyanarayan Reddy, and P. Nanda

Correction to:
Chapter "A Survey on Guiding Customer Purchasing Patterns in Modern Consumerism Scenario in India" in: D. S. Jat et al. (eds.), *Data Science and Security*, Lecture Notes in Networks and Systems 132, https://doi.org/10.1007/978-981-15-5309-7_12

In the original version of the book, in Chapter 12, the current affiliation (East Point College of Engineering, Bangalore, 49, Karnataka, India) of the author K. Satyanarayan Reddy is replaced with a revised affiliation "Cambridge Institute of Technology, Bangalore 36, Karnataka, India". The chapter and book have been updated with the change.

The updated version of this chapter can be found at
https://doi.org/10.1007/978-981-15-5309-7_12

D. S. Jat et al. (eds.), *Data Science and Security*, Lecture Notes in Networks and Systems 132, https://doi.org/10.1007/978-981-15-5309-7_35

Author Index

D. S. Jat et al. (eds.), *Data Science and Security*, Lecture Notes in Networks
and Systems 132, https://doi.org/10.1007/978-981-15-5309-7

Printed in the United States
by Baker & Taylor Publisher Services